NATO ASI Series

Advanced Science Institutes Series

A series presenting the results of activities sponsored by the NATO Science Committee, which aims at the dissemination of advanced scientific and technological knowledge, with a view to strengthening links between scientific communities.

The Series is published by an international board of publishers in conjunction with the NATO Scientific Affairs Division

A Life Sciences B Physics	Plenum Publishing Corporation London and New York
C Mathematical and Physical Sciences D Behavioural and Social Sciences E Applied Sciences	Kluwer Academic Publishers Dordrecht, Boston and London
F Computer and Systems Sciences G Ecological Sciences H Cell Biology I Global Environmental Change	Springer-Verlag Berlin Heidelberg New York London Paris Tokyo Hong Kong Barcelona Budapest

NATO-PCO DATABASE

The electronic index to the NATO ASI Series provides full bibliographical references (with keywords and/or abstracts) to more than 30000 contributions from international scientists published in all sections of the NATO ASI Series. Access to the NATO-PCO DATABASE compiled by the NATO Publication Coordination Office is possible in two ways:

- via online FILE 128 (NATO-PCO DATABASE) hosted by ESRIN, Via Galileo Galilei, I-00044 Frascati, Italy.
- via CD-ROM „NATO Science & Technology Disk" with user-friendly retrieval software in English, French and German (© WTV GmbH and DATAWARE Technologies Inc. 1989).

The CD-ROM can be ordered through any member of the Board of Publishers or through NATO-PCO, Overijse, Belgium.

Series I: Global Environmental Change, Vol. 10

The ASI Series Books Published as a Result of Activities of the Special Programme on Global Environmental Change

This book contains the proceedings of a NATO Advanced Research Workshop held within the activities of the NATO Special Programme on Global Environmental Change, which started in 1991 under the auspices of the NATO Science Committee.

The volumes published as a result of the activities of the Special Programme are:

Vol. 1: **Global Environmental Change.** Edited by R. W. Corell and P. A. Anderson. 1991.
Vol. 2: **The Last Deglaciation: Absolute and Radiocarbon Chronologies.** Edited by E. Bard and W. S. Broecker. 1992.
Vol. 3: **Start of a Glacial.** Edited by G. J. Kukla and E. Went. 1992.
Vol. 4: **Interactions of C, N, P and S Biogeochemical Cycles and Global Change.** Edited by R. Wollast, F. T. Mackenzie and L. Chou. 1993.
Vol. 5: **Energy and Water Cycles in the Climate System.** Edited by E. Raschke and D. Jacob. 1993.
Vol. 6: **Prediction of Interannual Climate Variations.** Edited by J. Shukla. 1993.
Vol. 7: **The Tropospheric Chemistry of Ozone in the Polar Regions.** Edited by H. Niki and K. H. Becker. 1993.
Vol. 8: **The Role of the Stratosphere in Global Change.** Edited by M.-L. Chanin. 1993.
Vol. 9: **High Spectral Resolution Infrared Remote Sensing for Earth's Weather and Climate Studies.** Edited by A. Chedin, M.T. Chahine and N.A. Scott. 1993.
Vol.10: **Towards a Model of Ocean Biogeochemical Processes.** Edited by G. T. Evans and M.J.R. Fasham. 1993.

Towards a Model of Ocean Biogeochemical Processes

Edited by

Geoffrey T. Evans

Department of Fisheries and Oceans
St. John's, Newfoundland
Canada A1C 5X1

Michael J. R. Fasham

James Rennell Centre for Ocean Circulation
Natural Environment Research Council
Chilworth Research Centre
Southampton SO1 7NS
U. K.

Springer-Verlag
Berlin Heidelberg New York London Paris Tokyo
Hong Kong Barcelona Budapest
Published in cooperation with NATO Scientific Affairs Division

Proceedings of the NATO Advanced Research Workshop Towards a Model of Ocean Biogeochemical Processes, held at Chateau de Bonas, France, May 3-9, 1992

ISBN-13:978-3-642-84604-5 e-ISBN-13:978-3-642-84602-1
DOI: 10.1007/978-3-642-84602-1

Library of Congress Cataloging-in-Publication Data

Towards a model of biogeochemical processes / edited by Geoffrey T. Evans, Michael J.R. Fasham. p. cm. — (NATO ASI series. Series I, Global environmental change; vol. 10) "Proceedings of the NATO Advanced Research Workshop Towards a Model of Ocean Biogeochemical Processes, held at Château de Bonas, France, May 3-9, 1992" — T. p. verso. "Published in cooperation with NATO Scientific Affairs Division."
Includes index.
1. Marine biology — Simulation methods — Congresses. 2. Biogeochemical cycles — Simulation methods — Congresses. I. Evans, Geoffrey T., 1948 - . II. Fasham, M. J. R. III. NATO Advanced Research Workshop Towards a Model of Ocean Biogeochemical Processes (1992: Château de Bonas, France) IV. North Atlantic Treaty Organization. Scientific Affairs Division. V. Series.
QH91.8.B5T68 1993 574.5'2636—dc20 93-23008 CIP

Softcover reprint of the hardcover 1st edition 1993

Typesetting: Camera ready by authors
31/3145 - 5 4 3 2 1 0 - Printed on acid-free paper

CONTENTS

THEMES IN MODELLING OCEAN BIOGEOCHEMICAL PROCESSES

Geoffrey T. Evans
Department of Fisheries and Oceans
PO Box 5667
St. John's, Newfoundland
Canada, A1C 5X1

Michael J. R. Fasham[1]

1. Introduction

This book is about how to write mechanistic models of marine ecology that describe and predict how concentrations and fluxes of biologically important elements (especially carbon) vary in space and time, throughout the ocean over many years, in response to physical forcing. Such models are needed because of widespread interest in the global carbon cycle and the ocean's place in it. The chances of achieving them are improving because computers are getting better, because satellite remote sensing offers much more information, and because many scientists worldwide are cooperating, through the Joint Global Ocean Flux Study (JGOFS), to collect relevant data and understand relevant processes. As different research groups undertake modelling projects, comparisons between them can become as much of an issue as comparisons with observations. It will help the whole modelling community if common approaches are used where possible, and if divergent approaches arise from clearly understood reasons instead of by accident. In the first week of May 1992, a NATO Advanced Research Workshop (with additional funding from the Scientific Committee on Oceanic Research) "Towards a Model of Ocean Biogeochemical Processes" was held at the Chateau de Bonas in Gascony,

[1]James Rennell Centre for Ocean Circulation, Natural Environment Research Council, Chilworth Research Centre, Southampton, SO1 7NS, U.K.

NATO ASI Series, Vol. I 10
Towards a Model of Ocean
Biogeochemical Processes
Edited by G T Evans and M J R. Fasham

France. It brought together 45 scientists from 16 countries to discuss the issues and to explore the extent to which a common approach is possible, emphasizing the model-building process rather than the results of particular models. This book is the record of invited lectures and working group discussions of that meeting.

This introductory chapter presents what the organizers see as the main themes in developing a model, elaborating most on those themes that subsequent chapters leave unresolved.

2. Biology in the ocean carbon cycle

To first order, the oceanic carbon cycle consists of water masses moving and carrying bicarbonate and carbonate ions with them. Temperature and alkalinity can influence the small fraction of carbon (~1%) that is dissolved CO_2 gas. If physics and chemistry were acting alone, then the large-scale circulation and mixing of the ocean would bring the carbon concentration to much the same value everywhere, with minor variations in surface waters due to exchange with the atmosphere modulated by temperature-dependent solubility of CO_2.

The distinctive role of biology in global element cycling in the ocean is to act against this homogenizing tendency: to separate substances and create particles that move through the water. Photosynthesis converts CO_2 to an organic, non-volatile form; the sinking of biologically-produced material, both organic tissue and carbonate shells, can induce gradients in the total carbon concentration. Such processes acting over thousands of years have led to total carbon concentrations in the deep North Pacific exceeding those in surface waters by 20%. Changes in these processes are implicated in large shifts in atmospheric CO_2 between glacial and interglacial periods. (The proximate cause may be changes in physical circulation, but without the separating effect of biology such changes would have no implications for CO_2) More rapid events, like a spring phytoplankton bloom, can reduce the local concentration of CO_2 in the surface water by 10% in a week or two. This only represents a reduction in total carbon of 0.1%, but it is comparable to the changes in atmospheric CO_2 that human activities have caused. Not that the natural biologically-induced variability has the potential to absorb the atmospheric increase. It is a local drawdown that will be cancelled, even locally, by events at other times of the year; and it cannot usefully be compared with a worldwide and unidirectional increase. We need to consider biogeochemical processes when contemplating a changed ocean carbon cycle, both because of their feedback potential and because of the variability they add locally to large-scale surveys.

It may be possible to set an upper bound on the possible effects of biology. Although biological production needs carbon, it is limited much more by the availability of other chemical elements (possible 'most limiting' elements in different circumstances include nitrogen, phosphorous, silicon and iron) or by light. If biological production goes as far as its other limiting factors allow - if the water in which there is enough light (say the top 100 m) is stripped of some essential nutrient (say nitrogen) and of the corresponding amount of carbon - then we know the greatest influence biology could possibly have on carbon concentrations. Shaffer (1993) makes calculations along these lines.

Such calculations would indicate a theoretical upper bound, were it not that biology acts to separate things. Nitrogen and carbon are combined, in an almost fixed ratio, into biological particles, some of which sink out of the well-lit water. Particles don't sink forever: the nitrogen and carbon are oxidized in deeper water to soluble forms that can be returned to the surface by physical transport processes. But they are not necessarily oxidized in the same fixed ratio in which they are combined. There is evidence that nitrogen is oxidized and dissolved somewhat faster (and therefore shallower) than carbon, and reaches the surface again sooner. Each nitrogen atom can only take 6 or 7 carbon atoms down with it on each trip, but it can possibly make more trips than the carbon it takes down.

This differential regeneration affects the rate of cycling; less obviously, it also affects equilibrium carbon concentrations. Calculating the size of the effect requires detailed knowledge of differential regeneration and vertical mixing rates (see Thingstad, this volume); but the principle can be demonstrated quite simply. Most carbon is regenerated at the same depth as its associated nitrogen; and so that part of the production and regeneration cycle is covered by the discussion above. Subtract it out, and consider a model of only that part of the new production in which nitrogen is regenerated significantly shallower than carbon. Exaggerate the difference in regeneration depths by considering that the ocean has three layers: all the nitrogen that is mixed into the top layer is immediately converted to organic particles that sink; the produced nitrogen sinks to be immediately regenerated in the middle layer; the produced carbon sinks to be immediately regenerated in the bottom layer. The nitrogen concentration in the top layer is thus always zero; and (because there is never any net flux between the top two layers) mixing will bring nitrogen to the same value, N say, in the lower two layers. Denote the carbon concentrations in the top, middle and bottom layers by C_T, C_M, C_B, and the exchange rates between the layers by K_{TM} and K_{MB}. Nitrogen is mixed into the top layer at a rate NK_{TM}; this is therefore also the rate of nitrogen production. The rate of carbon production is then RNK_{TM}, where R is the Redfield ratio. The equations that describe the changes in carbon concentrations are:

$$dC_T/dt = K_{TM}(C_M - C_T) - RNK_{TM} \quad (1)$$
$$dC_M/dt = K_{TM}(C_T - C_M) + K_{MB}(C_B - C_M) \quad (2)$$
$$dC_B/dt = K_{MB}(C_M - C_B) + RNK_{TM} \quad (3)$$

and at equilibrium, with $C_T + C_M + C_B = 3C$,

$$C_T = C - (2 + K_{TM}/K_{MB})RN/3 \quad (4)$$
$$C_M = C + (1 - K_{TM}/K_{MB})RN/3 \quad (5)$$
$$C_B = C + (1 + 2K_{TM}/K_{MB})RN/3 \quad (6)$$

The phenomenon of regeneration at different depths thus leads to a reduction on equilibrium surface carbon concentration. In principle there is no limit to how far this reduction can go, especially if K_{MB} is very small. In the extreme case where $K_{MB} = 0$, this is intuitively obvious: the bottom layer is an unlimited graveyard for carbon, with nitrogen as the undertaker. Sarmiento et al. (1988) have an identical argument concerning the separation of organic carbon and carbonate, and Walsh et al. (1981) commented on the importance of nitrogen cycling faster than carbon. If vertical mixing were to change in such a way that the ratio of K_{TM} to K_{MB} did not change, then there would be no effect on equilibrium carbon levels despite the change in new production.

In summary, the places where we should seek, and model, a distinctive role for biological processes in global carbon cycling are (i) the initial removal of carbon from inorganic, potentially volatile forms: primary production; (ii) the physical movement of particulate carbon through the water: sinking and perhaps swimming of organisms - especially vertical migration; (iii) the regeneration of the removed carbon and biologically associated elements in deeper water. We must also consider, in the first and third topics, the production and oxidation of carbon in non-biomass form: dissolved organic matter.

3. General modelling considerations

The biological part of the carbon cycle is made up of events that take place in a fraction of a second and a fraction of a cubic centimetre (the photosynthesis of a sugar molecule, the capture of a phytoplankton cell by a herbivore, the coalescence of two sinking particles), are typically studied in experiments lasting a few hours in a few litres of water (enough to collect measurable amounts and little enough to repeat many times under deliberately changed conditions), and must be extrapolated to apply to periods of a season or longer over an ocean basin or the whole world ocean - both to describe and assess flows now and to predict them in a changed climate. Moving among scales is therefore a key modelling issue. One would like to have a model that accurately represented processes in the ocean, whose structure and output at the scale of thousands of kilometres was understandable, and that made modest demands on computing resources. These wishes imply compromises that appear repeatedly in different guises. Although of course models of the very finest processes are of interest in their own right, often to the same people who are building the global models, the theme of this workshop was models that could be viewed at a scale orders of magnitude larger than the scale of the events that make it up. Models appropriate to this theme would go into more detail only for reasons of faithful representation or numerical accuracy, and only to the extent needed.

3.1 Accurate details

Even a gigantic model, with all the accurate detail we know how to put in, making no compromises in the direction of computing efficiency or ease of understanding, will simulate some aspect of reality poorly. How should we proceed when this happens? Do we add terms to the equations which, although there is no direct evidence for them, induce 'better' behaviour of the overall model? Can we in fact adduce the model behaviour as indirect evidence? It can be evidence that something is needed, but not that we have made the right choice. As a common example, models generally fail to produce the known diversity of species. Switching of predator functional response to feed more on the more abundant prey is a common move that allows more types of prey to coexist. But this move might create too easy coexistence, and overestimate the stability under changes in parameters.

It may be in principle wrong to start from the finest details - maybe there really are emergent properties and by taking account of them explicitly we will write better models than if we try to model the (partially unknown) details from which they emerge. If they exist, then they simplify both the attempt to model and the attempt to understand. (Indeed, the need for emergent properties may say as much about the process of human understanding as about the nature of the system to be understood.) But they may not exist, or if they do exist they may not emerge from the real details in quite the same way in a changed ocean. Is there any evidence one way or the other - for example, is it true that in fields where we have reliable emergent properties (stream functions, vortices) we also have models in which we can see them emerge?

Although the aim is a mechanistic model - a model in which all the terms are calculated (perhaps with deliberate intermediate simplifying assumptions) from known mechanisms, this aim will not always be achievable. There are details we simply do not know - effects of increased ultraviolet light, perhaps changes in pH. Much modelling will be based on an ad hoc parameterization and not a fundamental understanding. One can only guess how ad hoc parameters (or, indeed, parameterizations) will respond to climatic change - although guesses, if honestly labelled as such, can still be a valuable tool of investigation.

3.2 Time scales

Weekly model output would be sufficient for understanding and contemplating seasonal cycles, and so we will try to avoid smaller time intervals as much as possible. There are different ways to do this.

Even if we model detailed fast process, we can often avoid the need for a time step suited to the characteristic time of the fastest process, by using backwards difference methods for stiff equations (e.g. Press et al. 1992). The basic idea is easily explained: the solution of $dX/dt = f(X)$ can be approximated by $X(t+dt) = X(t)+dt.f(X(s))$ for s between t and $t+dt$. For decaying processes that are quick and have 'worked themselves out' or 'reached a quasi-equilibrium' within the time step we wish to use, it makes sense to choose $s = t+dt$ instead of the more natural $s = t$. For example, if $f(X) = -aX$ and a is positive, this prevents overshoot to negative answers for X even if dt is large.

There are also ways to formulate the equations themselves, to take account of the fact that 'fast' terms will have reached equilibrium during a long time step: separating variables into fast and slow, and writing algebraic, equilibrium equations for the fast variables to insert into differential equations for the slow variables. (The methods of the previous paragraph have the advantage that they can account for fast and slow interactions, not just fast and slow variables.)

One can view the search for emergent properties as an attempt to eliminate the fast interactions in the very formulation of the equations. Thus we seek equations that relate ambient nutrients and light to the population growth of phytoplankton cells, skipping over the steps of nutrient uptake and photosynthesis.

3.3 Space scales

As we deal with larger regions, some quantities that had been parameters can no longer be considered constant within a model run - either because they are literally position- (e.g. temperature-) dependent, or because they depend on the particular mix of organisms found in a place. As we increase the overall size of the modelled region or its spatial resolution, we are pulled in two directions. Demands on computing resources and on our understanding pull in the direction of simplifying the biological model; the actual spatial variability of the ecosystem argues for making the model more complicated. The complications have implications for the carbon cycle when we consider the geographical range of calcareous and siliceous organisms - coccolithophores and diatoms primarily - and how it may be affected by CO_2 concentrations affecting growth rates (see Raven, this volume). We would wish our models to explicitly predict the boundary between diatom and coccolithophorid areas and how it might respond to a changed climate. The Global Extrapolation Working Group (Murphy et al., this volume) considered the issues in more detail.

3.4 Trophic complexity

There are local issues here as well as the issues introduced by large-scale spatial structure. For reasons of computing needs and understanding one would want to make do with few types of phytoplankton, size classes of sinking particles, fractions of DOC: but how few? How are transfers between size classes of sinking particles and DOC to be modelled? Reducing the number of classes considered will increase the distortions in modelling transfers among them. See the Working Group reports of Tett et al. and Kirchman et al., this volume. Does size describe the differences adequately, or are more detailed and specific differences between organisms important? What about internal complexity, like storage pools or life history stages? The Trophic Resolution Working Group (Totterdell et al., this volume) considered the issues in more detail.

3.5 The use of multiple resources

The issue of how an organism responds to a choice of resources arises as soon as we move beyond the simplest model. Examples include light and nutrients, or nitrate and ammonium for phytoplankton, and different types of phytoplankton and bacteria for zooplankton. We seek a form of function for the reaction to a choice of prey items, that makes sense from the predator's point of view. The functional form should also behave well as we choose to aggregate or disaggregate the trophic level representing the prey. The effects of the form from the prey's point of view will then emerge. Queueing theory has the potential for conceptually unifying all of these treatments, and is worth examining in some detail.

The simplest queue has raw material arriving at a processor and then being processed. The time it takes in total is the time to arrive plus the time to be processed; if we write this in terms of rates instead of times we get a Michaelis-Menten hyperbola. This generalizes to any number of events that have to happen in a specific order. If two events can happen in either order, the total time required is reduced: this is the basis of the Poisson arrival time model of two nutrients of O'Neill et al. (1989). If there are two processes, things can get more complicated. One might think of a complicated single processor handing intermediates from one stage to the next; but in fact in organisms one processor often sends material to an internal pool from which the next processor takes them. The problem is to keep the internal pool from growing without limit. It is often the case that, for example, nutrient uptake can proceed faster than incorporation in tissue. Either the concentration of the internal pool suppresses the activity of the processor, or there is some other form of loss - say leakage of the internal pool materials out of the cell - so that the pool does not grow without limit. The attraction of queueing theory was that it had fewer arbitrary assumptions; and therefore arbitrary assumptions about the functional form of suppression rather spoil the point.

Let us illustrate the possible uses of queueing as a unifying treatment by considering the suppression of nitrate uptake by ammonium. The physiological story is that tissue formation needs reduced nitrogen, and therefore oxidized nitrogen must first be reduced before it can be used. If it has a lot of reduced nitrogen, why should the cell bother with the oxidation step? As mentioned in the previous paragraph, dealing with an internal pool of reduced nitrogen cannot be done easily (until physiologists are agreed about the form of suppression from the pool to the uptake step), and so as an illustration we shall consider what would happen if there were a single nitrogen processor with three states: idle, waiting for input; reducing oxidized nitrogen; using reduced nitrogen to make tissue. The key assumption of the model is that if an ammonium ion arrives when the processor is in the reducing state, then the processor rejects the nitrate it was reducing, accepts the ammonium, and moves to the using state.

The equations for the probabilities of moving between different states illustrate the reasoning that goes on throughout queueing theory; O'Neill et al. (1989) give simpler examples of the same thing.

Notation: P_I=probability of being in the idle state; P_R =probability of being in the reducing state; P_U=probability of being in the use state. N=rate of arrival of NO_3 , A=rate of arrival of NH_4 ; R=rate of reduction of nitrate; U=rate of using reduced N.

$$P_I(t+dt) = P_I(t)(1-(N+A)dt) + P_U(t)Udt \quad (7)$$
$$P_R(t+dt) = P_R(t)(1-(R+A)dt) + P_I(t)Ndt \quad (8)$$
$$P_U(t+dt) = P_U(t)(1-Udt) + P_I(t)Adt + P_R(t)(R+A)dt \quad (9)$$

Solving for steady state, when all the probabilities are unchanging:

$$P_I = 1 / [1 + \frac{N+A}{U} + \frac{N}{R+A}] \quad (10)$$

The rate of total nitrogen use, which is the rate of growth, is $P_U \cdot U = (N+A)/[\ ]$, where [ ] is the denominator of (10). The rate of NO_3 use is $P_I \cdot N - P_R \cdot A = RN/(R+A)\ [\ ]$. The rate of NH_4 use is $(P_I + P_R)A = A(A+R+N)/(R+A)[\ ]$.

Some limiting rates: as $A \to \infty, P_U \cdot U \to U$. As $N \to \infty$, we have a simple queue whose rate of arrival is $R+A$, and $P_U \cdot U \to \frac{U(R+A)}{U+R+A}$. As $A \to \infty$, NO_3 use $\to 0$ even if $N \to \infty$.

A queueing formulation of zooplankton grazing on several resources, including possible switching to the more abundant food type, will present challenges. The stage where items are waiting in queues is probably digestion rather than ingestion (e.g. Sjöberg 1980); but the discrimination between prey items must be made at ingestion.

Incorporating complications by ad hoc modifications of mechanistic equations is not recommended: problems can arise in the most innocuous-looking places. Consider, for example, the question of whether to parameterize a Michaelis-Menten hyperbola by its maximum value and half-saturation concentration: $VN/(K+N)$, or by its maximum value and initial slope: $V\alpha N/(V+\alpha N)$. There is no apparent problem, because a simple re-parameterization $\alpha = V/K$ makes the two forms identical. But suppose that the maximum value is a Michaelis-Menten function of some other resource, M. Should one consider

$$\frac{VM}{K_M+M}\,\frac{N}{K_N+N} = \frac{VMN}{K_MK_N+K_NM+K_MN+MN}$$

or

$$\left(\frac{V\alpha_M M}{V+\alpha_M M}\,\alpha_N N\right) \div \left(\frac{V\alpha_M M}{V+\alpha_M M}+\alpha_N N\right)$$

$$= \frac{V\alpha_M\alpha_N MN}{V\alpha_M M + V\alpha_N N + \alpha_M\alpha_N MN} = \frac{VMN}{K_MN+K_NM+MN}?$$

If M and N increase from zero together, this innocent choice suddenly becomes the difference between quadratic or linear dependence at low nutrient concentrations.

4. The use of data

Our remarks in this section concern the types of data available and how one might want to use them (topics that cannot be rigidly separated). We shall discuss data use only for determining the values of rate constants and initial conditions that are needed to define and simulate a specific instance of the model; we shall not address the issue of statistical tests to show (as they almost certainly will) that model and data are in some respects incompatible.

4.1 Types of data

One naturally thinks first of results of detailed seagoing programmes to measure concentrations and rates, like the North Atlantic Bloom Experiment (NABE) of JGOFS. This and other JGOFS process studies are mounted with the specific justification of collecting information that can refine our knowledge of model structure and parameter values. However, as Evans & Parslow (1985) pointed out, anecdotal observations of ecosystem persistence are also data. It is possible that the parameters estimated to best fit a single season of observations, when run in the model for many years, do not produce a plausible sustained ecosystem.

Data include not only open ocean observations but also experiments done to determine model parameters. What experiments are relevant? In principle, those that are conducted on the appropriate time scales. So, if we have built a model in which certain parameters implicitly or explicitly represent how cells or populations adjust their physiology to their environment (the number of NO_3 and NH_4 receptors they build when they have grown in different nutrient environments, for example) then experiments about how a cell responds to a short pulse of changed nutrients are not directly relevant. Although they may be interesting biology and may in the end lead someone to understand and model the cell's adaptation processes on the time scales we need, we cannot use the results directly.

4.2 Estimating rate constants

The mathematics of parameter estimation, or nonlinear regression, is hard and has received much study in other fields (e.g. Seber & Wild 1989). It does not need a specifically ocean biogeochemical slant. It may be worth investigating adjoint models, which can calculate the derivatives of the residuals with respect to the parameters rather more efficiently and speed up that part of the numerical solution. With the number of parameters to be estimated (over 25 for Fasham et al. 1990) the estimation is almost certain to be difficult and unreliable. Many different parameters can be competing to explain the same observations, and carefully designed auxiliary experiments to separate the effects of different parameters can be invaluable (see e.g. Hay et al. 1988).

4.3 Estimating initial conditions

A specific issue that will need some thought is the estimation of initial conditions. This is related to the requirement of persistence - perhaps a more important property than any values the model attains in its transient phase. One extreme is not to estimate initial conditions at all, but to let the model run to a steady annual cycle and compare its values with data. Fasham et al. (1990) took this approach with the parameter values that they estimated from Bermuda data. The opposite extreme would be to run the model only for a limited time - say April to July 1989 - and estimate rate constants and initial conditions that determine the model run that best matches observations from NABE at that time. An approach that includes both of the foregoing as extreme cases would be to run the model for, say, 10 years, and compare both the results in year 1 and the results in year 10 with the observations, giving the two comparisons different weights in the fitting process. Notice that if we are to model a steady annual cycle, then we need to specify a steady annual cycle of vertical mixing: either specified in advance or (unlikely) itself estimated from the data.

It is possible that a continuous measurement programme will move by accident into a different water mass - suddenly the data being collected are appropriate for a different model run. If we can believe that it is in fact the same model and same rate constants but different initial conditions, then we could simply choose to run the model in several chunks, each with its own set of initial conditions. This has great dangers of becoming ad hoc and multiplying unknowns beyond reason; but it may well represent what is going on. Similarly, if we knew that a salp swarm had passed through the area in the middle of the observational period, then it would make more sense to re-initialize the model afterwards instead of stubbornly insisting that one model run represent pre- and post-swarm populations.

The preceding remarks apply in principle to models with spatial structure as well as models of a single water column or spatially uniform region. In practice there is a tremendous difference because when there is spatial structure the number of initial values that must be determined far exceeds the number of rate constants. The problem of data assimilation to determine initial conditions has been faced more in models of physical oceanography (which have fewer unknown rate constants as well). Work on ecological models is just beginning (Ishizaka, this volume). We would guess that the 'initialization shock' that can plague physical models is less likely to be important for biological models, and that the influence of rate constants will be much longer lasting than the influence of initial conditions. Thus biological data assimilation is likely to be much more a problem of rate constant estimation.

4.4 Sensitivity analysis

This heading comprises sensitivity of the model to its parameters, sensitivity of the parameters to the available data, and sensitivity of the model output to the available data. Parameter sensitivity analysis must be done with care. It may not be sensible to test sensitivity to an independent change in just one parameter. Different parameters estimated from the same data set will not be independent, and parameter values may also not be independent if the parameters are derived quantities arrived at through abstracting from more basic concepts. For example, Slater et al. (this volume) choose not to vary the rates of creation and sinking of detritus independently. In short, before computing partial derivatives, think about what quantities can in fact be changed without changing any other quantities.

One use of a model is as a way to interpolate and smooth between observations. In this context, the parameter values are a necessary intermediate stage; but it may turn out that the model output is determined much more stably from the data than the parameter values are (this is true also of polynomial or other standard interpolation - see Press et al. 1992.) It will be of interest to compare the model performance with a more 'neutral' model, like Fourier series with the same number of parameters.

In considering experiments, either with organisms or with parameter values, one must be careful not to infer long-term changes from immediate responses. As a typical example, we examined a model of seasonal succession of phytoplankton at about the time of the phytoplankton peak.

The short-term effect of an increase in grazing half-saturation concentration is to delay the grazing down of the bloom, but the long term effect, after population feedbacks, is to advance it.

5. Model development

5.1 Respect the details, but only once

Recall the three goals of accuracy, understanding and efficiency. A strong consensus developed during the workshop that it was important, both for accuracy and for understanding, to base global biogeochemical models on the best available understanding of the details of the processes. Among the possibly less obvious ways in which this serves accuracy: the details represent more fundamental processes, and in making predictions for a changed ocean we tend to believe that the fundamental processes will not change as much as they way they are expressed. Understanding is served if the details are somehow kept in front of us and not hidden.

It need not follow that the actual simulation steps must incorporate all the detail that we know. Rather, we see a simulation as a two-stage process, in which the first stage is one of aggregation and contraction, following a set of explicit rules that reduces underlying detail to simplified summaries with which we will calculate. The first stage is performed once, at the beginning. Then there is the actual simulation stage in which the aggregated terms are evaluated repeatedly over time and space. The actual simulation stage follows a fairly fixed set of rules; all of the flexibility, all of the model development process, would occur in making changes within the initial aggregation and contraction phase.

5.2 Comparing submodels

Among the issues in investigating different candidate submodels for the same process is the question of how best to compare them, or how they should be parameterized to produce, as far as possible, the same model - so that differences between them can be ascribed to the functional form and not to the chosen parameters. Choosing two photosynthesis-light curves to have the same maximum value and initial slope is not the same as choosing them to have the same maximum value and half-saturation light intensity. It is worth remembering that almost anything about a function can serve as a parameter - for example, a n-parameter curve will in general be completely specified by its values at n distinct points. Requiring two n-parameter curves to agree at n points spaced within their typical working range is probably a good neutral way to compare them. Or fit each curve separately to the same data, either experimental data or (this is admittedly a bit circular) to data generated from the working range of the model.

5.3 Analysing models

A proper sensitivity analysis will test the sensitivity to the original mechanistic parameters rather than to the derived ones that might be used in the actual simulation. It is also useful to find the sensitivity of the model output to the data used to tune it, using the parameters only as not very important intermediaries. Sensitivity of model output to data, to parameter values, and to parameter estimates are three separate issues, all interesting but not to be confounded.

No rules can be laid down for how to analyse the output of a model from a scientific standpoint; but in any modelling project it is important to allot considerable time and thought to it. Analysis of the 3-dimensional ecosystem model of the North Atlantic presented in Sarmiento et al. (1993) lasted at least as long, and led to at least as many disagreements among its authors, as did creating and running the model.

One issue that may be relevant is predictability, for example how likely the model is to produce limit cycle or chaotic behaviour (and whether this is to be seen as a drawback or a reflection of reality).

5.4 The modelling community

A model of ocean biogeochemical cycles will not be constructed in its entirety by one researcher or closely-knit research group. The diverse interests, backgrounds and questions of modellers make a single community model neither desirable nor attainable. What is desirable and possibly attainable is a way to compare different ideas in a common context. This is valuable not only for modellers, but also for anyone who wants to be able to use models to learn about interactions between components in a complicated system. How can the modelling community's needs for diversity and intercomparability be served simultaneously? Investment in infrastructure to make the community function better would pay off, but who will see this as their task to invest?

5.5 A modelling workbench

A modelling workbench could form the basis of such a project. The idea is not to force people to take a common approach (this would never happen) but to make a common approach available for the bits they weren't interested in. Different modellers will have new ideas about different bits of the system. They cannot be tested and compared with older ideas in isolation, but only in the context of all the other ideas representing all the other processes in the sea. One purpose of the workbench is to provide a common set of defaults for all the ideas, so that as much as possible people can investigate their new contributions against a background that their colleagues are familiar with. (Perhaps there should be two defaults - one as good as we know how to build and one as cheap as we know how to make fairly good.)

We see three separate parts of the workbench. There is the collapsing part, in which the detailed physiological knowledge is summarized in pragmatic approximations down to a chosen level of computational requirement. The collapsing phase should be performed explicitly at the beginning of every serious simulation, so that the latest view of the 'real' physiology is automatically incorporated. The workbench might include rules for when to amalgamate: for example based on the range of variability to be expected in the parameters. It shouldn't be impossible to change the 'collapsed' parameters independently (think of debugging needs), but it shouldn't be easy either.

When there are different submodels to choose from, the workbench will take care of making sure that, to the extent possible, they are parameterized in the same way; i.e. the parameters will be derived from a more fundamental set on which we all agree, according to different aggregation rules.

Lest this collapsing phase seem a bit abstract, here is an example of the sort of thing it might contain. The function $\phi(y) = \ln(y+\sqrt{1+y^2}) - \frac{\sqrt{1+y^2}-1}{y}$ plays an important role in one model of calculating daily primary production integrated over some depth, used by Evans and Parslow (1985), Fasham et al. (1990), and Sarmiento et al. (1993). The dimensionless number y represents a combination of light and how plant cells use light to photosynthesize; it typically takes values between 0 and 30. The appropriate upper bound depends on the light regime and physiological parameters. $\phi(y)$ can be well approximated (to within 0.2% of its typical value) by a rational function of the form $(ay+by^2)/(1+cy)$, where the best values for a , b and c depend on the working range for y. The collapsing phase of the model could take quantities like the latitude, cloud cover, and initial slope of photosynthesis-light curve, and calculate the derived parameters of an accurate and economical approximation to $\phi(y)$, which would then be used in the simulation part.

Next there is the simulating part, about which nothing special need be said. (Assuming that there are no longer any people who write simulations without automatic numerical error checking and step size control.) Finally there is the data use part, which could in principle consist of parameter estimation, assimilation in the sense of estimating and updating initial conditions, and validation. This part could usefully include an experimental design component, that would let one investigate which types of data are most useful and influential for validation, and how errors travel through a foodweb.

A selection of community physical frameworks would be desirable. These could range from an eddy-resolving primitive equation coupled GCM, to a table of seasonal and vertical variation in vertical mixing rates for some typical locations. The selection should not be too wide, because we want to encourage people to try their different ideas in the same physical setting. It should be wide enough that even research groups with PCs can participate.

It will be a good idea to restrict the number of options. Even two choices made independently at each of 7 places gives over 100 models, and makes any systematic comparison impossible. Admittedly, the world is that complicated, but the natural tendency of the community is in any case in the direction of diversity, so when we have a choice we should pull somewhat in the other direction. In an ideal world, of course, everybody else would form a disciplined orchestra to whose accompaniment I would play my concerto.

The overall modelling project should not be tied to any one computer system. Inexpensive computer power is growing all the time, but one can take two different attitudes to that. One attitude says that the computer power available to a fixed person is growing so fast that there is no real need to worry about small and efficient implementations - the scientist need only wait for the power to run a big implementation to become routinely available. The other says that the number of people with access to a fixed level of computing is growing even faster, so that it makes a lot of sense to invest in doing clever things with that level. (In numerical linear algebra, a field where they really know what they are doing, improvements in algorithms have brought more orders of magnitude increase in speed over the past four decades than improvements in hardware.)

At least there should be a central registry of models, physical backgrounds, etc. The workbench is only one tool toward the more fundamental goal of giving people experience with how different models function. Communication of a model requires much more than can be written in a paper - it requires playing with the model, preferably at leisure. Another way to achieve this is to convene workshops at which different users can gain experience playing with a model and trying different approaches. It is not perfect because the time and leisure will be lacking, but it will have the advantage of a lively group of people including the originator to interact with. A series of workshops with working models devoted to subtopics of the whole Bonas workshop would be a sensible follow-up.

References

Evans GT, Parslow JS (1985) A model of annual plankton cycles. Biol. Oceanogr. 3:327-347.

Fasham MJR, Ducklow HW, McKelvie SM (1990) A nitrogen-based model of plankton dynamics in the oceanic mixed layer. J. mar. Res. 48:591-639.

Hay SJ, Evans GT, Gamble JC (1988) Birth, growth and death rates for enclosed populations of calanoid copepods. J. Plankton Res. 10:431-454

O'Neill RV, de Angelis DL, Pastor JJ, Jackson BJ, Post WM (1989) Multiple nutrient limitations in ecological modelling. Ecol. Model. 46:147-163.

Press WH, Teukolsky SA, Vetterling WT, Flannery BP (1992) Numerical Recipes in C, 2nd edn. Cambridge U. Press, Cambridge New York

Sarmiento JL, Toggweiler JR, Najjar R (1988) Ocean carbon-cycle dynamics and atmospheric pCO_2. Phil Trans Roy Soc A325:3-21

Sarmiento JL, Slater RD, Fasham MJR, Ducklow HW, Toggweiler JR, Evans GT (1993) A seasonal three-dimensional ecosystem model of nitrogen cycling in the North Atlantic euphotic zone. Global Biogeochemical Cycles, in press

Seber GAF, Wild CJ (1989) Nonlinear Regression. Wiley, New York.

Shaffer G (1993) Effects of the marine biota on global carbon cycling. In: M Heimann (ed) The Global Carbon Cycle. Springer, in press
Sjöberg, S (1980) Zooplankton feeding and queueing theory. Ecol. Model. 10:215-225.
Walsh JJ, Rowe GT, Iverson RL, McRoy CP (1981) Biological export of shelf carbon is a sink of the global CO_2 cycle. Nature 291:196-201

GLOBAL EXTRAPOLATION

Eugene J. Murphy
British Antarctic Survey
Natural Environment Research Council
High Cross, Madingley Road
Cambridge CB3 OET
United Kingdom

J. Field[1], B. Kagan[2], C. Lin[3], V. Ryabchenko[2], J. Sarmiento[4] and J. Steele[5]

1. Introduction

With the recognition that global-scale geochemical systems may be affected by anthropogenic perturbations (see Houghton *et al.* (1990) for recent discussions) there is now increased scientific awareness that the biological processes of the upper ocean are important in global geochemical cycles (Melillo *et al.*, 1990). A number of the key trace gases, for example, carbon dioxide (CO_2), nitrous oxide (N_2O) and dimethyl sulphide (DMS), show significant fluxes across the air-sea interface (Houghton *et al.*, 1990) and it is estimated that 18 to 40% of the anthropogenic CO_2 emissions are currently being absorbed by the oceans and 15 to 46% of the known N_2O sources are oceanic. Many of the pathways involved in these fluxes are biologically mediated and the interactions are complex and non-linear (see Fasham *et al.* (1990) for a summary of some aspects). Examples of the key role of the biological components include; the observed decrease in pCO_2 of the upper water column in association with high chlorophyll concentrations (Watson *et al.*, 1991) and the need for euphotic zone food web modelling for the prediction of the production of DMS (Wolfe *et al.*, 1991). Najjar (1992) and Sarmiento (1992) have recently produced detailed accounts of our current

---

[1]Zoology Department, University of Cape Town, 7700 Rondebosch, South Africa. [2]P.P. Shirshov Institute of Oceanology, USSR Academy of Sciences, St. Petersburg, USSR. [3]Department of Atmospheric and Oceanic Sciences, McGill University, Montreal, Quebec, Canada H3A 2K6. [4]Atmosphere and Ocean Sciences Program, Princeton University, Princeton, USA. [5]Woods Hole Oceanographic Institution, Woods Hole, MA 02542, USA.

NATO ASI Series, Vol I 10
Towards a Model of Ocean
Biogeochemical Processes
Edited by G T Evans and M J R Fasham

understanding of marine biogeochemistry and its role in the flux of radiatively important gases.

Until recently the biological aspects have generally not been included in studies of geochemical cycling in the oceans (Kagan *et al.*, 1986; Maier-Reimer and Hasselmann, 1987; Fokin, 1989; Toggweiller *et al.*, 1989a; Toggweiller *et al.*, 1989b; Bacastow and Maier-Reimer, 1990, 1991; Ryabchenko, 1990). Much of the impetus for the development of questions regarding the role of the oceanic biota in mediating material flux between the atmosphere and the oceans has been associated with the development of the JGOFS programme (JGOFS, 1990). Well defined aims are needed for modelling exercises which consider questions of global extrapolation. The full set of JGOFS objectives present clearly the reasons why global extrapolation is required (JGOFS, 1990). It is worth repeating here the two major objectives of JGOFS which are:

1 *To determine and understand on a global scale the processes controlling the time-varying fluxes of carbon and associated biogenic elements in the ocean, and to evaluate the related exchanges with the atmosphere, sea-floor, and continental boundaries (Fig. 1).*

2 *To develop a capability to predict on a global scale the response of oceanic biogeochemical processes to anthropogenic perturbations, in particular those related to climate change.*

One particular sub-objective of the JGOFS programme which encompasses the problems we are addressing here is:

2.2 *To develop coupled physical and biogeochemical models of the ocean for the purposes of testing our understanding and improving our ability to predict future climate-related change.*

The recognition of the importance of the biotic components in global systems has come at a time when realistic eddy resolving physical models of large scale oceanic circulation systems are being developed (Semtner & Chervin, 1988, 1992; Bryan & Holland, 1989; FRAM Group, 1991; Schott & Bøning, 1991). The development of regional, coupled, biological and physical models has occurred over the last two decades (see Wroblewski & Hofmann, 1989 for a review). As questions on the mechanisms underlying global biogeochemistry have been asked it has therefore been a logical step to include aspects of biotic influences in geochemical cycles into the large eddy resolving models of oceanic processes (Fasham *et al.*, in press; Sarmiento *et al.*, in press). Simultaneously the development of new methodologies for analysing remotely sensed data has begun to provide us with other ways of obtaining information on the role of the oceans in the global carbon budget (Platt & Sathyendranath, 1988). At this stage of the JGOFS programme it is appropriate to review the methods of extrapolating temporally and spatially restricted datasets and modelling exercises to investigate the role of the oceans in global biogeochemical cycles (see also USJGOFS, 1992).

The reports presented in this book reflect an attempt to consider the various aspects of the problems of including biotic components in ocean geochemical models. The chapters consider the various components of the marine biogeochemical system; the environment of the organisms (Physical Fluctuation), the major pathways of flux and the organisms involved in those fluxes (Phytoplankton, DOM, Sinking Particles, and Zooplankton) and the ways in which these biotic components can be aggregated to provide a practical description of ecosystem operations (Trophic Resolution). The problems and possible methods of linking these components together for global extrapolations are considered in this chapter. However, although this book considers dynamic modelling methods, we feel it is important to note other methods of obtaining global estimates as these often utilise the same datasets. Much of climate modelling of the physical systems of the atmosphere and oceans utilises a mixture of empirical relationships and dynamic process modelling and is now emphasising the need for a hierarchy of models to address climate change questions (Kiehl, 1992; Schneider, 1992). In modelling marine biogeochemistry we need to ensure that we do not restrict ourselves to one particular modelling strategy.

2. Extrapolation methods

There are various ways in which we can extrapolate to obtain larger scale estimates. The various approaches are complementary and include; box models or energy balance models (EBM), bio-optical models (dynamic biogeographical models (DBM)) and dynamic process models (DPM).

The EBMs are the main type of model that have been used to give global estimates of the role of the oceans in the carbon budget (see Henderson-Sellers & McGuffie (1987) for an introduction to this type of model and Sarmiento (1992) for a recent summary). These models usually assume a steady state and local tracer measurements are used to calibrate model parameters in larger domains (for example, Oescheger, *et al.*, 1975; Broecker & Peng, 1982; Shaffer, 1989; Peng & Broecker, 1991; Shaffer & Sarmiento, in press). These models have been used to consider anthropogenic CO_2 uptake by assuming that the CO_2 addition can be included as a perturbation. A recent development of this type of analysis is to use three-dimensional ocean models and perturb these with an addition of CO_2 (Sarmiento, 1992, Sarmiento *et al.*, 1992).

Although not yet a dynamic approach, the concept of bio-optical provinces defined on the basis of key characteristics mapped from satellite imagery has been proposed by Platt & Sathyendranath (1988), Mueller & Lange (1989) and Platt *et al.* (1992). According to this approach, the world oceans can be split into provinces based on bio-optical characteristics such as temperature, chlorophyll and surface roughness. The assumption is that there are particular regions that can be characterized and mapped from their surface fields to take account of major spatial and temporal variation. Justification for this assumption and discussion of related aspects is provided by Platt & Sathyendranath (1988), Platt *et al.* (1990; 1991; 1992), Sathyendranath & Platt (1988; 1989) and Sathyendranath *et al.* (1989) who have shown in several areas that shipboard experiments to derive primary production against light curves (P-I) give relatively constant values of parameters, alpha and the maximum rate of photosynthesis (P_{max}) within a province and season and that most of the variability in estimating depth-integrated primary production occurs in biomass (chlorophyll), estimated by satellite from ocean colour. Another assumption is that the depth of mixing can also be estimated from surface fields. The estimated production values in each province in each season (approximately 50 provinces are presently envisaged) are summed to give a global estimate of primary production, new production and, ultimately, carbon drawdown. At the moment extrapolations of this type are not dynamic. However, for global extrapolations the boundaries of the provinces may change in time and space generating more dynamic, predictive models.

This type of work requires shipboard process studies at particular times to provide chlorophyll profiles and regression relationships with depth that allow extrapolation from surface field measurements. These models generate bulk estimates of production and are likely to be extremely useful in developing ecosystem models.

Our main interest in this book is in dynamic process models (DPM). As described in the JGOFS (1990) science plan, our ability to model the ocean

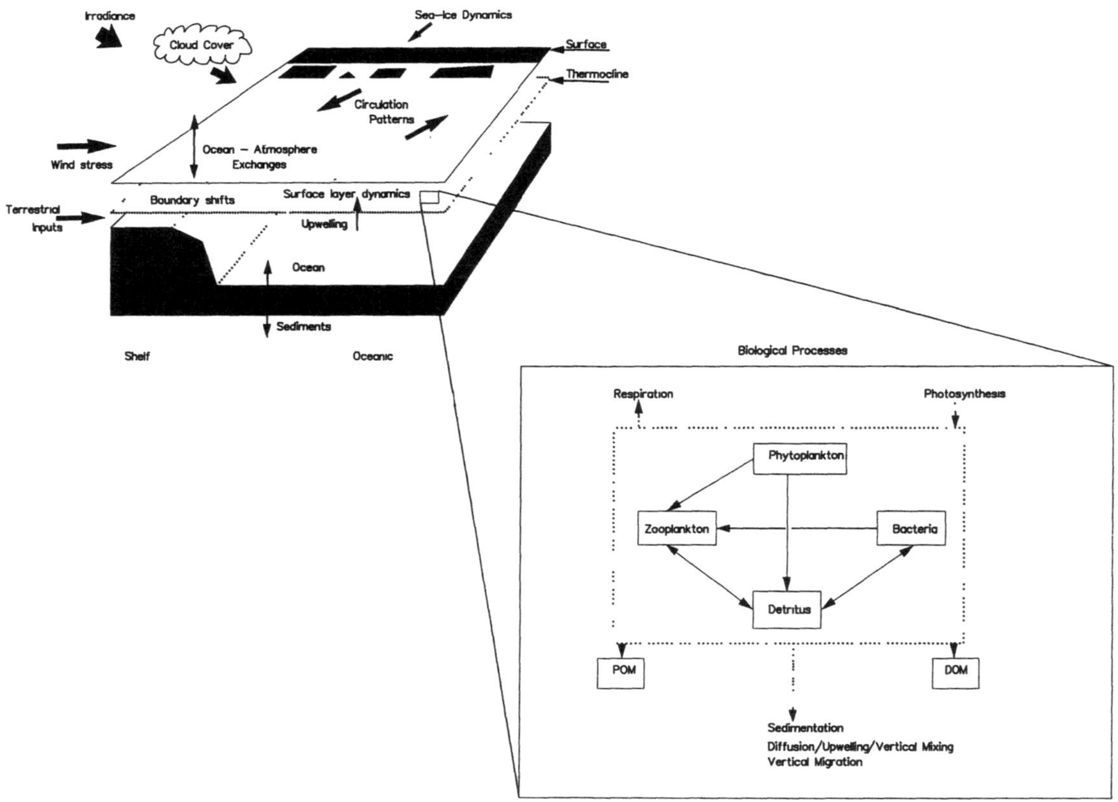

Fig. 1. Small-scale biological process models can be coupled with large-scale models of the physical regime to generate global extrapolations.

CO_2 system has until recently been very limited. One particular statement from the science plan is worth highlighting again as we consider the methods we should use to make global extrapolations using fully dynamic models.

> *"We must match the observations of the earth-ocean-atmosphere system with the basic laws of chemistry, physics and biology if we are to gauge the effects of human intrusion."*

This broad statement highlights that the biotic components of the models must be included in such a way that they allow a realistic and practical evaluation of climate change scenarios. Also, because of the high complexity, we cannot simply expect to include everything we observe into global models of the ocean-atmosphere system (Schneider, 1992). To be able to extrapolate dynamic models we need to couple models of physical and biological processes (Fig. 1). This raises questions of scale in which we have to consider the temporal and spatial scales over which the models were developed and scales to which we are extrapolating.

3. The spatial and temporal scale of processes and dynamic models

In coupling models the biological and hydrodynamic aspects may vary in complexity. An example of a model in which the physics used is simple is that of Fasham *et al.* (1990) who used an Eulerian biological model with aggregated physical driving variables (mixed layer depth) to simulate the biological-environment interactions. A much more complex small scale biological model with Lagrangian particle tracking coupled to a high resolution physical model of water column structure was used by Wolf and Woods (1988) to generate an ecological model.

The range of possible models that could be used and coupled is large and the models cover very different scales (Fig. 2). In the development of physical models the problems of scale are taken into account (Beckers, 1992). At fine scales (seconds-days, metres-kilometres) there are models which resolve the vertical structure and turbulent diffusion processes (Wolf & Woods, 1988). Regional models consider the physical processes applying over temporal and spatial scales of 1-100km and days (hours) to weeks (months) (see Wroblewski & Hofmann (1989) & Beckers (1992) for reviews of some of the more relevant aspects). The basic patterns of oceanic circulation can be reproduced by relatively crude models (Bryan and Lewis, 1979). The inclusion of mesoscale processes in large scale models results in much better description of oceanic heat budget and finer scale circulation patterns (Semtner & Chervin, 1988, 1992). General circulation models (GCMs) have grid sizes of approximately 100km and time steps of hours to days, thus they do not include much of the higher frequency variation of the environment. In addition climatology is usually entered as monthly values.

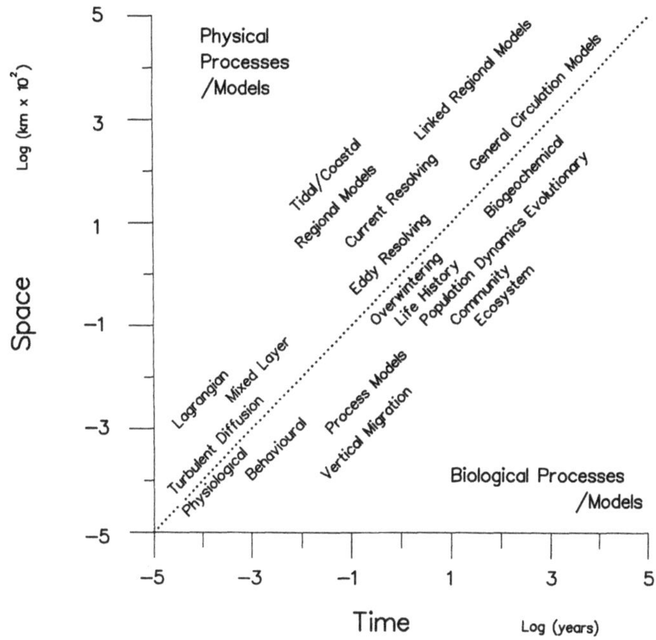

Fig. 2. The temporal and spatial scales over which particular physical and biological models apply.

Associated with this concept that different hydrodynamic processes are best studied and modelled over particular scales, the development of theory regarding scale and oceanic ecological processes (Haury *et al.*, 1978; Legendre & Demers, 1984; Harris, 1986; Murphy *et al.*, 1988; Powell, 1989; Steele, 1989a, b) has emphasized that different ecological processes are operating over different temporal and spatial scales. A hierarchy of ecological processes (Allen & Starr 1982; O'Neill *et al.* 1986; Murphy *et al.*, 1988) is operating; for example, the phytoplankton doubling times are of the order of hours to weeks; physiological scales are much faster over periods of seconds to hours, while shifts in community structure occur over longer time scales of weeks to months. In changing scales the non-linear nature of the interactions will need to be considered. There is no reason to expect that physiological process models alone will provide good models of community dynamics. The need for hierarchical development of models has been emphasized in the IGBP land cover discussions (Townsend, 1992) and in recent climate modelling reviews (Kiehl, 1992; Schneider, 1992). The need for an hierarchical approach to modelling ecological systems has not been emphasised by JGOFS although some of the basic principles have been applied.

4. Coupling physical and ecological models

The above indicates that there may be scientific problems with mismatching the scales of ecological and hydrodynamic models. These possible problems have been noted elsewhere; Platt & Sathyendranath (1991) suggested that temporally and spatially the scales of the environmental variability introduced into the models should match the scales of the ecological processes. So, where we are including ecological interactions which have time scales of the order of 1 to 10 days, such as the interactions affecting phytoplankton growth rates, we probably need to take account of variations in the physical environment over similar scales. How does this cause a problem? Examples given in some recent work by Sakshaug *et al.* (1991) and Sakshaug *et al.* (in press) have shown that the inclusion of storm events in models of bloom formation in the Southern Ocean can result in very different model conclusions with the retardation of the bloom. In modelling the Barents Sea primary production, these workers noted that as well as bloom retardation, enhanced total production occurred when storms of an approximate 10 day frequency were included in the model. Higher frequency variation in the environment may therefore be sufficiently important in the operation of the ecosystem models that, without it, misleading results will be produced. This problem may be even more significant in complex coupled models with a greater degree of non-linear feedbacks. Increasing complexity in the biology without increasing physical complexity may not be the most useful approach. Further work is required on the coupling of biological and physical models and, in particular, on the question of changing ecological complexity in such models.

As well as scientific problems, there are also technical problems associated with attempting to run very large coupled eco-hydrodynamic models. Computational constraints of running complex models has meant that their application has generally been restricted to small spatial scales. In the North Atlantic GCMs (Sarmiento, 1992) increasing the number of elements in the ecosystem models by a factor of 2 doubles the run time. However, to double the spatial resolution requires an order of magnitude increase in run time. A balance must therefore be achieved in the specification of detail in the ecological and hydrodynamic models. Currently the eddy resolving North Atlantic GCMs can only be run practically for approximately 10 simulations a year. The ability to explore the models is, therefore, constrained by the computer time, not the specification and run of the problem. This situation is changing rapidly with the advent of parallel computing so computational limits should not limit what we think we would like to do in the next decade as more powerful computers become available (Chervin & Semtner, 1990). Possibly a more serious constraint is the ability of scientists to analyze the voluminous output produced by one run, restricting the amount of experimentation that is practical with GCMs.

The effect of these scientific and technical problems is such that the complexity of the models that are required will vary depending on the scale appropriate to the questions being asked. In extrapolating ecological process models we are considering the combination of temporal and spatial scales of process operation and biological complexity. Complexity is defined here as the number of components included in ecological process models. In this 3-dimensional representation we can consider the scales over which the models are constructed against those for which we must extrapolate (Fig. 3). Process models of high ecological complexity will probably require high resolution in time and space of the environment. Regional models have a lower temporal and spatial resolution and hence deal with a reduced ecological complexity. Low complexity, highly aggregated models will be associated with low resolution physical models such as GCMs.

5. The development of extrapolations

If we are to understand ocean biogeochemical processes we will need a range of model approaches so that different questions are addressed and inter-calibration comparisons carried out. The models within the hierarchy could then be used to allow definition of inputs and outputs from one level to the

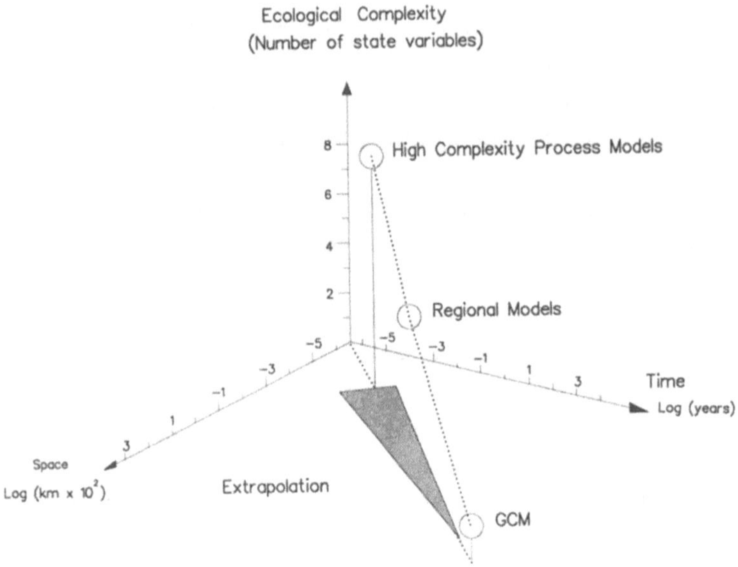

Fig. 3. Extrapolation in time and space will require a reduced complexity in the ecosystem models for examining biogeochemical processes. The large ocean models do not include the high-frequency environmental variation required to operate high complexity ecologically based biogeochemical models.

next. Access to supercomputers is limited, so a variety of approaches to solving the problems, particularly those relating to the physical models will be needed. These include linking together many localised models of very high complexity, linking regional models of medium resolution, linking large GCM models of coarse resolution, generating a single GCM for the global ocean, or using box models. Other methods involve combining empirically based model methods with remotely sensed data. In carrying out these types of studies a range of different levels of detail will probably need to be tested. At the most basic level generic models with the same parameter values could be applied widely. Using the generic models with different parameter values could then be applied in different regions. Thereafter, a more complex scenario could be tested using different models for different regions.

Clearly, in any extrapolations an ideal solution for including the biotic components would be to utilise a generic model. In such models a standard set of parameter relationships would be defined and the basic structure of the system and the parameterisations would not have to change in the extrapolation. Any parameter changes required should be a feature of the model rather than dependent on reparameterisation for a range of sites at different times. Changes in ecosystem structure or parameter values in the model should therefore be dependent on changes in the forcing variables.

To take such models and extrapolate to larger scales we need to know what aspects of the larger scale driving variables cause the flux rates and key parameter values to vary. It will be much simpler to model and understand the system if we can extrapolate on the basis of forcing variable-parameter relationships or some other derived secondary parameters. Sensitivity analyses are often carried out on such process models (e.g Fasham *et al.*, 1990). If these analyses indicate that particular parameters are sensitive, effort can be increased to define the driving variable:parameter relationships of these parameters.

In some areas the form of the model may have to change fundamentally. The ecosystems of polar regions are characterised by a number of elements different from open ocean regions, such as the swarming krill of the Antarctic and primary production systems of the marginal ice zone, so these may require very different model specifications (e.g. Lancelot *et al.*, 1991; Murphy & Priddle, 1992). Developing models which are sufficiently flexible to allow structural changes in the ecosystem will be a major challenge for the next few years.

The ecological models available for extrapolation to address questions of marine biogeochemistry can be highly complex biologically. They are often validated and applied over small spatial scales but extrapolated in time to yield a temporal sequence, with the usual assumption that this represents some spatial average. These process models require relatively high resolution physical models in the vertical dimension, at least in terms of how aggregated variables such as mixed layer depth affect the ecosystem. Often these models are considered to give some average of the temporal sequence when higher frequency variation is not introduced. The range of ecological models developed and applied over these scales is large, for example: Woods & Onken (1982), Wolf & Woods, (1988), Simonot *et al.* (1988), Fasham *et al.*

(1990), Moloney & Field (1991), Moloney *et al.* (1991), Lancelot *et al.* (1991) and Sakshaug (1991). The need for an adequate model of the mixed layer was recognised as a key component of coupling biological-physical models over these small to medium scales. Calibrating such models at one site without the inclusion of horizontal advection and vertical mixing may give misleading results in areas of complex physics (Bryan, 1986). In extrapolating to larger models with lower vertical resolution but greater horizontal range, such high resolution models will be useful in correctly specifying the parameter relationships required. Such models are often used to study detailed mechanisms and consider the match of the model output to high resolution observations. There is little objective basis for considering the relative merits of the different models for extrapolation purposes. Fasham *et al.* (1990) have generated a model which appears to capture many of the key elements of marine biogeochemical cycles. This volume has addressed many questions

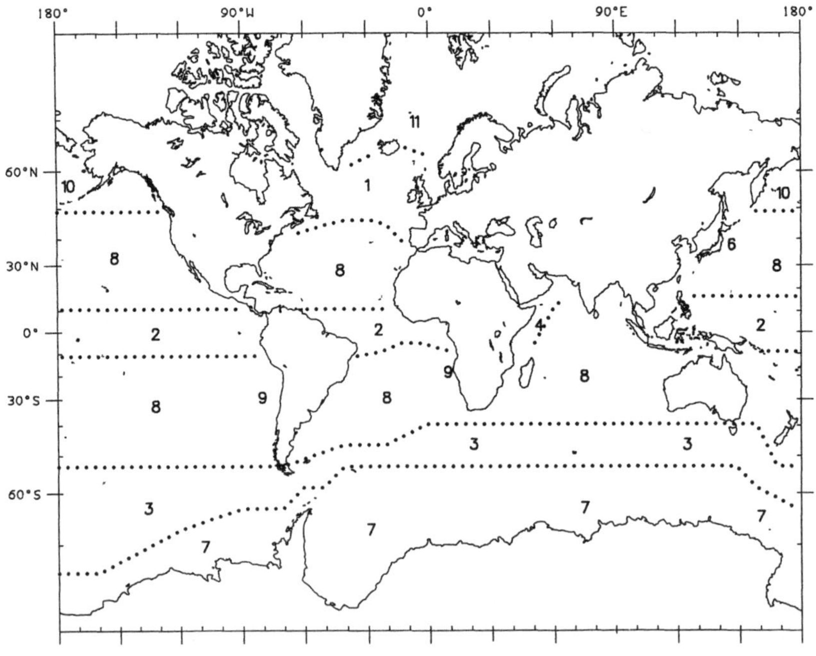

Fig. 4. Biogeochemical provinces can be defined to allow global extrapolation (Mercator projection; based on JGOFS, 1990).

relating to the number of components of different groups required in a generic model. The structural sensitivity of the models will probably need to be tested and the effects of increasing ecological complexity included. This may need to be carried out at the same time as an assessment on best methods for both biological and physical variable aggregation in such models.

6. Current extrapolations using dynamic models

6.1 GCM based models

The need to begin the extrapolation exercises at this time means that decisions are required now on how to extrapolate and that these should be based on current knowledge of the system operation. However, at this stage it is only possible to carry out high resolution sampling of ocean biogeochemistry over very restricted temporal and spatial scales (for example Fasham *et al.*,1990, in press; Ducklow & Harris, 1992). The basic principle upon which many of these studies have been developed is that biogeochemical provinces can be defined within which the biogeochemical cycle can be considered to be structurally homogeneous (JGOFS, 1990) (Fig. 4). Thus, the assumption is that if process studies are carried out over restricted scales within these provinces, extrapolation to the larger areas on longer time scales will be possible by utilising a set of constant parameters.

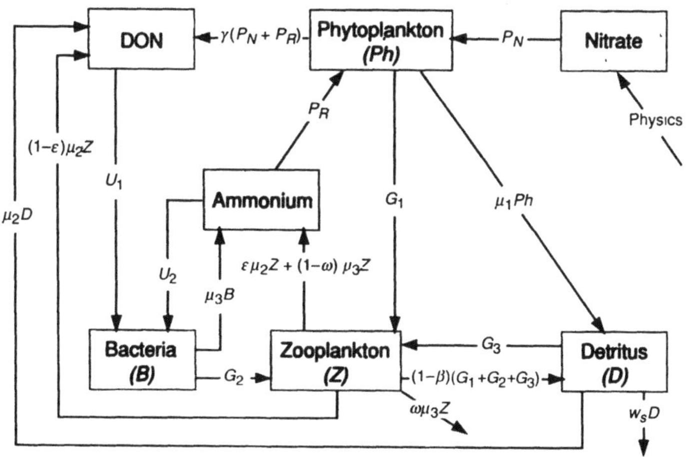

Fig.5. The Fasham *et al.* (1990) ecosystem model. See Fasham *et al.* (1990) for a discussion of the model and the parameters. Reproduced from Sarmiento (1992).

Linking ecological models into large ocean models is at an early stage. So far this has involved direct extrapolation from small scale, reasonably high resolution ecological models, to large scales by coupling them with regional or global GCMs. A good example of the state of the art is the North Atlantic coupled model described in Fasham *et al.* (in press) and Sarmiento *et al.* (in press). A seven component ecological model (Fasham *et al.*, 1990) has been developed (Fig. 5) which includes some of the key biological processes involved in ocean biogeochemistry. The model has a phytoplankton, zooplankton and bacterial class and four forms of nitrogen; nitrate, ammonium, dissolved organic and particulate nitrogen. The model can thus separately quantify new nitrate based production and "regenerated" ammonium-based production. The model has been incorporated into a 2° horizontal resolution, 25 vertical level, seasonal ocean general circulation model of the North Atlantic (Sarmiento, 1986).

A range of aspects of ocean biogeochemistry have been investigated, for example quantifying the sources of nitrate to the upper water column through upwelling, vertical mixing and convective overturn. The large scale hydrodynamic systems maintaining the nitrogen levels have been studied and various aspects of the biogeochemistry considered. The results indicate that during the early part of the season the bloom is nitrate based while later ammonium, produced by zooplankton and bacteria, becomes the more important source of nitrogen for production (Fig. 6). Many of the problems encountered in examining the biogeochemical aspects of the model are the result of problems in its hydrodynamic element. This large scale exercise has emphasised the importance of stressing processes, rather than specific organisms, in the development of coupled models. Various questions regarding the controlling mechanisms in food webs affecting nutrient distributions have been highlighted by such studies (Sarmiento, 1992).

Problems in applying the GCMs remain when we are trying to consider questions of the role of the oceans in the global carbon budget. Possibly the most difficult to deal with are the boundary conditions, both at the arbitrary latitudinal boundaries and more particularly at the continental margins. These are some of the most ecologically important regions, but they require high resolution physical models to produce realistic hydrodynamic regimes. These coastal shelf regions are of major importance in the biogeochemistry of the oceans. Recent calculations indicate that whilst approximately 20% of the primary production occurs on the continental shelves, 80-90% of the carbon burial into the sediments occurs in these regions (Emerson & Hughes, 1988; Tromp, 1992). The residence times of carbon in the sediments is also very much shorter (approximately 50 times shorter) on the shelf than off the shelf (Tromp, 1992).

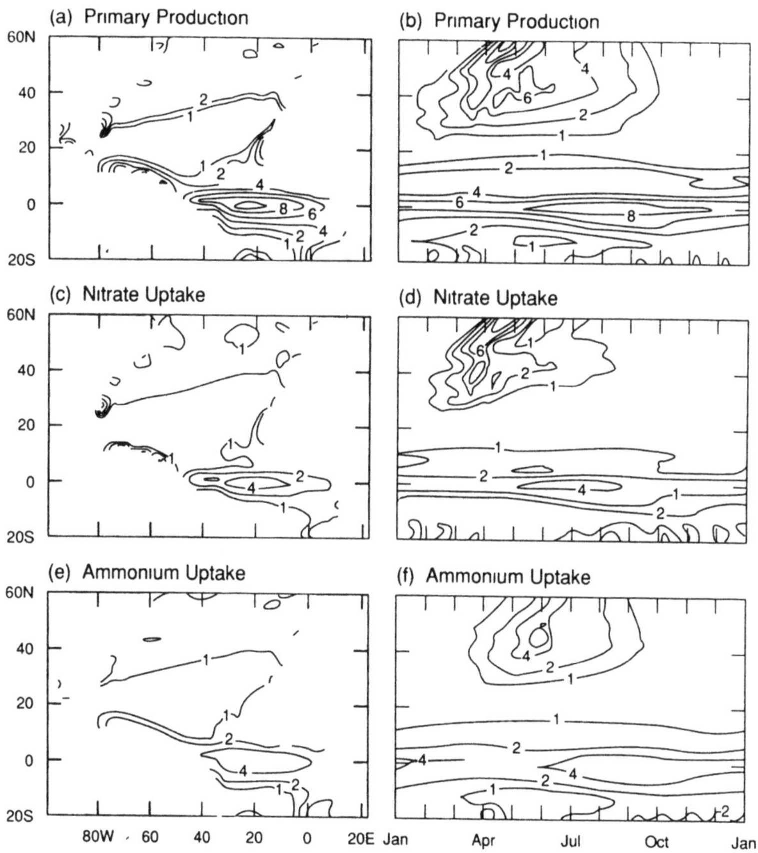

Fig. 6. Coupled-model simulations of the biogeochemistry of the North Atlantic. The left hand panels show the distributions of annual means over the top 123m. The right hand panels show zonal integrals over the top 123m as a function of time. Reproduced from Sarmiento (1992) (based on Sarmiento *et al.*, in press).

Other studies are also underway to couple ecological models into large scale oceanic models in the UK (FRAM) and in France. The large scale extrapolations carried out so far have highlighted a number of lines of biogeochemical research such as: the need for good data on air-sea gas exchange, the need to quatify fluxes into the oceans in the boundary regions, the importance of sedimentary burial, what are the important control mechanisms in aquatic food webs? and what is the role of DOM in the transfer from surface to deep waters?

6.2 Regional models

Regional models usually have reduced biological complexity in terms of numbers of process components, but may introduce more large scale ecological aspects such as life history associated changes in ecosystem structure (Böning & Cox, 1988; Hofmann 1988; Hofmann & Ambler, 1988; Ishizaka & Hofmann, 1988; Ishizaka, 1990a, 1990b, 1990c; Walsh *et al.*, 1988; Wroblewski, 1989; Niebauer & Smith, 1989; McClain *et al.*, 1990; Spall, 1990; Betello & Bergamasco, 1992; Gregg & Walsh, 1992; also see Wroblewski & Hofmann (1989) and Beckers (1992) for reviews). These models may be appropriate for increasing the complexity of aspects relevant to larger organisms where the scale of the physical models are compatible with the ecological processes. Such models can include high frequency variation in their physics. These models may be the most satisfactory compromise in terms of spatial and temporal resolution and ecological complexity for addressing many JGOFS related questions. For extrapolation to larger scales, models of this type could be linked together into a mosaic to generate a global estimate of production, interpolating to fill gaps where necessary. Such models could also be used to specify particularly important regions and then be linked together within cruder GCMs (Wroblewski & Hofmann, 1989; Giorgi & Mearns, 1991).

7. Other considerations

Linking different types of models of differing resolution is one way in which we can generate a global budget. Other methods also need to be considered, such as the use of models which have adaptive grids, or nested grids which allow changes in resolution in different regions (Voltzinger *et al.*, 1989). Other ways of coupling ecological and hydrodynamic models may be to include explicitly the different scales of process operation. For example time scales of various processes in surface and deep water can be treated separately (Gorchakov *et al.*, in press; Kagan & Ryabchenko, 1992). Complex models based on other aggregated ecosystem properties, such as models which switch between the type of ecosystem (e.g. recycling or producing), may be suitably embedded in larger GCM-type models. The question of the level of complexity required in such models is fundamental. There are various techniques that can be used to consider the aggregation of food webs (see Powell this volume; Jørgensen, 1992; Rastetter *et al.*, 1992). However, many questions of higher level ecosystem properties have yet to be addressed.

The majority of effort being expended on modelling for JGOFS has been aimed at modelling the present day situation. Questions of how other aspects of the system may evolve have by default been left largely to the second stage. Now that we are in the situation where large scale extrapolations are being developed we need to consider other aspects of oceanic systems that may change under various climate change scenarios and how these may affect ecosystems. Appropriate models need to be developed to address scenarios of future atmospheric and chemical conditions.

Some of the possible factors that need to be included are changed temperature, CO_2, pH, wind stress patterns, sea-ice, circulation patterns, nutrient supply shifts, irradiance changes associated with cloud cover effects, UV-B radiation selection pressure on planktonic species, trace element effects and secondary feedback effects. Some of these aspects are being concentrated upon, such as temperature and CO_2 changes. Many of the others are poorly understood and the physical models available so far are insufficient to resolve many of the problems relative to biogeochemical cycling. Of the list above, possibly the pH changes and the wind stress pattern changes affecting upper water column structure are aspects which should be addressed in the near future.

The effects of changes in these various factors (temperature, CO_2, pH etc.) on different ecosystems will require development of different types of ecological models. As indicated above, mismatching scales of ecological and physical models could result in incorrect scenarios being developed. The change of structure of ecosystems is a likely consequence of environmental change, so aspects of evolution in ecological systems need to be considered. This may apply particularly when sub-systems (or provinces) are used as the basic extrapolation unit.

Specific large scale aspects of modelling projects that require further development include those relating to processes of carbon sequestration into the sediments and deep ocean, and their validation. These processes include formation of refractory dissolved organic matter (see Dissolved Organic Matter report in this volume) (sequestration of carbon in euphotic or aphotic zone) and processes resulting in sinking of particles of phytoplankton, faeces, etc. (see Sinking Particles report in this volume). Some of the more specific aspects of model development have been discussed by USJGOFS (1992).

8. Validation and analyses of extrapolations

Many data sets are now available which can be used to parameterise and validate coupled ecological-physical models. Of recent developments, possibly the most important relate to aspects of automated remote sensing technologies. These include the development of fixed station arrays for physical and ecological sampling, the improvement of hydroacoustic systems for environmental and biological analyses, improvement of towed body technologies (BATFISH/Sea-Soar/UOR) and new innovations such as the free roving AUTOSUB. At the very large scale the new generation of satellite (SeaWiFS and MODIS) data promises to be particularly useful in generating global estimates of primary production (Platt *et al.,* 1991). While global validation of surface fields such as ocean colour, sea-surface temperature and surface roughness from satellite imagery are likely to be very useful, it is also very important to validate models using depth profile fields, since the sequestration of carbon from the atmosphere into the ocean for long periods is an important focus of JGOFS.

As part of the analyses and model validation, there is a need for inter-calibration exercises with comparisons of different types of models using the same datasets. This structural sensitivity analysis is missing from most marine ecological modelling. Data assimilation methodologies have been considered by USJOGFS (1992) and food web models which can assimilate satellite ocean colour data may be required (Sarmiento, 1992). Many of the problems we are likely to encounter over the next decade relate to the large volumes of data becoming available. To carry out global extrapolations and develop global climate models which include biotic aspects, the relevant datasets must be widely available. Protocols of handling, access and visualisation must be rapidly developed to allow us to produce realistic global extrapolations.

9. Conclusions and recommendations

1) In order to address questions relating to the role of the oceans in global biogeochemical cycles, we need to develop a range of models operating over different spatial and temporal scales.

a) Detailed high resolution models rich in vertical structure and biological detail are important to synthesize our understanding of processes of biogeochemical cycling and carbon transport. These can also be used to estimate local fluxes of carbon under particular conditions. They are also needed to intercalibrate with less detailed models operating over longer time scales and larger spatial scales.

b) Medium spatial and temporal resolution models are needed to simulate the interaction of physical and biological processes at the regional scale and to provide seasonal temporal resolution. These can be calibrated against the high resolution models (a) for finer scale closure and against basin scale GCMs to set large scale boundary conditions.

c) The present generation of basin-scale GCMs is criticised as needing better physics to describe the vertical processes which are required to address the important ecological questions. The ecological components are being refined and appear adequate for the present level of physics that can be included. These general models can be calibrated against the more detailed regional models and extrapolated over larger areas and longer periods.

2) These dynamic simulation models may need to be linked across scales to generate global estimates of carbon fluxes. Models at one scale could generate integrated outputs to act as inputs to different levels. Different resolution models can be linked together to generate global estimates.

3) One difficulty with most of the present GCMs is that they tend to have problems with boundary conditions, both at the arbitrary latitudinal boundaries and particularly in dealing with continental margins. The latter can be improved by adopting variable adaptive grids with a curvilinear relationship to the contours, providing much more detail over the slope and shelf regions where fluxes tend to be greater.

4) Intercalibration exercises between different types of models applied over different scales will help to elucidate the operation of the systems.

5) There is a need to develop models which address the various aspects of the climate change scenarios affecting biotic components such as temperature, chemical changes, wind stress effects and circulation changes.

6) The ecological models that are being applied to consider present day fluxes cannot yet be used to investigate how the various factors (e.g temperature, CO_2, pH, circulation patterns etc.) that may be affected by climate changes will affect the ecosystems. Ecological models will be required which address ecological processes operating over the scales affected by climate change scenarios.

7) Many of the problems that modellers are likely to encounter over the next decade in attempting carry out global extrapolations relate to the questions of data acquisition, access, handling and management.

References

Allen TFH, Starr TB (1982) Hierarchy: perspectives for ecological complexity. University of Chicago Press Chicago. 310p

Bacastow R, Maier-Reimer E (1990) Ocean-circulation model of the carbon cycle. Clim Dynam 4:95-125

Bacastow R, Maier-Reimer E (1991) Dissolved organic carbon in modelling oceanic new production. Global Biogeochemical Cycles 5:71-85

Beckers JM (1992) The modelling and simulation of ocean's circulation on supercomputers. In: Melli P, Zannetti P (eds) Environmental modelling. Elsevier London. p 319-343

Bettello G, Bergamasco A (1992) A multilevel model for the study of the dynamics of the North Adriatic Sea. In: Melli P, Zannetti P (eds) Environmental modelling. Elsevier London. p 307-317

Böning CW, Cox MD (1988) Particle dispersion and mixing of conservative properties in an eddy-resolving model. J Phys Oceanogr 18:320-338

Broecker WS, Peng T-H (1982) Tracers in the sea. Lamont-Doherty Geological Observatory Palisades New York. 690 pp

Bryan F (1986) High latitude salinity effects and interhemispheric circulations. Nature 323: 301-304

Bryan K, Lewis KJ (1979) A water mass model of the world ocean. J Geophys Res 84: 2503-2517

Bryan FO, Holland WR (1989) A high resolution simulation of the wind- and thermohaline-driven circulation in the North Atlantic. In: Muller P, Henderson D (eds) Parameterization of small-scale processes. Proceedings of the A 'Aha Huliko'a Hawaiian workshop. University of Hawaii, Honolulu. p 99-115

Chervin RM, Semtner AJ (1990) An ocean modelling system for the supercomputer architectures of the 1990's. In: Schlesinger ME (ed)

Climate-ocean interactions. Kluwer Academic Press Boston. p 87-95
Ducklow HW, Harris RP (eds) (1992) Joint Global Ocean Flux Study North Atlantic Bloom Experiment. Deep-Sea Res Suppl Issue 20
Emerson S, Hughes JI (1988) Processes controlling the organic carbon content of open ocean sediments. Paleocean 3:621-634
Fasham MJR, Ducklow HW, McKelvie SM (1990) A nitrogen-based model of plankton dynamics in the oceanic mixed layer. J Mar Res 48:591-639
Fasham MJR, Sarmiento JL, Slater RD, Ducklow HW, Williams R (in press) Ecosystem behaviour at Bermuda station "S" and Ocean Weather Station "India": a GCM model and observational analysis. Global Biogeochemical Cycles (in press)
Fokin SA (1989) The use of bomb-tritium for testing the performance of a global ocean model. In: Modelling of global climate change and variability. Proceedings of an International Conference, 11-15 Sept. Hamburg University Hamburg. p 55
FRAM Group (1991) An eddy-resolving model of the southern ocean. Eos Trans 72:174-175
Giorgi F, Mearns LO (1991) Approaches to the simulation of regional climate change : a review. Rev Geophys 29:191-216
Gorchakov VA, Maslova NB, Ryabchenko VA (in press) Application of the averaging method for solution to the nonstationary ocean hydrothermodynamic equations. Ocean Modelling (in press)
Gregg WW, Walsh JJ (1992) Simulation of the 1979 spring bloom in the mid-Atlantic bight: a coupled physical/biological/optical model. J Geophys Res 97:5723-5743
Harris GP (1986) Phytoplankton ecology: structure, function and fluctuation. Chapman and Hall London. 384pp
Haury LR, McGowan JA, Wiebe PH (1978) Patterns and processes in the time-space scales of plankton distributions. In: Steele JH (ed) Spatial pattern in plankton communities. Plenum Press New York. p 277-328
Henderson-Sellers A, McGuffie K (1987) A climate modelling primer. John Wiley & Sons Chichester. 217pp
Hofmann EE (1988) Plankton dynamics on the outer southeastern U.S. continental shelf, part III, A coupled physical-biological model. J Mar Res 46:919-946
Hofmann EE, Ambler JW (1988) Plankton dynamics on the outer southeastern U.S. continental shelf, part II, A time-dependent biological model. J Mar Res 46:883-917
Houghton JT, Jenkins GJ, Ephraums JJ (1990) Climate change: The IPCC scientific assessment. Cambridge University Press Cambridge. 365pp
Ishizaka J (1990a) Coupling of coastal zone color scanner data to a physical-biological model of the Southeastern U. S. continental shelf

ecosystem. 1. CZCS data description and lagrangian particle tracing experiments. J Geophys Res 95:20167-20181

Ishizaka J (1990b) Coupling of coastal zone color scanner data to a physical-biological model of the Southeastern U.S. continental shelf ecosystem. 2. An eulerian model. J Geophys Res 95:20183-20199

Ishizaka J (1990c) Coupling of coastal zone color scanner data to a physical-biological model of the Southeastern U.S. continental shelf ecosystem. 3. Nutrient and Phytoplankton fluxes and CZCS data assimilation. J Geophys Res 95:20201-20212

Ishizaka J, Hofmann EE (1988) Plankton dynamics on the outer southeastern U.S. continental shelf, Part I, Lagrangian particle tracing experiments. J Mar Res 46:853-882

Jørgensen SE (1992) Recent and future development in environmental modelling. In: Melli P, Zannetti P (eds) Environmental modelling. Elsevier London. p 351-372

JGOFS (1990) The Joint Global Ocean Flux Study -JGOFS- Science Plan. SCOR Halifax Canada. 61pp

Kagan BA, Ryabchenko VA, Fokin SA (1986) A two-layer circulation model of the world ocean with detailed description of the seasonal evolution of the upper mixed layer. Dkl Akad Nauk USSR. 291:669-703

Kagan BA, Ryabchenko VA (1992) The biogeochemical cycles in the ocean. JGOFS meeting, 3-9 May 1992, Chateau de Bonas France.

Kiehl JT (1992) Atmospheric general criculation modeling. In: Trenberth KE (ed) Climate system modeling. Cambridge University Press Cambridge. p 319-369

Lancelot C, Billen G, Veth C, Becquevort S, Mathot S (1991) Modelling carbon cycling through phytoplankton and microbes in the Scotia-Weddell Sea area during sea ice retreat. Mar Chem 35:305-324

Legendre L, Demers S (1984) Towards dynamic biological oceanography and limnology. Can J Fish Aquat Sci 41:2-19

Maier-Reimer E, Hasselmann K (1987) Transport and storage of CO_2 in the ocean - an inorganic ocean-circulation carbon cycle model. Climate Dynamics 2:63-90

McClain C, Ishizaka J, Hofmann EE (1990) Estimation of the processes controlling variability in phytoplankton pigment distributions on the Southeastern U.S. continental shelf. J Geophys Res 97:20213-20235

Melillo JM, Callaghan TV, Woodward FI, Salati E, Sinha SK (1990). Effects on ecosystems. In: Houghton JT, Jenkins GJ, Ephraums JJ (eds.) Climate change: The IPCC scientific assessment. Cambridge University Press Cambridge. p 283-310

Moloney CL, Field JG (1991) The size-based dynamics of plankton food webs. I. A simulation model of carbon and nitrogen flows. J Plankton Res 13(5):1003-1038

Moloney CL, Field JG, Lucas, MI (1991) The size-based dynamics of plankton food webs. II. Simulations of three contrasting southern Benguela food webs. J Plankton Res 13(5):1039-1092

Mueller JL, Lange RE (1989) Bio-optical provinces of the Northeast Pacific ocean: a provisional analysis. Limnol Oceanogr 34:1572-1586

Murphy EJ, Morris DJ, Watkins JL, Priddle J (1988) Scales of interaction between Antarctic Krill and the environment. In: Sahrhage,D (ed) Antarctic ocean and resources variability. Springer-Verlag Berlin. p 120-130

Murphy EJ, Priddle JP (1992) Processes determining primary production in the Southern Ocean. Report of JGOFS modelling workshop October 1991. British Antarctic Survey (BAS) Cambridge UK. 62pp

Najjar RG (1992) Marine biogeochemistry. In: Trenberth KE (ed) Climate system modeling. Cambridge University Press Cambridge. p 241-280

Niebauer HJ, Smith WO (1989) A numerical model of mesoscale physical-biological interactions in the Fram Strait marginal ice zone. J Geophys Res 94:16151-16175

Oescheger, H, Siegenthaler, U, Schotterer, U, and A Gugelmann (1975) A box diffusion model to study the carbon dioxide exchange in nature. Tellus 27: 168-192

O'Neill RV, DeAngelis DL, Waide JB, Allen TFH (1986) A hierarchical concept of ecosystems. Monographs in population biology 23. (Series ed: May RM). Princeton University Press Princeton. 253 pp

Peng T-H, Broecker WS (1991) Dynamical limitations on the Antarctic iron fertilization strategy. Nature 349:227-229

Platt T, Sathyendranath S (1988) Oceanic primary production; estimation by remote sensing at regional and larger scales. Science 241:1613-1620

Platt T, Sathyendranath S (1991) Biological production models as elements of coupled, atmosphere-ocean models for climate research. J Geophys Res 96:2585-2592

Platt T, Caverhill C, Sathyendranath S (1991) Basin-scale estimates of oceanic primary production by remote sensing: the North Atlantic. J Geophys Res 96:15147-15159

Platt T, Sathyendranath S, Ulloa O, Harrison WG, Hoepffner N, Goes J (1992) Nutrient control of phytoplankton photosynthesis in the Western North Atlantic. Nature 356:229-231

Platt T, Sathyendranath S, Ravindran P (1990) Primary production by phytoplankton: analytic solutions for daily rates per unit area of water surface. Proc R Soc Lond B 241:101-111

Powell TM (1989) Physical and biological scales of variability in lakes, estuaries and the coastal ocean. In: Roughgarden J, May RM, Levin SA (eds) Perspectives in ecological theory. Princeton University Press Princeton. p 157-176

Rastetter EB, King AW, Cosby BJ, Hornberger GM, O'Neill RV, Hobbie (1992) Aggregating fine-scale ecological knowledge to model coarser-scale attributes of ecosystems. Ecological Applications 2(1):55-70

Ryabchenko, V A (1990) Seasonal variability of primary production in the world ocean (numerical results). Dokl Akad Nauk USSR. 313: 441-445

Sakshaug E, Slagstad D, Holm-Hansen O (1991) Factors controlling the development of phytoplankton blooms in the Antarctic Ocean - a mathematical model. Mar Chem 35:259-271

Sakshaug E, Slagstad D (in press) Sea ice and wind: effects on primary productivity in the Barents Sea. Atm Ocean (in press)

Sarmiento JL (1986) On the north and tropical Atlantic heat balance. J Geophys Res 91: 11677-11689

Sarmiento JL (1992) Biogeochemical ocean models. In: Trenberth KE (ed) Climate system modeling. Cambridge University Press Cambridge. p 519-551

Sarmiento JL, Orr JC, Siegenthaler U (1992) A perturbation simulation of CO_2 uptake in an ocean general circulation model. J Geophys Res 97: 3621-3645

Sarmiento JL, Slater RD, Fasham MJR, Ducklow HW, Toggweiler JR, Evans GT (in press) A seasonal three-dimensional ecosystem model of nitrogen cycling in the North Atlantic euphotic zone. Global Biogeochemical Cycles (in press)

Sathyendranath S, Platt T (1988) The spectral irradiance field at the surface and in the interior of the ocean: a model for applications in oceanography and remote sensing. J Geophys Res 93:9270-9280

Sathyendranath s, Platt T (1989) Computation of aquatic primary production: extended formalism to include effect of angular and spectral distribution of light. Limnol Oceanogr 34:188-198

Sathyendranath S, Platt T, Caverhill CM, Warnock RE, Lewis MR (1989) Remote sensing of oceanic primary production: computations using a spectral model. Deep-Sea Res 36:431-453

Schneider SH (1992) Intorduction to climate modeling. In: Trenberth KE (ed) Climate system modeling. Cambridge University Press Cambridge. p 3-26

Schott FA, Bøning CW (1991) The WOCE model in the western Atlantic: upper layer circulation. J Geophys Res 96:6993-7004

Semtner AJ, Chervin RM (1988) A simulation of the global ocean circulation with resolved eddies. J Geophys Res 93:15502-15522

Semtner AJ, Chervin RM (1992) Ocean general circulation from a global eddy-resolving model. J Geophys Res 97:5493-5550

Shaffer G (1989) A model of biogeochemical cycling of phosphorus,

nitrogen, oxygen and sulphur in the ocean: one step towards a global climate model. J Geophys Res 94:1979-2004
Shaffer G, Sarmiento J (in press) Biogeochemical cycling in the global ocean 1. A new analytical model with continuous vertical resolution and high latitude dynamics. Deep-Sea Res (in press)
Simonot J-Y, Dollinger E, Le Treut H (1988) Thermodynamic - biological -optical coupling in the oceanic mixed layer. J Geophys Res 93:8193-8202
Spall MA (1990) Circulation in the Canary Basin: a model/data analysis. J Geophys Res 95:9611-9628
Steele JH (1989a) Discussion: scale and coupling in ecological systems. In: Roughgarden J, May RM, Levin SA (eds) Perspectives in ecological theory. Princeton University Press Princeton. p 177-180.
Steele JH (1989b) The ocean 'landscape'. Landscape Ecology 3:185-192
Toggweiller JR, Dixon K, Bryan K (1989a) Simulations of radiocarbon in a coarse-resolution world ocean model, 1. Steady-state prebomb distributions. J Geophs Res 94:8217-8242
Toggweiller JR, Dixon K, Bryan K (1989b) Simulations of radiocarbon in a coarse-resolution world ocean model, 1. Distributions of bomb-produced carbon 14. J Geophys Res 94:8243-8264
Townsend, JR (ed) (1992) Improved Global Data for Land Applications. Global Change Report Number 20. 87pp
Tromp TK (1992) Global carbon cycles: a coupled ocean-atmosphere model. PhD Dissertation. Princeton University Princeton. 140pp
USJGOFS (1992) Modelling and data assimilation. U.S. JGOFS Planning Report Number 14. 28pp
Voltzinger, N.E. Klevanny, K.A., Pelinovsky, E.N. (1989). Long-wave dynamics of the coastal zone. Gidrometeoizdat Leninigrad. 300pp
Walsh JJ, Dieterle DA, Meyers MB (1988) A simulation analysis of the fate of phytoplankton within the mid-Atlantic bight. Cont Shelf Res 8:757-788
Watson AJ, Robinson C, Robinson JE, Williams PJ le B, Fasham MJR (1991) Spatial variability in the sink for atmospheric carbon dioxide in the North Atlantic. Nature 350:50-53
Wolf KU, Woods JD (1988) Lagrangian simulation of primary production in the physical environment - the deep chlorophyll maximum and nutricline. In: Rothschild BJ (Ed.) Towards a theory of biological-physical interactions in the world ocean. Kluwer Press Dordrecht. p 55-70
Woods JD, Onken R (1982) Diurnal variation and primary production in the ocean - preliminary results of a Lagrangian ensemble model. J Plankton Res 4:735-756
Wroblewski JS (1989) A model of the spring bloom in the North Atlantic and its impact on ocean optics. Limnol Oceanogr 34:1563-1571

Wroblewski JS, Hofmann EE (1989) U.S. interdisciplinary modelling studies of coastal-offshore exchange processes: past and future. Prog Oceanog 23:65-99

Wolfe GV, Bates TS, Charlson RJ (1991) Climatic and environmental implications of biogas exchange at the sea surface: modelling DMS and the marine biolgic sulfur cycle. In: Mantoura RFC, Martin, J-M, Wollast, R (eds) Ocean margin processes in global change. John Wiley & Sons Chichester. p 383-400.

FLUCTUATIONS : A TASK PACKAGE FOR THE PHYSICISTS

Véronique C. Garçon
UMR39, GRGS/CNRS
18 Avenue E. Belin
31055 Toulouse Cedex
France

H. Baumert[1], W. Schrimpf[2] , J. D. Woods[3]

This paper aims to give a list of questions that the biogeochemical modelling community will pose for physicists in the context of the design and development of a "Model of Ocean Biogeochemical Processes". It has no pretensions to being all-encompassing, rather it surveys the physical topics we see as the most relevant to the global ocean flux study and which as a whole address all time and space scales. Oceanographic phenomena are expressed over scales ranging across many orders of magnitude, from seconds to millennia, and from microns to megameters (see Table 1 below), and variation is observed on all scales.

In order to address the basic difficulties and modelling techniques for the consideration of fluctuations in biogeochemical ocean models as well as the variety of specific fluctuations in the global ocean, we chose to discuss the following problems:

1. Fluctuations and particles with memory: difficulties and techniques
2. Fluctuations and particles without memory: sub-grid scale (SGS) heterogeneities
3. Seasonal boundary layer
4. The permanent thermocline : ventilation, gyres and eddies
5. Ocean-atmosphere coupled models

---

[1] Universität Hamburg, Institut für Meereskunde, Troplowitzstr. 7, 2000 Hamburg 54, Germany
[2] Institute for Remote Sensing Applications, Joint Research Center, Commission of the European Communities, I-21020 Ispra (VA), Italy
[3] Linaere College, University of Oxford, Oxford OX1 3JA, United Kingdom

NATO ASI Series, Vol I 10
Towards a Model of Ocean Biogeochemical Processes
Edited by G T Evans and M. J. R Fasham

6. Geostrophic turbulence
7. Deep water circulation
8. Mesoscale turbulence
9. Ocean-ice interactions
10. Slope/open ocean exchanges

This division is functional rather than unique and the tasks are far from being independent, yet it does provide a convenient demarcation for the present purpose. This list involves most of the critical issues of the World Ocean Circulation Experiment (WOCE). For each of these tasks, we intend to provide the relevant references rather than a complete review, and to identify weaknesses and future avenues of development.

Table 1. Scales of oceanographic variability

Scale	Dimension	
	Time	Space
Small-scale	Min - Hours	cm - 100 m
Local	Hours - Days	1 - 10 km
Mesoscale	Days - Months	10 - 100 km
Regional	Weeks - Years	100- 1000 km
Basin	Seasons - Years	1 000 km
Global	Decades and longer	10 000 km

1. Fluctuations and Particles With Memory: Difficulties and Techniques

The implications of the fluctuations of the marine environment and of the biological non-linearities for primary production estimates and biodynamics are well established (Woods, 1988, Yamazaki and Osborn, 1988). This situation is in principle also known from classical theories of non-linear chemical reactions, where the relaxation of the reacting particles with respect to environmental fluctuations is assumed to be fast. Unfortunately, the internal relaxation time of microorganisms is in many cases not small compared with the temporal characteristics of the fluctuations of the marine environment:

some microorganisms have a significant "memory". In contrast to classical non-linear chemistry, this has the fundamental consequence that a pure continuum description of an ecosystem in terms of concentrations and concentration-based rate constants is useful only as a first approximation: when two phytoplankton cells exchange their place on the vertical coordinate due to turbulence, the concentration distribution of cells remain unchanged while each of the two cells experiences different events: the lower cell moves from the dark into the light, and the upper *vice versa*. Due to their internal inertia, both cells are averaging the experienced light on their path. If the light intensity were a linear function of the vertical coordinate, the result in the temporal average (e.g., after a new exchange of places) would be the same for both cells and the same as when they were at rest. Unfortunately, light is non-linearly (exponentially) distributed in the vertical and therefore an effect remains which cannot be covered by the classical continuum picture.

Therefore, biogeochemical ocean modelling is faced with the following specific features:

- non-linear interactions of the particles with each other and with the environment,
- irregular fluctuation of the environment in space and time,
- internal inertia of the reacting particles.

There is only one way to overcome all three above difficulties: to give up the continuum picture of the population and to replace it by a discrete-cell description, i.e., to write down dynamic equations with fluctuating forces (Langevin equations) for each cell and to tackle them statistically.

In a first approximation, this leads to Fokker-Planck equations (FPE) which are parabolic partial differential equations. The "classical" concentration distributions can be found as special (simplified) solutions while more general solutions contain the above sophisticated effects.

There are several approximation techniques to solve FPEs in such cases. Besides finite difference and finite element methods, the Lagrangian ensemble method pioneered for marine ecological problems by Woods (Woods and Onken, 1982, Woods and Barkmann, 1986) is a robust numerical simulation method of high efficiency when implemented on modern computers. By this technique Wolf and Woods (1988) simulated successfully the complicated dynamical interplay between mixed layer physics, upwelling events, phytoplankton production and nutricline with special emphasis on the deep chlorophyll maximum.

For special boundary conditions of the FPEs, analytical approximations were found by Baumert (1988) who derived a new formula for the column phytoplankton production considering the effects of cell motion due to "coloured" turbulence, exponential light distribution and inertia of the photosynthetic apparatus including photoadaptation. In the limit of low turbulence, the formula approaches *Talling's* primary production integral. In the limit of strong turbulence, it approaches the production formula of *Riley*.

2. Fluctuations and Particles without Memory: Sub-Grid Scale (SGS) Heterogeneities

Even if the memory of the assimilating cells can be neglected, the mathematical description of the interaction of environmental fluctuations and non-linearities of the reaction rates is still a non-trivial problem: today's large-scale to global-scale numerical models of biogeochemical processes consist of many discrete spatial boxes. Computer limitations make it unlikely that the grid spacing in models will be less than several tens of kilometers. Because such large boxes are not homogeneous either in their biogeochemical or their hydrophysical properties and due to the non-linear character of many biogeochemical interactions, it is crucial to be able to take the sub-grid scale heterogeneities or fluctuations in space into account (see the discussion of phytoplankton patchiness and production in section 8: Mesoscale turbulence below). Similar arguments can be applied to the fluctuations in time: numerical models use discrete time steps with an order of magnitude from minutes to months, during which the physical and biogeochemical variables and the external forces are assumed to be constant. Fluctuations with small characteristic times in relation to the time step can not be resolved. These phenomena have also to be considered properly.

In the purely physical momentum balance equations the consideration of fluctuations is an old problem: turbulence is considered there by means of the (infinite) Friedman-Keller expansion and closure assumptions of different order (see, e.g., Mellor and Yamada, 1974, Rodi, 1987, Aldama, 1990). Similar approaches are used in other physical sciences such as kinetic theory of gases and plasmas. In geoscience these methods are mainly applied in stochastic subsurface hydrology (e.g., McLaughlin and Wood, 1988). This approach can also be used for biogeochemical equations.

For illustrating the method by an example, the following biogeochemical rate equations for an internally homogeneous and isolated box model shall be considered:

$$\frac{dX}{dt} = \Phi(X,Y),\ \frac{dY}{dt} = -\Phi(X,Y),\ X(t) + Y(t) = E = \text{const.} \qquad (1)$$

Here Y is the concentration of a dissolved limiting nutrient, X the particulate concentration of the nutrient (i.e., the biomass), and E is its total concentration. In the above example, E is assumed to be a constant.[1] Equations (1) can be interpreted as a general model for phytoplankton spring blooms in isolated mixed layers where horizontal exchanges are neglected. For simplicity, turbulent fluctuations are not considered.

Usually the specific form of Φ in (1) is derived from laboratory experiments with homogeneous distributions of X and Y through the box (e.g., Bannister, 1979, Kiefer and Reynolds, 1992). If Y fluctuates spatially within the box, then X will also fluctuate in space accordingly and both variables become functions of the space coordinate r. In this case (1) must be rewritten as follows:[2]

$$\frac{\partial X}{\partial t} = \Phi(X,Y),\ \frac{\partial Y}{\partial t} = -\Phi(X,Y),\quad X(r,t) + Y(r,t) = E(r). \qquad (2)$$

Small spatial deviations from the spatial mean can be introduced by

$$\delta X = X - \overline{X},\ \ \delta Y = Y - \overline{Y},\ \ \delta E = E - \overline{E},$$

where the overbar denotes the average over the box, i.e., over the spatial fluctuations, and a Taylor series expansion of (2) reveals the following:

---

[1] In mathematical terms, equations (1) are ordinary differential equations and Φ is an ordinary function of the real variables X and Y.

[2] Now, equations (2) are partial differential equations. In the particular case that Φ contains a spatial operator like $\partial/\partial r$ (e.g., describing advective or diffusive exchanges with neighbor boxes), Φ is a real-valued functional.

$$\frac{\partial(\overline{X}+\delta X)}{\partial t} = \Phi(\overline{X},\overline{Y}) + \delta X\Phi_x + \delta Y\Phi_y + \frac{1}{2}\delta X^2\Phi_{xx} + \delta X\delta Y\Phi_{xy} + \frac{1}{2}\delta Y^2\Phi_{yy} \quad (3)$$

$$+ \text{ higher order terms}$$

where $\Phi_x = \partial\Phi/\partial x$ etc. denotes a partial derivative of the biogeochemical reaction function Φ evaluated at $X=\overline{X}$, $Y=\overline{Y}$.[3] When averaging (3) over the box, all terms linear in the perturbations become zero and (3) may be rewritten as ordinary differential equations for the box-averaged variables $\overline{X}$ and $\overline{Y}$,

$$\frac{d\overline{X}}{dt} = \Phi(\overline{X},\overline{Y}) + \frac{1}{2}\varphi, \quad \frac{d\overline{Y}}{dt} = -(\Phi(\overline{X},\overline{Y}) + \frac{1}{2}\varphi), \quad \overline{X} + \overline{Y} = \overline{E}, \quad (4)$$

where φ represents a new additional term due to the joint effect of spatial fluctuations and non-linearities:

$$\varphi = \overline{\delta X^2}\Phi_{xx} + 2\overline{\delta X\delta Y}\Phi_{xy} + \overline{\delta Y^2}\Phi_{yy} + \text{ higher order terms} \quad (5)$$

The function φ includes biogeochemical aspects (Φ_{xx},...) as well as correlators of the fluctuations, e.g. $\overline{\delta X^2}$. Unfortunately, the time evolution of these correlators and higher order terms are not given by the solution of (4). Mathematically speaking, the system (4,5) is not closed. This is in analogy to the turbulence closure problem for the momentum equations where it is solved by the use of assumptions and/or empirical information. In the present example the system (3,4) is closed assuming that all terms higher than second order can be neglected. It is easy to show that the system (3,4) closed by this assumption reads

$$\frac{d\overline{X}}{dt} = \Phi(\overline{X},\overline{Y}) + \underbrace{\frac{1}{2}\overline{\delta X^2}(\Phi_{xx} - 2\Phi_{xy} + \Phi_{yy}) + \overline{\delta E\delta X}(\Phi_{xx} - \Phi_{yy})}_{=\varphi}, \quad (6.1)$$

[3] When Φ contains a spatial differential operator (see footnote 2), the expansion method used here is of the Taylor-Frechet type and the partial derivatives are Frechet derivatives in the sense of functional analysis (see, e.g., Milne, 1980 or McLaughlin and Wood, 1988, Appendix A).

$$\frac{d(\overline{\delta X^2})}{dt} = 2(\ \overline{\delta X^2}(\Phi_x\ - \Phi_y) + \overline{\delta E \delta X}\Phi_y) \qquad (6.2)$$

$$\frac{d(\overline{\delta E \delta X})}{dt} = \overline{\delta E^2}\Phi_y + \overline{\delta E \delta X}\ (\Phi_x - \Phi_y). \qquad (6.3)$$

Note that the difference between the equations for the spatially homogenous case (1) and the equations (6) for the averaged variables in the spatially heterogeneous case may lead even to a qualitative different behavior of the solutions. The additional term φ might be interpreted in a popular fashion as "feeding by fluctuations" or "starvation by fluctuations", depending on the sign of φ. Temporal fluctuations of "external" variables such as light can be treated in a similar way, leading analogously to additional terms in the dynamic equations like φ in (6).

The question arises whether the term φ and the dynamic closure equations for $\overline{\delta X^2}$ and $\overline{\delta E \delta X}$ (in the above example) are of relevance or not, whether they should be included in the simulation algorithm or not. This can only be decided in cooperation between biogeochemists and physical oceanographers: The task of the biogeochemist is to estimate the second derivatives (Φ_{xx},...) of "his" trophic interaction function Φ.The physical oceanographer is asked to provide estimates of the above correlators by means of observations (remotely sensed data, ship observations,etc...) and/or numerical models.

3. The seasonal boundary layer

This can be defined as the portion of the water column which is vertically mixed at the end of the cooling season through the action of the winds and wintertime overturning. It includes the mixed layer, the diurnal thermocline and the seasonal thermocline. The mixed layer is a buffer zone between the higher frequency scales of solar and atmospheric forcing and the slow, large scale circulation of the thermocline and deep waters of the ocean. Here the turbulent kinetic energy (TKE) plays the role of a main control variable for the mixed layer depth, etc.

This buffering action is critical for the chemical and biological components, as for instance in the regulation of the partial pressure of carbon dioxide in seawater. Plankton consume some of the dissolved CO_2 in the mixed layer and sequester carbon into the thermocline and deep waters away from the immediate influence of high frequency processes at the ocean-atmosphere interface. It must be noted that also for these assimilation processes the turbulence intensity can play a major role (Riebesell et al., 1992). The seasonal layer water gradually leaks into the permanent thermocline.

For biogeochemical modelling it is a difficult but crucial task to resolve the dynamics of the seasonal boundary layer explicitly by models and to couple them with biological mixed layer models.

Due to the complicated nature of mixed layer dynamics including turbulence, these processes are mainly described by means of vertical one-dimensional models. This corresponds to the experience that observations of local biological production and local physical variables such as the vertical distributions of currents, shear, stability, TKE, temperature and eddy diffusivity exhibit strong correlations in important regions of the world ocean (see, e.g., Wroblewski, 1989).

Today, the one-dimensional mixed layer models in broader use can be described by the conceptual approaches they follow. Integral or bulk models (Kraus and Turner, 1967, Niiler and Kraus, 1977, Garwood, 1977, Miyakoda and Rosati, 1984, Martin, 1985 and Gaspar, 1988) rely on the idea that the water column can be divided on the vertical into two (three) areas separated by movable membranes which are permeable only for light: the turbulent mixed layer at the surface (ML), the viscous transit layer below the ML (and, for shallow seas, the bottom layer where tidal friction generates bottom turbulence). The variables within the ML and the ML depth are governed by balance equations of heat, TKE and salinity for the whole ML. The fluxes of these quantities through the sea surface, and -in the interior of the ML- the production of heat via light absorption and the generation and dissipation of TKE, are empirical functions of meteorological and ML variables. In these models, entrainment and detrainment at the bottom of the ML are consequences of the motions of the above "membranes" at the bottom of the ML and at the top of the bottom layer.

The second class of models are differential models. There are several model groups within this wide class. One classification criterion is the level of turbulence description in the sense of Mellor and Yamada (1974). It may serve as a guideline but does not cover all models.The Mellor-Yamada classification generally assumes that the turbulent length scale can be described by algebraic expressions or by an integral of the TKE while some models use also transport equations for the length scale or a corresponding quantity, e.g., the k-ε models used by Rahm and Svensson (1989) and by Baumert and Radach (1992).

Generally, this second class of mixed layer models resolves the vertical distributions of the state variables in detail. A first model group uses vertical turbulent diffusion equations only for momentum, heat and salinity (Mellor and Durbin, 1975, Kochergin, 1987, Frey, 1991). In a second model group the vertical diffusion equations are also applied to TKE (Klein and Coantic, 1981, Mellor and Yamada, 1982, Richards, 1982, Martin, 1985, Price et al., 1986). These models are called the Second Moment Closure models (SMC, Mellor, 1989). The model of Price et al. (1986) represents an interesting combination of the bulk and the vertically resolved description.

A key problem in all SMC models is the adequate parameterization of how the coefficient of vertical diffusion depends upon the vertical density gradient, mostly done via the Richardson or flux Richardson number. Vertical convection is not explicitly described by this model class. Another general problem for all SMC models is the formulation of the TKE boundary conditions at the water surface which is strongly controlled by wave action. Important contributions on this topic were given by Woods (1980) and Kitaigorodskij and Lumley (1983).

A great deal of work remains to be done in order to take into account mixing generated remotely (by storms for instance) but effective locally through the breaking action of internal waves. Similarly we can no longer ignore the biological feedbacks into the physics: phytoplankton reduce the penetration of solar radiation into the upper ocean thereby modifying the heating profile and the vertical stability of the water column.

At high latitudes, sea ice becomes an important control variable for the dynamics of the seasonal boundary layer. One-dimensional boundary layer models which include sea-ice thermodynamics are discussed in section 9: Ocean-ice interactions below.

4. The permanent thermocline : ventilation, gyres and eddies

A zone of enhanced vertical temperature and density gradient exists in the upper kilometer of the world ocean; this is called the permanent thermocline. Since the velocity field of wind-driven gyres is closely linked to the structure of the thermocline, the study of the three-dimensional structure of the wind-driven circulation is also called the thermocline problem. Observing and modelling the thermocline circulation are impeded by the relatively weak signal of the permanent flow compared to the energetic ocean eddies. Eddies contain around 99% of the ocean's kinetic energy. In addition most of the potential-kinetic energy interchange occurs at the smaller scales. Physical and chemical properties of the warm water sphere are determined by the gradual leakage from the seasonal boundary layer over many years. Western boundary currents, large-scale mid-latitude gyres and mesoscale eddies are all part of the thermocline circulation.

The main characteristic of the thermocline problem is that the corresponding circulation is no longer dominated by vertical mixing but is in near geostrophic balance, the perturbations being induced by convergences and divergences due to surface wind and buoyancy forcing and inherent instabilities in the mean flow.

There is a whole hierarchy of model approaches. The low vertical resolution models limit the physics and the vertical resolution : the reduced gravity model in the context of equatorial modelling is an example. Numerical efficiency of such models enables truly eddy resolving horizontal resolutions (Kindle and Thompson, 1989).

The quasi-geostrophic (QG) models are based on the principle that thermocline motions are in near geostrophic balance and conserve potential vorticity. Numerical models consistent with this principle have been developed (Holland, 1978, Holland et al., 1983, McWilliams et al., 1978). The QG models have made valuable contribution to the understanding of mesoscale and gyre-scale processes, a knowledge which is now being successfully applied to eddy-resolving OGCMs (The FRAM Group, 1991, Böning et al., 1991).

Gent and McWilliams (1984) have derived a set of balanced model equations by combining an exact heat equation with truncated vorticity and divergence equations, reducing the degrees of freedom of the primitive equations by 1/3. The model resolves mesoscale turbulence and diabatic processes. It can be a viable alternative for climate studies, for instance in the study of sub-grid scale parametrization.

Another class, the isopycnic models, make explicit use of the fact that the preferred plane of flow is along isopycnic surfaces; the fixed vertical grid of the primitive equation models is replaced by a set of isopycnic surfaces, in essence a Lagrangian approach to the vertical coordinate (see Huang, 1991). Parsons (1969) first used a two-layer model to describe the separation of the Gulf Stream. Bogue et al. (1986) have compared the Parsons model solution and a one-layer model for isopycnal outcropping with good agreement. Bleck and Boudra (1981) have used an isopycnic model to study the dynamics of the Agulhas retroflection and mesoscale frontogenesis. Huang and Bryan (1987) used an isopycnic model to study wind driven gyres in a non eddy resolving climate framework with substantial isopycnal outcroppping. This model is able to produce the basic features of thermocline ventilation and water mass formation. Woods (1985) examined the connection between subducted waters and penetration of wintertime mixing. Oberhuber et al. (1991) developed an isopycnic model in a coupled atmosphere-ocean-cryosphere context. Smith et al. (1990) carried out experiments with a seasonal wind-driven isopycnic coordinate model. They found the model capable of reproducing the major features of the North Atlantic circulation, including western boundary current separation and a Labrador current.

Much remains to be learnt of the fundamental processes by which water masses acquire their essential characteristics. Temporal effects and subsurface buoyancy forcing are two aspects to investigate further. Uncertainty in the lateral open boundary conditions, and in the initial conditions remains a problem for regional models.

5. Ocean-atmosphere coupled models

We just mention here ocean-atmosphere coupled models because information from the interface with the atmosphere is very important in defining the ocean's role in climate change and also because global air-sea flux estimates are needed with the best accuracy. Meehl (1990) gives an excellent review of the state of coupled climate modelling. Presently coupled model studies are concerned with issues of climate change. Deficiencies in the oceanic and atmospheric components have an immediate impact on the way such climatic integrations reach equilibrium. The separate components differ in their simulations of the surface heat and freshwater fluxes and differ from the "best" observed estimates so that the coupled system tends towards an unrealistic climate. The usual solution is to adjust the surface exchanges to account for this mismatch, usually referred to as "flux correction" or "flux adjustment " (Manabe and Stouffer, 1988). This mismatch might be due, on the oceanic side, to poor resolution, uncertainty in eddy parameterization, poor upper ocean physics and poor representation of the deep thermohaline field.

6. Geostrophic turbulence

Oceanic motions span a large range of spatial and temporal scales. The small-scale motions such as microstructure turbulence and locally generated gravity waves are not affected by the rotation of the earth. While some motions with time scales near the inertial period, such as tidal waves and inertial oscillations are affected by the rotation of the earth, the relative acceleration is unimportant to their dynamical balance. Between these motions and the basinwide circulation is the domain of the so-called mesoscale variability, which exists on length scales on the order of or greater than the baroclinic radius of deformation, and on time scales larger than the inertial period. Rossby waves, mesoscale ocean eddies and fronts fall in this class of motions.

The major characteristic of mesoscale motions is the near-geostrophic balance whereby the Coriolis and pressure gradient forces dominate and almost balance each other. The variability of ocean eddies, namely their generation, drift, merging and breaking and their interaction with Rossby waves and fronts, defines geostrophic turbulence (Charney, 1971). More generally, geostrophic turbulence refers to chaotic, nonlinear motions subject to a state of near-geostrophic balance. At the root of the quasi-geostrophic (QG) formalism is the approximation of near-geostrophic balance, resulting from a relatively small Rossby number. Mathematically, the number is made small by assuming weak vertical displacements of isopycnals compared to the vertical scale of the stratification.

The study of geostrophic turbulence has traditionally been limited to the use of the QG equation (Rhines, 1975, Salmon, 1980, McWilliams, 1984). Recently Cushman-Roisin and Tang (1990) were the first to use a generalized geostrophic equation to study geostrophic turbulence. Their study revealed certain phenomena previously missed by the QG studies, namely the equilibration of the geostrophic turbulence at some intermediate scale and the preferred emergence of anti-cyclones. Their study, however, was restricted to the reduced gravity system (one moving layer), which excludes baroclinicity, a key element in geostrophic turbulence. Tang and Cushman-Roisin (1991) extended this study to the two-layer ocean using the generalized geostrophic equations. These equations offer certain advantages in the study of geostrophic turbulence. On one hand, unlike the original primitive equations, they filter out gravity-inertia motions, on the other hand, in addition to the QG regime, they also cover a wide variety of other regimes. They are thus suitable as a model of geostrophic turbulence where a flow field may contain motions on several amplitudes and length scales or may evolve in time from one regime to another.

7. Deep water circulation

The motivation for considering the deep waters as a separate component is provided by their role in long-term climate change where they are considered as the memory of the climate system. It is important to keep in mind that the large-scale circulation is subject to forcing by the faster time-scale components considered previously. In general the deep ocean is the domain of Ocean General Circulation Models (OGCMs) and inverse models since simulation and understanding of the "equilibrium" state is the primary consideration. In fact, the deep ocean is in a continuous state of slow change.

The beginning of numerical modelling for large scale ocean circulation started with the work of Bryan (1963, 1969). Cox (1975) first attempted to model the baroclinic response of ocean models to wind forcing in realistic domains. Resolution was necessarily coarse (2°), integration periods were short and horizontal eddy parameterizations were over-dissipative. However, the main features of the circulation were captured, such as western boundary currents and the Antarctic Circumpolar Current. Bryan and Lewis's (1979) work remained a benchmark for stand-alone global ocean simulations for many years. Cox (1985) made readily available the GFDL (Geophysical Fluids Dynamics Laboratory, Princeton) primitive equation code within the ocean modelling community.

Nowadays it is important that a variety of models be used if only to ensure that models intercomparisons are meaningful and not simply a self-congratulatory exercise. The Semtner and Chervin (1988), Hasselman (1982), Maier-Reimer et al. (1982) and Haidvogel et al. (1990) models are good examples of such a diversity.

The WOCE Community Ocean Modelling Effort (CME) aims to establish a series of baseline calculations of the wind and thermohaline driven large-scale ocean circulation and to critically evaluate the performance of the state-of-the-art ocean circulation models through direct comparisons with observations and identify those areas where improvements are needed. The initial focus is an eddy-resolving, thermodynamically active basin-scale model of the North Atlantic circulation and the distribution of passive tracers within it. The GFDL primitive equation code (Bryan-Semtner-Cox model) is used in both medium (1°, 20 levels) and fine (1/3°, 30 levels) resolution configurations.

The Fine Resolution Antarctic Model (The FRAM Group, 1991) is a primitive equation model of the Southern ocean, extending from 24°S to the Antarctic continent and is developed from the Cox (1985) version of the GFDL code. The horizontal resolution is 1/2° longitude by 1/4° latitude, there are 32 levels thus enabling an explicit resolution of the eddies.

The Semtner and Chervin (1988) model is the first global eddy-resolving study with a resolution of 1/2° by 1/2° globally. The model provides a global map of the mean and eddy circulations in quite good agreement with existing observations (e.g. SEASAT altimeter data).

Cox (1989) and England (1992) used the GFDL code to investigate the processes by which the major water masses of the world ocean are formed.

All the initiatives discussed above derived from the GFDL model, although each included enhancements and modifications in both the physics and numerical design. Alternate OGCMs configurations exist.

In Hasselman (1982) and Maier-Reimer et al. (1982) models, the circulation is divided into a barotropic component with associated surface displacement and a baroclinic component which includes prognostic equations for temperature, salinity and other tracer fields of interest. It is computationally very effective and reveals a good insight into the causality of ice-age periods (Heinze and Maier-Reimer, 1991).

Haidvogel et al. (1990) developed a semi-spectral primitive equation ocean circulation model. The equations of motion are the hydrostatic primitive equations, much as in the GFDL code, with an explicit Price et al. (1986) style mixed layer at the surface. There is a sigma coordinate system in the vertical and a horizontal orthogonal curvilinear coordinate system for improved handling of lateral boundaries. Hofmann et al. (1991) used it to generate circulation fields in the coastal transition zone off the coast of California. This model has not been applied yet on a basin or global scale.

The further development of models for the deep ocean is undoubtedly linked to progress in each of the previous listed tasks : surface forcing (air-sea fluxes of momentum, heat and water), seasonal boundary layer physics and ventilation of the thermocline.

8. Mesoscale turbulence

Understanding the physical processes that influence horizontal variation in primary production on scales of kilometers to megameters is a key issue in models of climate change associated with variation of atmospheric CO_2 : e.g., primary production is dominated by patchiness on scales of 1-100 km (Steele, 1978). Principal oceanographic features in this mesoscale range are internal waves and fronts (Woods, 1980). The influence of variation on scales smaller than the model grid (a few tens of kilometers) must be parameterized in terms of the larger structure (see section 2: Fluctuations and Particles without memory: SGS Heterogeneities). In particular, it is necessary to parametrize the effects on atmospheric CO_2 of primary production patches with smaller scale.

Recently there have been several attempts at understanding the way in which mesoscale upwelling influences primary production. MacVean and Woods (1980) have modelled vortex stretching in the mixed layer without entrainment, showing how the latter thins on the anti-cyclonic side of an intensifying mesoscale jet. As upwelling raises particles and nutrients towards the surface, the mixed layer draws away from them, delaying the onset of the extremely high surface concentration of chlorophyll in an otherwise oligotrophic regime. Wolf and Woods (1988) have used the one-dimensional Lagrangian ensemble method pioneered by Woods and Onken (1982) to model the response to upwelling of plankton growing in a seasonally oligotrophic regime. They showed that, if upwelling is slower than the sinking speed of the particles, their enhanced growth is sufficient to prevent the nutricline from rising into the mixed layer, and the result is an intensification of the deep chlorophyll maximum. Woods (1988) presented a dynamical theory for phytoplankton patchiness, which explains many features of the observed fine structure of chlorophyll concentration and opens the way to a statistical approach to predicting the large scale distribution of primary production and its interannual variability. The theory extends classical one-dimensional treatment of the role of density stratification in controlling primary production to three dimensions by adopting the isopycnic gradient of potential vorticity as the dynamic variable. A similar method has been used by Dippner (1991) to study the phytoplankton dynamics of the Adriatic sea.

9. Ocean-ice interactions

For climate scales, it is clearly important to take into account ocean-ice interactions at high latitudes. Sea ice effectively insulates the ocean from incoming solar radiation and moderates the exchange of heat between the ocean and the atmosphere. Exchange is felt most strongly in the seasonal boundary layer but is also important for the rates of formation of deep and bottom water masses. Let us summarize the state of the art for ice modelling. Sea ice models usually consist of four parts :

- a surface energy balance which considers incoming and outgoing radiation, atmospheric heating, and upward conduction of heat through the ice to compute the surface temperature,
- an ice thermodynamic model for determining the rate of heat conduction through the ice,
- a momentum balance: atmospheric and ocean stresses, Coriolis forcing, sea surface tilt and internal ice strain are used to predict the ice velocity,
- a water budget which takes into account surface accumulation, internal movement, and melting to determine the spatial variability of ice thickness and concentration.

Models of Hibler (1979), Parkinson and Washington (1979) and Owens and Lemke (1990) are typical of ice models being used in coupled ocean-ice studies. Sophisticated sea ice models have so far only been applied to limited regions. To simulate sea-ice conditions, also the one-dimensional model type is in use (Ikeda, 1986, McPhee et al., 1987, Martinson, 1990, for a review see Gade, 1991). These models include also turbulence closures of higher order because turbulence is relevant for the heat transport across the water-ice interface. A drawback of the presently used SCM models is that these closures do not cover the so-called transient and viscous layer at the very interface where the flow becomes viscous. Here a progress can be made by application of special turbulence models for low Reynolds numbers (Patel et al., 1985).

Global models usually adopt a much simpler thermodynamic configuration and include a very simple advection of sea ice. Experiments with ice models coupled to mixed layer models clearly indicate that ocean-ice coupling is important for climate studies. Purely thermodynamical configurations, for global studies, are probably extra-sensitive to changes in boundary conditions; dynamic sea-ice models have reduced sensitivity because of thermodynamic-dynamic interactions which provide a stabilizing effect. Significant progress is needed towards a better understanding of ice dynamics and thermodynamics.

10. Slope/open ocean exchanges

The recognition that the ocean is a major factor in the global biogeochemical cycling of carbon has resulted in the need to quantify horizontal fluxes of biogenic material from continental margins to the ocean interior and vertical fluxes of this material from the surface to the ocean bottom. This last pathway may be critical to the dynamics of climate change on the time scale of decades (Longhurst and Harrison, 1989). Continental margins may be a significant source of organic carbon in mass balance carbon budgets constructed for the open ocean (Walsh et al., 1981, Walsh, 1983). A significant portion of the primary production in the ocean occurs in coastal regions. The key question is: what fraction of this organic production ultimately enters the subthermocline waters of the deep ocean? The answer of course depends on the dynamic interchange between coastal and oceanic waters and the transfer of dissolved and particulate matter from the shoreline, across the continental shelf over the continental slope into the open ocean, and *vice versa*. In particular upwelling areas like the East Atlantic off Northwest Africa have to be considered, as cool and nutrient rich subsurface waters cause high biological productivity on the shelf (Minas et al., 1982, Mittelstaedt, 1983). A summary of the physical characteristics of the upwelling areas in the world oceans is found in Sündermann (1986). Tides and local winds are the major driving forces over the shelf whereas the offshore part over the deep sea along the continental slope is governed by the open ocean circulation, which depends not so much on local winds and mesoscale eddies.

The interactions of open ocean and shelf sea processes, as well as the spatial and temporal variability of coastal winds, irregularities of the bottom topography and of the coastline, generate features, including upwelling, which are transient, three-dimensional mesoscale processes with variations along the coast. Due to the complex processes in the shelf/open ocean area, hydrodynamic and biogeochemical models should preferably be three-dimensional. The combination of shelf and deep sea in a model can require specific numerical treatment as the transition from deep to coastal waters can occur within a few kilometers.

Recently, several initiatives to study the exchange fluxes from shelves into the open ocean have been prepared for launching. LOICZ is an activity of the International Geosphere-Biosphere Programme on Land-Ocean Interactions in the Coastal Zone (IGBP, 1990). NOWESP (North West European Shelf Programme) is a CEC initiative to quantify the biogeochemical fluxes across the shelf-break (e.g., nutrients, plankton, carbon), to estimate their variability and to understand the nature of shelves as a link between land and ocean. Both programmes have a strong focus on modelling and will provide valuable boundary conditions for future global biogeochemical model studies.

First attempts at quantifying the flux of carbon from continental margins to the open ocean have already been carried out. Some of the physical mechanisms (and models) through which this carbon supply occurs will now be briefly discussed. Physical processes include shelfbreak upwelling, shelf-slope interactions, rings and mid-ocean eddies. Recently, Ishizaka and Hofmann (1988), with a Lagrangian particle tracking model, and Hofmann (1988), with a coupled biological-physical model, investigated the effect of Gulf Stream frontal eddies and bottom intrusions on biological production of the southeastern U.S.A. shelf. For shelf-break areas not directly influenced by intense currents like the Gulf Stream, horizontal exchanges between the shelf and slope waters can occur through low-frequency eddy fluxes (Houghton et al., 1978). Gulf Stream rings are energetic mesoscale eddies formed by the closure of a Gulf Stream meander. Formation of rings is a major mechanism of transport of water from one side of the Gulf Stream to the other side, in both directions. Ring-like eddies are associated with any major boundary current such as the Benguela current, the Agulhas current, etc. The Warm Core Rings Experiment (WRCE) and the Cold Core Rings Program were multidisciplinary programs in the early 80's to study the physics, chemistry and biology of a Gulf Stream warm-core ring over its lifetime and of cyclonic eddies in the Northwest Atlantic, respectively. Long-lived mesoscale eddies have been observed to cross ocean basins, carrying remnants of original water parcels thousands of kilometres before they dissipate by turbulence. Exchange processes also occur where fronts separate oceanic water masses with different physical, chemical and biological characteristics. Modelling the contribution of rings to property exchanges across ocean basins requires circulation models capable of resolving these mesoscale features such as those mentioned in section 7.

A significant contribution to the understanding of biogeochemical processes on the continental shelf and adjacent deep waters (and in general to the marine environment) can be obtained through information collected by remote sensing. The synoptic pictures provided by these data are well suited to discover and monitor spatial and temporal variability of state variables and processes in the surface layer of the world oceans (e.g. Van Camp et al., 1991). While remotely sensed data have been used mainly as "stand-alone" data up to now, the combination with biogeochemical models can improve and widen application and interpretation of these data (Gregg and Walsh, 1992).

As a concluding remark, we would like to stress the issue of models validation. One of the major goals in marine ecosystem modelling is to devise suitable techniques for parameter fitting and validation of the models and to collate suitable data sets for testing the models. Suitable data mean observations of parameters (physical and/or biogeochemical) whose evolution is simulated by the models.

Acknowledgements

Thanks are due to all participants of the NATO Advanced Research Workshop in Chateau de Bonas for their valuable discussions. The work is financially supported by a grant from the Centre National de la Recherche Scientifique (FRANCE-JGOFS, Modelling Program) to V.G. and by the Programme "Monitoring the Marine Environment" (Activity Sheet No. 73012/2AE1) at the Joint Research Centre of the Commission of the European Communities to W.S. H.B. thanks NATO for financial support with respect to the travel costs.

References

Aldama AA (1990) Filtering techniques for turbulent flow simulation.In: Lecture Notes in Engineering, Vol 56, Springer Verlag

Bannister TT (1979) Quantitative description of steady-state, nutrient saturated algal growth, including adaptation. Limnol Oceanogr 24:76-96

Baumert H (1988) Dr Sc Nat Thesis, Techn Univ of Dresden, Germany
Baumert H, Radach G (1992) Hysteresis of turbulent kinetic energy in nonrotational tidal flows : a model study. J Geophys Res 97 : 3669-3677
Bleck R, Boudra DB (1981) Initial testing of a numerical ocean circulation model using a hybrid (quasi-isopycnic) vertical coordinate. J Phys Oceanogr 11:755-770
Bogue NM, Huang RX, Bryan K (1986) Verification experiments with an isopycnal coordinate ocean model. J Phys Oceanogr 16:985-990
Böning CW, Döscher R, Budich RG (1991) Seasonal transport variations in the western subtropical North Atlantic : Experiments with an eddy-resolving model. J Phys Oceanogr 21:1271-1289
Bryan K (1963) A numerical investigation of a nonlinear model of a wind-driven ocean. J Atmos Sci 20:594-606
Bryan K (1969) A numerical method for the study of the world ocean. J Comput Phys 4:347-376
Bryan K, Lewis LJ (1979) A water mass model of the World Ocean. J Geophys Res 34: 2503-2517
Charney JG (1971) Geostrophic turbulence. J Atmos Sci 28:1087-1095
Cox MD (1975) A baroclinic numerical model of the World Ocean. In: Numerical models of Ocean Circulation. National Academy of Sciences, Washington DC, p 107
Cox MD (1985) An eddy resolving numerical model of the ventilated thermocline. J Phys Oceanogr 15:1312-1324
Cox MD (1989) An idealised model of the World Ocean. Part I. The global-scale water masses. J Phys Oceanogr 19:1730-1752
Cushman-Roisin B, Tang B (1990) Geostrophic turbulence and emergence of eddies beyond the radius of deformation. J Phys Oceanogr 20:97-113
Dippner JW (1991) A Lagrangian model of phytoplankton growth dynamics - a sensitivity analysis. CEC Report EUR 13371 EN, CEC Joint Research Centre Ispra, p 63
England MH (1992) On the formation of Antarctic Intermediate and Bottom Water in ocean general circulation models. J Phys Oceanogr 22:918-926
Frey H (1991) A three-dimensional, baroclinic shelf sea circulation model, part 1: The turbulence closure scheme and the one-dimensional test model. Cont Shelf Res 11:365-395
Gade HG (1991) When ice melts in sea water; a review of recent contributions to the study of dynamics and thermodynamics of melting ice sheets in seawater. Rep No 69, University of Berger, Norway
Garwood RW (1977) An ocean mixed layer capable of simulating cyclic states. J Phys Oceanogr 7:455-468
Gaspar P (1988) Modeling the seasonal cycle of the upper ocean. J Phys Oceanogr 18:161-180
Gent PR, McWilliams JC (1984) Balanced models in isentropic coordinates and the shallow water equations. Tellus 36A:166-171

Gregg WW, Walsh JJ (1992) Simulation of the 1979 Spring Bloom in the Mid-Atlantic Bight : A Coupled Physical/Biological/Optical Model. J Geophys Res 97:5723-5743

Haidvogel DB, Wilkin JL, Young R (1990) A semi-spectral primitive equation ocean circulation model using vertical sigma and orthogonal curvilinear horizontal coordinates. J Comput Phys 94:151-185

Hasselmann K (1982) An ocean model for climate variability studies. Prog Oceanogr 11:69-92

Heinze C, Maier-Reimer E (1991) The Hamburg oceanic carbon cycle circulation model. Dtsch. Klimarechenzentrum Hamburg Techn. Rep. No 2

Hibler WD III (1979) A dynamic thermodynamic sea ice model. J Phys Oceanogr 9:815-846

Hofmann EE (1988) Plankton dynamics on the outer southeastern U.S. continental shelf. Part III: A coupled physical-biological model. J Mar Res 46:919-946

Hofmann EE, Hedström KS, Moisan JR, Haidvogel DB, Mackas DL (1991) Use of simulated drifter tracks to investigate general transport patterns and residence times in the coastal transition zone. J Geophys Res 96(C8):15041-15052

Holland WR (1978) The role of mesoscale eddies in the general circulation of the ocean: Numerical experiments using a wind-driven quasi-geostrophic model. J Phys Oceanogr 8:363-392

Holland WR, Harrison DE, Semtner AJ Jr (1983) Eddy-resolving models of large-scale ocean circulation. In: Robinson AR (ed.) Eddies in Marine Science. Springer-Verlag, Berlin and New York, p 379

Houghton RW, Smith PC, Fournier RO (1978) A simple model for cross-shelf mixing on the Scotian shelf. J Fish Res Bd Can 35:414-421

Huang RX, Bryan K (1987) A multilayer model of the thermohaline and wind-driven ocean circulation. J Phys Oceanogr 17:1909-1924

Huang RX (1991) The three-dimensional structure of wind-driven ventilation and subduction. Rev Geophys (Suppl) 29:590-609

IGBP (1990) Coastal Ocean Fluxes and Resources. In: Holligan P. (ed.) International Geosphere Biosphere Programme : A study of the Global Change. Rept No 14, Ad Hoc Workshop, Tokyo

Ikeda M (1986) A mixed layer beneath melting sea ice in the marginal ice zone using a one-dimensional turbulent closure model. J Geophys Res 91:5054-5060

Ishizaka J, Hofmann EE (1988) Plankton Dynamics on the outer southeastern U.S. continental shelf. Part I : Lagrangian particle tracing experiments. J Mar Res 46:853-882

Kiefer DA, Reynolds R (1992) A general steady-state model for light absorption and growth rate of Skeletonema costatum. In: Proceedings of the NATO Advanced Research Workshop "Towards a Model of Ocean Biogeochemical Processes, Chateau de Bonas, Springer- Verlag

Kindle J, Thompson JD (1989) The 26- and 50-day oscillations in the western Indian Ocean : model results. J Geophys Res 94:4721-4736

Kitaigorodskij SA, Lumley JL (1983) Wave-turbulence interactions in the upper ocean. Part I: The energy balance of the interacting fields of surface wind waves and wind-induced three-dimensional turbulence. J Phys Oceanogr 13:1977-1999

Klein P, Coantic M (1981) A numerical study of turbulent processes in the marine upper layers. J Phys Oceanogr 11:849-863

Kochergin VP (1987) Three-dimensional prognostic models. In: Heaps NS (ed.) Three-dimensional coastal ocean models, Am Geophys Union, Washington D.C., p 201

Kraus EB, Turner JS (1967) A one-dimensional model of the seasonal thermocline, II. The general theory and its consequences. Tellus 19:98-106

Longhurst AR, Harrison WG (1989) The biological pump : Profiles of plankton production and consumption in the upper ocean. Prog Oceanogr 22:47-123

MacVean MK, Woods JD (1980) Redistribution of scalars during upper ocean frontogenesis: a numerical model. Q J roy meteor Soc 106:293-311

Maier-Reimer E, Hasselman K,Olbers D, Willebrand J (1982) An ocean circulation model for climate studies. Tech Rep, Max-Planck-Institut für Meteorologie

Manabe S, Stouffer RJ (1988) Two stable equilibria of a coupled ocean-atmosphere model. J Climate 1:841-866

Martin PJ (1985) Simulation of the mixed layer at OWS November and Papa with several models. J Geophys Res 90: 903-916

Martinson DG (1990) Evolution of the Southern Antarctic mixed layer and sea-ice; ocean deep water formation and ventilation. J Geophys Res 95:11641-11654

McLaughlin D, Wood E (1988) A distributed parameter approach for evaluating the accuracy of groundwater model predictions, Part I. Water Resources Res 7:1037-1047

McPhee MG, Mayku GA, Morison JH (1987) Dynamics and thermodynamics of the ice/upper ocean system in the marginal ice zone of the Greenland Sea. J Geophys Res 92:7017-7031

McWilliams JC (1984) The emergence of isolated coherent vortices in turbulent flow. J Fluid Mech 146:21-43

McWilliams JC, Holland WR, Chow J (1978) A description of numerical Antarctic circumpolar currents. Dyn Atmos Oceans 2:213-291

Meehl GA (1990) Development of global coupled ocean-atmosphere general circulation models. Clim Dyn 5:19-33

Mellor GL, Durbin PA (1975) The structure and dynamics of the ocean surface mixed layer. J Phys Oceanogr 5:718-728

Mellor GL, Yamada T (1974) A hierarchy of turbulence closure models for planetary boundary layers. J Atmos Sci 31:1791-1806

Mellor GL, Yamada T (1982) Development of a turbulence closure model for geophysical fluid problems. Rev Geophys Space Phys 20:851-875
Mellor GL (1989) Retrospective on oceanic boundary layer modeling and second moment closure. In: Müller P, Henderson D (eds.) Proceedings of the 'Aha Huliko'a Hawaiian Winter Workshop on Parameterization of Small-scale Processes,Honolulu,Hawaii,Hawaii
Institute of Geophysics Special Pub. p 251
Milne RD (1980) Applied Functional Analysis, Pitman, London
Minas HJ, Codispoti LA, Dugdale RC (1982) Nutrients and primary production in the upwelling region off Northwest Africa. In: Rapports et Procès-Verbaux des Réunions, Conseil International pour l'Exploration de la Mer 180:148-183
Mittelstaedt E (1983) The upwelling area off Northwest Africa- a description of phenomena related to coastal upwelling. Prog Oceanogr 12:307-331
Miyakoda K, Rosati A (1984) The variation of sea surafce temperature in 1976 and 1977,2. The simulation with mixed layer models. J Geophys Res 89:6533-6542
Niiler PP, Kraus EB (1977) One dimensional models of the upper ocean. In: Kraus EB (ed.) Modelling and Prediction of the Upper Layers of the Ocean, Pergamon Press, p 143
Oberhuber JM, Lunkeit F, Sausen R (1991) CO_2 doubling experiments with a coupled global isopycnic ocean-atmosphere circulation model. MPI Rept, Max-Planck Institut für Meteorologie, Hamburg
Owens WD, Lemke P (1990) Sensitivity studies with a sea ice-mixed layer-pycnocline model in the Weddell Sea. J Geophys Res 95: 9527-9538
Parkinson CL, Washington WM (1979) A large-scale numerical model of sea ice. J Geophys Res 84: 311-337
Parsons AT (1969) A two-layer model of Gulf Stream separation. J Fluid Mech 39:511-528
Patel VC, Rodi W, Scheuerer G (1985) Turbulence models for near-wall and low Reynolds number flows: a review. AIAA Journal 23 9:1308-1319
Price JF, Weller RA, Pinkel R (1986) Diurnal cycling: observations and models of the upper ocean response to diurnal heating, cooling, and wind-mixing. J Geophys Res 91:8411-8427
Rahm L, Svensson U (1989) Dispersion in a stratified benthic boundary layer. Tellus 41A:148-161
Richards KJ (1982) Modelling the benthic boundary layer. J Phys Oceanogr 12:428-439
Riebesell U, Wolf-Gadrow EA, Smetacek V (1992) Carbon dioxide limitation of marine phytoplankron growth. Nature, submitted
Rhines PB (1975) Waves and turbulence on a beta-plane. J Fluid Mech 69:417-443
Rodi W (1987) Examples of calculation methods for flow and mixing in stratified fluids. J Geophys Res 92:5305-55328
Salmon R (1980) Baroclinic instability and geostrophic turbulence. Geophys Astrophys Fluid Dyn 15:167-211

Semtner AJ Jr, Chervin RM (1988) A simulation of the global ocean circulation with resolved eddies. J Geophys Res 93:15502-15522

Smith LT, Boudra DB, Bleck R (1990) A wind-driven isopycnic coordinate model of the North and Equatorial Atlantic Ocean. J Geophys Res 95:13105-13128

Steele J (1978) Spatial patterns in plankton communities, Plenum, New York

Sündermann J (ed) (1986) Landolt-Börnstein, Numerical Data and Functional Relationships in Science and Technology, Group V:Geophysics and Space Research Vol 3 Oceanography, Springer-Verlag, Berlin Heidelberg New York London Paris Tokyo

Tang B, Cushman-Roisin B (1991) Two-layer geostrophic dynamics. Part II:Geostrophic turbulence. J Phys Oceanogr 22:128-138

The FRAM Group (1991) Initial results from a fine resolution model of the Southern Ocean. EOS, Trans Amer Geophys Union 72(169) :174-5

Van Camp L, Nykjaer L, Mittelstaedt E, Schlittenhardt P(1991) Upwelling and boundary circulation off Northwest Africa as depicted by infrared and visible satellite observations. Prog Oceanogr 26:357-402

Walsh JJ (1983) Death in the sea:enigmatic phytoplankton loss. Prog Oceanogr 12:1-86

Walsh JJ, Rowe GT, Iverson RL, McRoy CP (1981) Biological export of shelf carbon is a sink in the global CO_2 cycle. Nature 291:196-201

Wolf U, Woods JD (1988) Lagrangian simulation of primary production in the physical environment- the deep chlorophyll maximum and nutricline. In: Rothschild BJ (ed.) Towards a Theory on Biological-Physical Interactions in the World Ocean. Kluwer Academic Publishers, Dordrecht, p 51

Woods JD (1980) Do waves limit turbulent diffusion in the ocean? Nature 288:219-224

Woods JD, Onken R (1982) Diurnal variation and primary production in the ocean-preliminary results of a Lagrangian ensemble model. J Plank Res 4:735-756

Woods JD (1985) The physics of thermocline ventilation. In: Nihoul JCJ (ed.) Coupled Ocean-Atmosphere Models. Elsevier Oceanogr Ser 40, Elsevier, New York, p 543

Woods JD, Barkmann W (1986) A lagrangian mixed layer model of Atlantic 18°C water formation. Nature 319:574-576

Woods JD (1988) Mesoscale upwelling and primary production. In: Rothschild BJ (ed.) Toward a Theory on Biological-Physical Interactions in the World Ocean. Kluwer Academic Publishers, Dordrecht, p 7

Wroblewski JS (1989) A model of the spring bloom in the North Atlantic and its impact on oceanic optics. Limnol Oceanogr 34:1563-1571

Yamazaki H, Osborn TR (1988) Review of oceanic turbulence: Implications for biodynamics. In: Rothschild BJ (ed.) Toward a Theory on Biological-Physical Interactions in the World Ocean. Kluwer Academic Publishers, Dordrecht, p 215

TROPHIC RESOLUTION

Ian J. Totterdell
James Rennell Centre for Ocean Circulation
Gamma House
Chilworth Research Centre
SOUTHAMPTON SO1 7NS
United Kingdom

Robert A. Armstrong[1], Helge Drange[2], John S. Parslow[3],
Thomas M. Powell[4], Arnold H. Taylor[5]

1. Objectives and overview

1.1. General principles

The Working Group was asked to consider the level of complexity (in terms of the number of state variables and the processes included) suitable for use in models of oceanic biogeochemical cycles. The Group adopted the following principle in their deliberations:

'A model should be only as complex as demanded by the question to be addressed'

The Group felt that modellers would be justified in using this approach for two reasons:

(i) The demands for economy of computer storage and computation time when biogeochemical models are embedded in ocean general circulation models (GCMs). A model with a large number of state variables in a GCM will also produce an immense volume of data for modellers to analyse. It was recognised however that not all models would be governed by the

[1] AOS Program, Princeton University, P.O. Box CN710, Princeton, NJ 08544-0710, USA
[2] Nansen Environmental and Remote Sensing Centre, Edvard Griegsvei 3A, N-5037 Solheimsviken, Bergen, Norway
[3] CSIRO Div. of Fisheries, GPO Box 1538, Hobart, Tasmania 7001, Australia
[4] Division of Environmental Studies, University of California-Davis, Davis, CA 95616, USA
[5] Plymouth Marine Laboratory, Prospect Place, West Hoe, Plymouth PL1 3DH, United Kingdom

NATO ASI Series, Vol I 10
Towards a Model of Ocean
Biogeochemical Processes
Edited by G T Evans and M J R Fasham

requirements of GCMs, and that there was also a need for regional models and for models restricted to a small area where there was data coverage good enough to test predictions.

(ii) It is expected that as the model complexity increases, confidence in the predictions will eventually decrease (Ludwig & Walters, 1981). This might be for several reasons: the interactions of the various processes will become more difficult to understand and spurious ones less easy to identify, while there may be increased uncertainty in the mathematical forms of functions and the parameter values required.

1.2. The questions to be addressed

Having adopted the principle enunciated above, it is necessary to define the questions the biogeochemical models seek to answer. The Working Group listed these as the identification and quantification of biochemical pools and fluxes, and also their responses to changes in global climate forcing. The pools included both organic and inorganic, and both dissolved and particulate, forms of carbon, nitrogen, phosphorous, silicon and calcium, and possibly iron, sulphur and other trace elements. The fluxes would be those on time-scales varying from seasonal to inter-annual, and those operating from the scale of particular regions up to the global scale. The Group recognised that some models will be constructed to study the cycles of all these elements for their own sake, while others will include them only to enable accurate prediction of the carbon cycle.

There will continue to be substantial interest in the response of the pools and fluxes to changes in the physical and climatic forcing, with particular reference to global climate change. The Group felt that there was a definite need for the inclusion of some biological community structure in the models dealing with this topic, because fluxes to depth can depend on which species dominate at different times. For example, if species which have a high sinking rate are abundant, it would be expected that the remineralization products will re-appear deeper in the water column than if species with low sinking rates were dominant. It is important that models addressing global change do not fix communities in their present-day patterns by prescribing them by regions, but instead allow them to be continuously predicted as climate forcing changes.

Related to this point is the question of how accurately the timing of seasonal cycles have to be replicated in models primarily concerned with long-term fluxes and changes. For such models, it is probably more important to get the community composition correct at roughly the right time of year than to force annual cycles to fit a particular set of data at the cost of distorted parameter values and invalid trophic relationships. The meteorological data used to drive the models are usually averages taken over many years, while any observation set will have been recorded in one particular year. On the other hand, it will be necessary to compare the models' predictions of the annual cycles with available data, for if they are significantly in error confidence in the accuracy of predictions of the fluxes will be reduced. It is recommended that modellers should be aware of this situation and decide what their priorities are in each case.

1.3. The basic approach

The Working Group were unanimous in deciding that the most effective approach to minimising model complexity lies in considering the biogeochemical functions of each trophic level, and then choosing suitable functional groups to represent them. In order to be able to authentically model trophic relationships and find values of parameters each of these groups will probably correspond approximately to a collection of species in the case of the biotic components (bacteria, phytoplankton and zooplankton), rather than being variables with just one function.

The Group recognised that other approaches, for example that of size structure, are possible for modelling marine biogeochemical systems, but did not consider them in any detail.

1.4. Trophic levels

The Group restricted its discussion, for the sake of simplicity, to the lower trophic levels. The categories for which functions and functional groups were considered were phytoplankton, zooplankton, bacteria and the non-biotic components detrital particles and dissolved organic matter (DOM).

The forms of nutrients were not examined, although some existing models (e.g. Walsh, 1975; Wroblewski, 1977; Hofmann & Ambler, 1988; Fasham *et al.*, 1990) separate ammonium (reduced nitrogen) and nitrate (oxidised nitrogen) as substrates for autotrophic growth while others (e.g. Evans & Parslow, 1985; Frost, 1987; Parsons & Kessler, 1987; Andersen & Nival, 1989; Moloney & Field, 1991; Taylor *et al.*, 1991) combine them into one component. It was recognised that there are questions to be resolved for this category, especially relating to the ability to model the possibly important process of ammonium inhibition of nitrate uptake (Fasham *et al.*, 1990), but in general it seemed that the components to use are fairly self-evident, and correspond closely to the chemical elements included in the model.

It may be possible to represent some of the functional groups implicitly rather than as a separate state variable. For example, the remineralization and breakdown of detrital particles is mediated by bacteria which live on the particle as it sinks, but in models the process is often modelled as a rate constant, because the growth of these attached bacteria is strongly dependent on the size and composition of the particle and independent of most of the external processes. Anderson *et al.* (this volume) discuss when implicit modelling can be justified.

The Working Group proceeded to consider each of the categories in turn, listing the functions and discussing the most suitable functional groups. The results of these discussions are presented in the following four sections, after which model closure, the role of formal aggregation techniques and other problems are considered.

2. Phytoplankton

2.1. Functions

The biogeochemical functions of phytoplankton were identified as:

(i) Primary production and growth - the uptake of chemical species from the water and their incorporation into tissue. The fixing of carbon is an especially notable process, both because of the importance of the carbon cycle and because of the high proportion of carbon in the plankton tissue.

(ii) Nutrient dependence - because the growth rates of the plankton are functions of the availabilities of various nutrients, the cycles of the different chemical species are linked.

(iii) Being grazed - the ultimate fate of the planktonic tissue is often not to return to solution immediately or in the same spatial location but instead to form the body tissue of other trophic levels.

(iv) Sinking or swimming - the majority of phytoplankton in the ocean are non-motile, and their lateral movements are governed by currents and tides; this is of course also the case for the dissolved chemical species from which their cell tissues have been made. Unlike the dissolved species however the plankton may sink, and although (at least in eutrophic waters) the effects can be minimised by the cells' control of their buoyancy (Eppley *et al.*, 1967) there is usually a significant downward flux of material from the surface waters to the deep. Some species of phytoplankton do possess motility, which can lead to a lateral flux of biogeochemical components not solely dependent on the movements of the water masses.

2.2. Functional groups

Having defined the functions performed by the phytoplankton, a set of functional groups was chosen. The groups were generally selected to correspond to the most common types of phytoplankton, and are listed below. How each group fulfils the functions is also described.

(i) Diatoms: these are dominant in the spring blooms that are observed in certain areas of the oceans. Because they form shells, they are linked to the silicate cycle, and can be silicate-limited. Diatoms can have high sinking rates. They are grazed by meso- rather than micro-zooplankton. It has been suggested that in some areas, where the available nutrients are not completely taken up, the diatoms' growth rate is limited by lack of iron (Martin & Fitzwater, 1988).

(ii) Coccolithophores: these produce liths or shells of calcium carbonate which are very important in biogeochemical terms. This group is linked to the calcium cycle, and is instrumental in the flux of calcium carbonate to deep waters. The formation of the liths affects the carbon cycle in that bicarbonate is used (rather than carbonate), a process which increases the concentration of dissolved CO_2 gas in the surface waters. The coccolithophore blooms which occur in the summer at high latitudes can strongly scatter light, significantly affecting its penetration into the water column and hence the depth of the upper mixed layer. It has also been suggested that dimethyl-sulphide produced towards the end of a bloom (Turner *et al.*, 1988; Iverson *et al.*, 1989) can, having diffused into the atmosphere, encourage cloud formation.

(iii) Nitrogen fixers: the total amount of nitrogen available for growth in the form of nitrate or ammonium in some areas of the ocean may be affected by the rate at which atmospheric nitrogen is fixed by certain species of phytoplankton. This is sometimes accomplished by symbiotic intracellular bacteria (Carpenter & Romans, 1991). This process is not easy to model however as little is understood about the controlling factors. See also Section 4.

(iv) Picoplankton (including autotrophic bacteria): these have very high growth rates, but their main importance is that they form part of the microbial loop, being grazed by zooflagellates which also graze heterotrophic bacteria. The inclusion of this group will therefore be necessary if the dynamics of the microbial loop are to be correctly modelled.

(v) Phytoflagellates: the growth of this group is not limited by the availability of silicate or iron, and the population peaks around the time of the spring bloom. They are grazed by microzooplankton.

(vi) Dinoflagellates: these are large and have relatively slow growth rates, frequently forming the bulk of the summer population. They are motile, and can practise vertical migration. They may be facultative heterotrophs.

It should be noted that the groups can be split into those that have an important role in particular chemical cycles (e.g. of silicate, iron and/or calcium carbonate), and those that are necessary in order to correctly model the seasonal pattern of the carbon and nitrogen fluxes, taking into account processes such as grazing, sinking, etc. The latter role will be important in areas where the physics of the ocean is strongly seasonal and coupled to the biological fluxes.

2.3. Areas of uncertainty

Two linked questions must be considered regarding the completeness of the description of phytoplankton using the six functional groups listed above:

are there any other species which play an important role in biogeochemical fluxes and which cannot be included in one of the groups above, and, within the groups, is further discrimination between types or sub-species required to describe the seasonal cycles adequately. Both these questions address the possibility that the actual species representing a particular functional group in one area of the ocean will have values of parameters (e.g. the half-saturation constant for nitrate uptake) different from those for the species representing the same group in another area. For large scale models, especially those to be included in GCMs, it is undesirable to require a single state variable to have parameters which are dependant on geographical position. If the variation in the parameters is mainly due to a change in latitude however it may be possible to model the parameter values as a function of temperature, which itself varies strongly with latitude. This correlation has been suggested by Eppley (1972).

Overwintering can be a major problem with many models of phytoplankton annual cycles, especially in the higher latitudes and where there is deep convection. Taylor *et al.* (1991) set thresholds below which the population densities may not fall. The thresholds set are not well-constrained by data, since the population densities in winter are at or below the limit of detectability, and very little survey work is done at that time of year. Unfortunately the timing of the spring blooms of the various phytoplankton groups can vary by as much as a week for different but justifiable threshold values, and care has to be taken that the model results are not an artifact of the values chosen if this approach is adopted. Modelling 'tricks' are only necessary because the methods by which real phytoplankton survive the winter are poorly understood. Suggested mechanisms by which diatoms are available at the end of winter in the deep ocean to initiate a bloom include forming resting spores through the worst conditions, and reseeding from coastal populations. Both of these mechanisms provide problems for modellers, in the former case because of the complete lack of knowledge about the environmental triggers that cause the cells to enter or leave the spore state, and in the latter because of the implied need to correctly model the coastal ecosystem to predict the deep ocean succession adequately. The Group were unanimous in recognising the need for more study of winter populations.

3. Zooplankton

3.1. Functions

In this study the Working Group concentrated on herbivorous, rather than carnivorous, zooplankton. Some of the herbivores discussed however may display carnivorous or even cannibalistic behaviour in certain situations. The modelling of the pure carnivores and higher trophic levels is discussed in Section 7.

Further discussions on the modelling of zooplankton and their biogeochemical functions may be found in the report of the Zooplankton Working Group by Anderson *et al.* (this volume).

The biogeochemical functions of the herbivorous zooplankton were identified as:

(i) Grazing and control of populations. Zooplankton prey on various species of phytoplankton. Some also graze detrital particles and/or bacteria. This process not only transfers material to a higher trophic level but also can be the major factor in limiting the growth of the autotrophs in some areas of the oceans (Colebrook, 1979).

(ii) Defecation. Unassimilated material is formed into fecal pellets which are released. The larger of these particles can have high sinking rates and form a major part of the flux of material to depth, while smaller ones are attacked by bacteria. Section 5 has a fuller discussion.

(iii) Vertical migration. Some of the larger zooplankton spend a part of the day (usually the daylight hours) in the aphotic zone where they are relatively safe from their predators, and come to the surface to feed at other times. This behaviour will affect the rates of grazing and mortality, and can lead to a downward flux if food is consumed in the surface layers and waste defecated or excreted at depth (Longhurst & Harrison, 1988; Longhurst *et al.*, 1990, 1991).

(iv) Life history strategy. Many of the larger zooplankton pass through several distinct juvenile stages before appearing as fully mature adult organisms. The importance of such stages is discussed in Subsection 3.4., and Anderson *et al.* (this volume).

(v) Particle scavenging and repackaging. Some zooplankton also feed on detrital particles. This topic is discussed more fully in Section 5.

(vi) Shell formation. Certain species of zooplankton, e.g. foraminifera and pteropods, produce shells of calcium carbonate and, like the coccolithophores (Section 2), can constitute an important carbonate flux to depth when they sink.

(vii) Mortality. Herbivores can die for a number of reasons, the main ones being starvation and being grazed by predators, usually from higher trophic levels. The latter process leads to the bulk of the transferred material being remineralized on time-scales of months to years, and probably at depth, whereas much remineralization due to the former process will take place on a timescale of days, and in the surface layers. The higher trophic levels are discussed in Section 7.

3.2. Functional groups

There was generally much less confidence in the assignment of functional groups for this category than for that of the phytoplankton, and it was suggested that a better approach might be to have a single grazer which can represent different zooplankton as required; this idea is discussed in Subsection 3.3 below. The following groups were however proposed:

(i) Zooflagellates. These form part of the microbial loop (Azam *et al.*, 1983), grazing free-living bacteria and picoplankton and having high rates of growth; if the microbial loop is modelled implicitly (see Section 9), then this group will not be required.

(ii) Microzooplankton. These also have fast growth dynamics. They graze the zooflagellates and phytoflagellates, and excrete ammonia. Their primary function in an ecosystem context is the control of various phytoplankton groups.

(iii) Mesozóoplankton. These graze larger phytoplankton groups, such as diatoms and (possibly) dinoflagellates, and may control their populations in many areas of the oceans. The members of this group include several species with cohort structures, but because these strategies are different it is not possible to model them in one group. Fast-sinking fecal pellets are produced.

(iv) Salps. These were suggested tentatively for their large fecal pellets and hence their participation in the export flux mechanism. They have high growth rates.

The Working Group also considered how and if any of these functional groups could be combined. It was suggested that:

- zooflagellates and microzooplankton could be combined because they both have high rates of growth
- microzooplankton and salps could be combined for similar reasons
- salps and mesozooplankton could be combined because they both produce fecal pellets with high sinking speeds

The Group's consensus was that there is scope for reducing the number of zooplankton functional groups in this way, but that care should be taken in case the seasonal succession is sensitive to the zooplankton structure.

3.3. Switching grazers

It is possible to represent the activities of all the important herbivores by one single grazer, if it is allowed to switch between different types of prey. The grazer will therefore imitate different zooplankton at different times, at least in terms of their feeding behaviour (fecal pellet production may have to be more consistent through the year). The switching can be determined by the time of year (which will tend to prescribe the system) or by the availability of the possible prey. For the latter approach, the response of the grazer (in terms of preference for a particular prey) could be proportional to, or a non-linear function of, the concentration of that prey. Fasham *et al.* (1990) use a grazer which switches between detritus, bacteria and phytoplankton as prey, and Pace *et al.* (1984) also allow their grazers to vary their preferences as the prey concentrations change. Fasham *et al.* (1990) discuss the suitability of three different functions of the prey concentrations for calculating feeding preferences.

One problem with a switching grazer is that spurious interactions may be introduced between the grazer being simulated at one time and that at another. For example, in a model which represents the meso- and microzooplankton separately the mesozooplankton population may peak by preying on a diatom bloom, and then decrease as the bloom dies off. Phytoflagellates will begin to bloom as the diatom concentration falls, and for a time they will not be significantly grazed because their predators, the microzooplankton, will not have had time to grow and so will still be at a low density. If a switching grazer is used however it will begin by imitating the mesozooplankton and graze on the numerous diatoms, but subsequently, as the phytoflagellates become dominant, it will start to imitate the microzooplankton; the difference is that the population density will already be high, and might prevent the phytoflagellates from reaching their peak concentration. Concern was also expressed that the model predictions can be very sensitive to the parameter values used in the switching function (Evans, 1988), while there is very little data on zooplankton feeding preferences to provide such values.

3.4. Cohort structure and overwintering

The Working Group considered whether it is necessary to model the cohort structure of zooplankton which feature distinct juvenile stages. Hofmann & Ambler (1988), for example, represent copepods by five size categories, i.e. eggs, three juvenile stages and the adult stage. The disadvantage in terms of model complexity and extra state variables is obvious, and even this is a simplification from the observed five juvenile stages. Another problem if cohorts are going to be modelled, particularly for GCMs, is that zooplankton occupying a similar position in the food chain but living in different geographical locations have developed different life history strategies, possibly in response to local physical forcing (Evans & Parslow, 1985).

There are several ways in which the existence of cohorts may be important. It is possible that some account will have to be taken of the different prey and predators of the separate stages if the seasonal succession is to be simulated accurately (however if microzooplankton, which lack cohort structure, are the primary means by which phytoplankton are controlled (Burkill *et al.*, 1993) then this reasoning will be less valid). Also, the existence of several development stages with separate trophic links may enable the species to survive the winter more readily.

Given that including cohort structure in biogeochemical models will mean a lot of extra complexity and variables, the Group recommends that the life history strategies should only be modelled explicitly if their effects are indispensable for the results and predictions sought, and they cannot be reproduced by some implicit formulation. Evidence that such effects are not indispensable comes from the success of those models already running that treat zooplankton in a simple way (e.g. Fasham *et al.*, 1990).

4. Heterotrophic bacteria

This section considers only heterotrophs, since autotrophic bacteria were included (as part of the picoplankton functional group) in the phytoplankton category.

4.1. The functions of heterotrophic bacteria

Heterotrophic bacteria were identified as having three main functions in biogeochemical cycles in the ocean:

(i) Respiration and remineralization. These two process have been grouped together because for bacteria they both involve the consumption of complex substrates and the release of simple compounds. Bacteria play a very important role in this stage of the recycling of nutrients and carbon, feeding on dissolved and particulate organic matter and excreting carbon dioxide and ammonia (the latter of which may be further converted to nitrate by the action of other bacteria).

(ii) Being grazed. Bacteria form part of the microbial loop (Azam *et al.*, 1983) and are grazed by microzooplankton.

(iii) Nitrogen fixing (strictly a chemotrophic rather than a heterotrophic process). Some bacteria have the ability to fix gaseous nitrogen, and so may be important in controlling the total nutrient content of regions of the ocean (Carpenter & Romans, 1991). The resulting global flux of nitrogen into the ocean is likely to be small compared to other fluxes however and since, as with the nitrogen-fixing phytoplankton, the factors which control this process are poorly understood it seems prudent to omit it from the models at present. Howarth *et al.* (1988a, 1988b) discuss the importance and possible controls of nitrogen fixation in different aqueous environments.

4.2. Functional groups

Two groups were proposed, free-living bacteria and those attached to particulate matter. Although the organisms in these two groups may be of the same species and have identical parameters for growth, etc, it is still necessary to group them separately because of the significant vertical motion of the particulate matter.

For each of the groups, it was considered whether it was necessary to model the bacteria explicitly, or if their functions could be adequately described using an implicit representation. This is, of course, a particular example of the choice considered in the introduction. The decision of the Working Group was that, at least for large-scale models, the attached bacteria should be modelled implicitly as a rate of breakdown of the particulate matter which might vary with temperature and/or depth. One reason for this was that the growth of such bacteria is only weakly determined by factors (other than pressure - i.e. depth - and temperature) outside the particles, even the grazing being performed by attached flagellates.

In the case of free-living bacteria, no final decision was reached, but a few points were raised that might help modellers come to a decision for their particular model(s). The bacteria form part of the microbial loop and are grazed by the same micro-grazers which also feed on the picoplankton. If the microbial loop is included explicitly, the bacteria must be modelled explicitly as well (the possibility that the microbial loop as a whole can be modelled implicitly is discussed in a later section). If however the microbial loop is not modelled explicitly, then the grazing function of the bacteria is not so important and it may be acceptable to model the remineralization function implicitly. Another factor that needs to be considered is whether the bacterial biomass is controlled by the concentrations of the substrate on which the bacteria feed, or by grazing or other factors.

The consensus of the working group was that because the bacteria have such a short doubling time and hence are generally tightly-coupled to their substrate, if a need existed to reduce the number of state variables in a model then the bacteria would be the most suitable component to model implicitly, either on their own or as part of an implicitly-modelled microbial loop. However an assessment of the errors introduced should be made first.

5. Detritus

5.1. Composition and functions

Detrital particles include the fecal pellets of zooplankton, fragments of phytoplankton and zooplankton cells, and some complete cells that have suffered natural death. These three types of detritus frequently aggregate to form larger particles.

The 'function' of detritus was perceived to be the flux of material from the compartments representing predators and their prey to either the deep ocean or to dissolved components in the surface layers. This could be split up into various parts:

(i) Production. Detritus is produced by the defecation of zooplankton (and unmodelled higher trophic levels), and also by the death of phytoplankton and zooplankton cells, either by being grazed or by natural mortality. For example, grazing can lead to fragments of cells being released, while at the end of diatom blooms cells that have succumbed to viral infection or environmental stress from nutrient depletion sink out as passive particles, often in clumps or sheets.

(ii) Sinking. The particles either float or sink with one of a wide range of speeds, depending on their size and relative buoyancy. There is a strong correlation between size and sinking rate, and if small particles aggregate the larger particles formed will sink faster. Particles which sink faster will, for the same time-rates of breakdown and remineralization, be less affected by these processes than particles with lower sinking speeds.

(iii) Breakdown/remineralization. Attached bacteria feed on at least some components of detrital particles (see Section 4) releasing CO_2 and ammonium. The removal of this material can cause the undigested parts to fragment, decreasing their sinking speed; these fragments may be remineralized over longer time-scales.

(iv) Scavenging/repackaging. Grazers from the mixed or surface layers will often feed on detrital particles; this feeding may take place while the grazers are sheltering at depth during the daylight hours if they practise vertical migration. Specialised detritivores can also be found at mid-depths. Because some of the detritus grazed will be indigestible, a proportion will be returned to the water as new fecal pellets, having been repackaged.

5.2. Groups

With a large range of sinking speeds - from zero to a few hundred metres per day - it is not obvious how detritus should be represented in a model. Geochemists performing radio-tracer studies have found it adequate to split the particles into just two classes, those which sink and those which do not (Clegg & Whitfield, 1990, 1991). The problem is to decide which sinking speed to use; this will probably depend on the system being modelled. The appropriate speed may also vary with time if the sources of the particles change through the year; for example, at the end of a diatom bloom the sinking particles might be mainly aggregations of diatom cells with a high sinking rate, while later on in the year fragments of smaller cells and fecal pellets might dominate and give a much lower rate. This may be difficult to model in a simple but accurate way.

Another possible division of detritus is to link particles to their source. If this method were adopted then it would be possible to separate occasionally dominant sources, such as the diatom blooms, from general production which occurs at a more constant rate through the year and features a lower sinking speed. This gain would however be bought at the cost of extra state variables in the model, and was not thought by the Working Group to be a fruitful approach.

5.3. Interactions with other components

In Section 4 it was recommended that the attached bacteria responsible for the remineralization of detritus should be modelled implicitly as a decay rate. Although zooplankton graze detritus down to depths of several hundred metres, and this process will probably be modelled explicitly in the surface layers, it would be inconsistent to include the process in that way at depths where remineralization is modelled implicitly. A more consistent way would be to include another decay rate for this zooplankton-mediated removal process, applicable below a certain depth (or the base of the upper mixed layer, whichever was the lower).

6. Dissolved organic matter

DOM is now regarded as an important component of marine biogeochemical cycles (Toggweiler, 1989). The longer-lived (refractory) fractions have decay timescales of thousands of years, and hence may be a sizeable sink for carbon, while some fractions are labile enough to be taken up as food by bacteria. A fuller discussion is given by Kirchman *et al.* (this volume).

DOM is produced by cell lysis and by the release of cell contents during grazing. Certain types of cell produce it actively, for example *phaeocystis*, and it might be necessary to add this as another functional group for the phytoplankton.

The Working Group recognised that great uncertainty exists about the concentration of DOM in the oceans, its chemical composition and the timescales for breakdown of the various fractions. In view of this, the Group only felt justified in recommending that the most suitable division of DOM was into labile and refractory groups.

7. Closure

The zooplankton category was assumed to consist of herbivores, with carnivores and higher trophic levels to be modelled implicitly by the closure term (although some of the herbivores might also behave as carnivores in certain circumstances). It was reasoned that a closure term will be necessary at some level in the model, and that cutting the trophic links at a relatively low level enables the number of state variables to be kept small and means that the closure term is not trying to model behaviour patterns that are very complicated. The term used for closure is therefore the mortality term for the herbivores.

This term was considered in some detail by the Working Group because previous work (Steele & Henderson, 1992) has shown that the dynamics of the system can be very sensitive to the form of the closure. Herbivore (and carnivore) mortality terms are also discussed by Anderson *et al.* (this volume).

The use of a constant *per capita* rate of mortality a leads (at least in simple models) to the equilibrium phytoplankton population being set by the zooplankton parameters with no relation to the zooplankton numbers. Several other forms of the mortality rate were discussed, and it was suggested (J. Steele, pers. comm.) that it can be useful for diagnostic purposes if the term can be split into a numerical response (of the predator population) and a functional response (the predator behaviour). For example, a mortality term of aH (giving a loss of herbivore biomass per unit time of aH^2, H being the density of herbivores) might be explained in terms of a constant rate of feeding per carnivore a times a carnivore density which increases in proportion to the prey available; while it is usually unreasonable to suppose that the total carnivore biomass can track that of the herbivores, when spatial heterogeneity is considered it can be quite reasonable that a local patch of high herbivore density will attract predators from a larger area and so make the relation locally accurate.

Another possible approach is to prescribe a rate of mortality as a function of the geographical position and time of year. This rate can be set from observations (although it was recognised that there is a paucity of such data in most regions) or to match the distribution of phytoplankton through the year in each area (for which there is much more complete data coverage). The Group was not in favour of such an approach however, especially for models whose aim is to predict the response of the biota to changes in physical and climatic forcing, because it might preclude the very changes looked for.

The consensus of the Group was that making the mortality rate proportional to the herbivore biomass is a promising and reasonable formulation, although many other forms (some including time lags) are possible (see Steele & Henderson, 1992, or Holling, 1965 and 1966, for a wider discussion). Whichever form of closure is chosen, it will be advisable to undertake a detailed sensitivity analysis of the effects of the closure assumption on model behaviour.

8. Formal aggregation

8.1. The role of formal aggregation

The Working Group considered how the mathematical techniques of formal aggregation (Iwasa *et al.*, 1987, 1989) might be most usefully applied to the construction and diagnostics of marine biogeochemical models.

The Group identified the following questions where formal aggregation techniques might be helpful:

(i) Aggregating different species with the same biogeochemical function(s) into one functional group; particularly defining the parameters of that group:

(ii) Aggregating different functional groups, to reduce the number of model state variables. This might be possible, for example, for the phytoplankton functional groups that are primarily concerned with the timing of the seasonal succession.

(iii) Aggregation of detrital particles of different sizes (and hence with different sinking rates.

(iv) Aggregation of zooplankton cohorts.

(v) Parameterization of sub-gridscale variability and effects by aggregation over space.

8.2. Practical application

As previously stated, much aggregation of variables is undertaken already by those constructing models, usually in an ad-hoc way. If the formal aggregation techniques can be applied simply and without excessive restrictions on the type of system considered then they could be a powerful means of testing the effectiveness of aggregation. Perfect aggregation (Iwasa *et al.*, 1987) will usually not be possible, so choices will have to be made as to which aspects of the models' output are most important. Iwasa *et al.* (1989) consider various different criteria.

Some members of the group expressed reservations about how easy it will be to apply these techniques, developed for economics, chemical engineering and other fields, to complicated ecosystem models, and whether they will deliver a great improvement on less formal methods of aggregation.

9. Discussion

9.1. The microbial loop

The Group discussed whether the microbial loop, consisting of bacteria, picoplankton, zooflagellates and microzooplankton, (Azam *et al.*, 1983) can be represented implicitly, probably as a rate of flow to the microzooplankton compartment. The benefits are that the number of state variables will be reduced, and the structure of the model simplified. There are several supporting arguments. For example, all the components have fast growth rates, and so should be able to respond quickly to changes in other variables (e.g. the concentrations of DOM or detrital particles) and forcings, although there are in practice short time lags which could be important. Also, model studies have shown (Ducklow *et al.*, 1986; Taylor & Joint, 1990) that the flow of material from the microbial loop to higher trophic levels is small. Lastly, in Section 2 the main reason for including picoplankton as a functional group was to enable the grazing on bacteria to be calculated correctly, while in Section 4 one argument for modelling free-living bacteria explicitly was so that the grazing on the picoplankton would not be in error; if the time lags in the responses of the components do not lead to significant effects then this circular reasoning can be dispensed with. The Group suggests that implicit representation of the microbial loop should be examined.

9.2. Summary

In conclusion, therefore, the Working Group makes the following recommendations:

(i) Modellers should be clear what questions are to be addressed and construct their models accordingly. The models should only be as complex as the questions demand; any extra complexity may decrease understanding of, and confidence in, the models' results.

(ii) Modellers should investigate the dependence of results on their choice of variables and processes. This is particularly important for grazing and model closure, and for other processes which might be modelled implicitly.

(iii) Functional groups are a promising and effective approach to choosing state variables within a trophic level. However size structured models, which were not considered, may also be worthy of further study.

(iv) In ocean GCMs, a four-fold increase in the number of biological components is equivalent to a 2-fold change in the horizontal spatial resolution, and the former would allow significantly more effects to be modelled. It is suggested that effort expended on aggregating biological components should be balanced by that expended on parameterizing the effects of the sub-gridscale physics.

(v) In order to validate regional and global models, much improved data coverage is required. This will ideally be in the form of time-series of the biogeochemical components, rather than intermittent surveys.

References

Andersen V, Nival P (1989) Modelling of phytoplankton population dynamics in an enclosed water column. J mar biol Ass UK 69:625-646

Anderson TR, Andersen V, Fransz HG, Frost BW, Klepper O, Rassoulzadegan F, Wulff F (1993) Modelling zooplankton. (this volume)

Azam F, Fenchel T, Field JG, Gray JS, Meyer-Reil LA, Thingstad F (1983) The ecological role of water-column microbes in the sea. Mar Ecol Prog Ser 10:257-263

Burkill PH, Edwards ES, John AWG, Sleigh MA (1993) Microzooplankton and their herbivorous activity in the north eastern Atlantic Ocean. Deep-Sea Res II 40:479-493

Carpenter EJ, Romans K (1991) Major role of the cyanobacterium *Trichodesmium* in nutrient cycling in the north Atlantic Ocean. Science 254:1356-1358

Clegg SL, Whitfield M (1990) A generalized model for the scavenging of trace metals in the open ocean - I. Particle cycling. Deep-Sea Res 37:809-832

Clegg SL, Whitfield M (1991) A generalized model for the scavenging of trace metals in the open ocean - II. Thorium scavenging. Deep-Sea Res 38:91-120

Colebrook JM (1979) Continuous plankton records: Seasonal cycles of phytoplankton and copepods in the north Atlantic Ocean and the North Sea. Mar Biol 51:23-32

Ducklow HW, Purdie DA, Williams PJLeB, Davies JM (1986) Bacterioplankton: a sink for carbon in a coastal marine plankton community. Science 232:865-867

Eppley RW (1972) Temperature and phytoplankton growth in the sea. Fish Bull 70:1063-1085

Eppley RW, Holmes RW, Strickland JDM (1967) Sinking rates of marine phytoplankton measured with a fluorometer. J Exp Mar Biol Ecol 1:191-208

Evans GT (1988) A framework for discussing seasonal succession and coexistence of phytoplankton species. Limnol Oceanogr 33:1027-1036

Evans GT, Parslow JS (1985) A model of annual plankton cycles. Biol Oceanogr 3:327-347

Fasham MJR, Ducklow HW, McKelvie SM (1990) A nitrogen-based model of plankton dynamics in the oceanic mixed layer. J Mar Res 48:591-639

Frost BW (1987) Grazing control of phytoplankton stock in the open subarctic Pacific Ocean: a model assessing the role of mesozooplankton, particularly the large calanoid copepods *Neocalanus* spp. Mar Ecol Prog Ser 39:49-68

Hofmann EE, Ambler JW (1988) Plankton dynamics on the outer southeastern U.S. continental shelf. Part II: A time-dependent biological model. J Mar Res 46:883-917

Holling CS (1965) The functional response of predators to prey density and its role in mimicry and population regulation. Mem Entomol Soc Canada No. 45

Holling CS (1966) The functional response of invertebrate predators to prey density. Mem Entomol Soc Canada 48:1-85

Howarth RW, Marino R, Cole JJ (1988) Nitrogen fixation in freshwater, estuarine and marine ecosystems. 2. Biogeochemical controls. Limnol Oceanogr 33:688-701

Howarth RW, Marino R, Lane J, Cole JJ (1988) Nitrogen fixation in freshwater, estuarine and marine ecosystems. 1. Rates and importance. Limnol Oceanogr 33:669-687

Iverson RL, Nearhoof FL, Andreae MO (1989) Production of dimethylsulfonium and dimethylsulfide by phytoplankton in estuarine and coastal waters. Limnol Oceanogr 34:53-67

Iwasa Y, Andreasen V, Levin S (1987) Aggregation in model ecosystems. I. Perfect aggregation. Ecol Modelling 37:287-302

Iwasa Y, Levin SA, Andreasen V (1989) Aggregation in model ecosystems. II. Approximate aggregation. IMA Journal of Mathematics Applied in Medicine and Biology 6:1-23

Kirchman DL, Lancelot C, Fasham M, Legendre L, Radach G, Scott M

(1993) Dissolved organic matter in biogeochemical models of the ocean. (this volume)

Longhurst AR, Bedo A, Harrison WG, Head EJH, Horne EP, Irwin B, Morales C (1991) NFLUX: a test of vertical nitrogen flux by diel migrant biota. Deep-Sea Res 38:1705-1719

Longhurst AR, Bedo A, Harrison WG, Head EJH, Sameoto DD (1990) Vertical flux of respiratory carbon by oceanic diel migrant biota. Deep-Sea Res 37:685-694

Longhurst AR, Harrison WG (1988) Vertical nitrogen flux from the oceanic photic zone by diel migrant zooplankton. Deep-Sea Res 35:881-889

Ludwig D, Walters CJ (1981) Measurement errors and uncertainty in parameter estimates for stock and recruitment. Can J Fish 38:711-720

Martin JH, Fitzwater SE (1988) Iron deficiency limits phytoplankton growth in the north-east Pacific subarctic. Nature 331:341-343

Moloney CL, Field JG (1991) The size-based dynamics of plankton food webs. I. A simulation model of carbon and nitrogen flows. J Plankton Res 13:1003-1038

Pace ML, Glasser JE, Pomeroy LR (1984) A simulation analysis of continental shelf food webs. Mar Biol 82:47-63

Parsons TR, Kessler TA (1987) An ecosystem model for the assessment of plankton production in relation to the survival of young fish. J Plankton Res 9:125-137

Steele JH, Henderson EW (1992) The role of predation in plankton models. J Plankton Res 14:157-172

Taylor AH, Joint I (1990) A steady-state analysis of the 'microbial loop' in stratified systems. Mar Ecol Prog Ser 59:1-17

Taylor AH, Watson AJ, Ainsworth M, Robertson JE, Turner DR (1991) A modelling investigation of the role of phytoplankton in the balance of carbon at the surface of the North Atlantic. Global Biogeochem. Cycles 5:151-171

Toggweiler JR (1989) Is the downward dissolved organic matter (DOM) flux important in carbon transport? In: Berger WH, Smetacek VS, Wefer G (eds) Productivity of the Oceans: Present and Past. J Wiley & Sons, Chichester, p 65

Turner SM, Malin G, Liss PS, Harbour DS, Holligan PM (1988) The seasonal variation of dimethyl sulfide and dimethylsulfoniopropionate concentrations in nearshore waters. Limnol Oceanogr 33:364-375

Walsh JJ (1975) A spatial simulation model of the Peru upwelling ecosystem. Deep-Sea Res 22:201-236

Wroblewski JS (1977) A model of phytoplankton plume formation during variable Oregon upwelling. J Mar Res 35:357-394

MODELLING GROWTH AND LIGHT ABSORPTION IN THE MARINE DIATOM *SKELETONEMA COSTATUM*

Dale A. Kiefer
Department of Biological Sciences
University of Southern California
University Park
Los Angeles, CA 90089-0371
U.S.A.

Introduction

In this paper I consider the relationship between light absorption and growth in marine phytoplankton. In particular, I will derive a mathematical description of the daily carbon-specific rate of photosynthesis and the photosynthetic quantum yield of *Skeletonema costatum* as a function of four environmental factors: temperature, nutrient concentration, light intensity, and photoperiod. The formulations describe laboratory measurements of the growth of this marine diatom when the cells are fully acclimated and in continuous culture. Among other goals it is hoped that this study will aid in interpreting global ocean color imagery, which provides synoptic maps of the concentration of chlorophyll *a* at the sea surface in regions where the four environmental parameters differ.

My approach to examining the relationship between light absorption and photosynthetic growth is based upon phenomenology. The instantaneous rate of photosynthesis of a suspension of cells (F_c, e.g. units of gm-at carbon m^{-3} h^{-1}) is by definition the product of the instantaneous rate of light absorption by the suspension, $F_a(\lambda)$ (e.g. mol quanta m^{-3} h^{-1} nm^{-1}), and the quantum yield of photosynthesis, $\Phi_c(\lambda)$, (e.g. gm-at carbon $quanta^{-1}$ nm^{-1}),

$$F_c = \int_{400}^{700} F_a(\lambda)\, \Phi_c(\lambda)\, d\lambda$$

NATO ASI Series, Vol I 10
Towards a Model of Ocean
Biogeochemical Processes
Edited by G T Evans and M J R Fasham

Symbol	Description	Example Units
λ	light wavelength	nm
PAR	photosynthetically available radiation (400 - 700 nm)	mol quanta m^{-2} s^{-1}
E_o	scalar irradiance	mol quanta m^{-2} s^{-1}
C	concentration of cellular carbon	gm-at C m^{-3}
Chl	concentration of chlorophyll a	mg Chl a m^{-3}
θ	ratio of cellular carbon-to-chlorophyll *a*	mg-at C (mg Chl a)$^{-1}$
a_x	absorption cross section normalized to unit x	m^2 (x)$^{-1}$
$\sigma(\lambda)$	absorption cross section of PSU	m^2 (mol PSU)$^{-1}$
F_a	instantaneous rate of light absorption	mol quanta m^{-3} h^{-1}
F_c	instantaneous rate of carbon fixation	gm-at C m^{-3} h^{-1}
g	daily carbon-specific rate of photosynthesis	d^{-1}
p^m	max. instantaneous carbon-specific photosynthetic rate	h^{-1}
μ	specific growth rate	d^{-1}
r	specific respiration rate	d^{-1}
Φ_c	quantum yield of carbon fixation	gm-at C (mol quanta)$^{-1}$
Φ_e	quantum yield of electron flow	mol e (mol quanta)$^{-1}$
D	photoperiod (illuminated fraction of a day)	dimensionless (0-1)
η	carbon-specific number of photochemical units	mol PSU (gm-at C)$^{-1}$
$x(\lambda)$	photochemical unit rate of light absorption	mol quanta (mol PSU)$^{-1}$ s^{-1}
γ	mean PSU photon processing rate	s^{-1}
Ψ	transfer efficiency to reaction center	dimensionless (0-1)

Table 1. Definition of symbols and example units used within the text.

The instantaneous carbon-specific rate of photosynthesis is defined as the ratio F_c to the concentration of cellular carbon in the suspension, C, (e.g. gm-at carbon m^{-3}), and the daily carbon-specific rate of photosynthesis, g (d^{-1}), is conveniently described as the product of the mean instantaneous carbon-specific rate of photosynthesis and the photoperiod, D:

$$g = \frac{D}{\text{off} - \text{on}} \int_{\text{on}}^{\text{off}} \frac{F_c}{C}(t)\, dt$$

The integral is evaluated over the photoperiod when the cells are undergoing photosynthesis; photosynthesis begins at time on and ends at time off. D, the photoperiod (units of hours of light day^{-1}), is of course equal to the difference between off and on. The specific growth rate of the cell suspension, μ, is simply the difference between the carbon-specific rate of photosynthesis and the carbon-specific rate of respiration, r:

$$\mu = g - r$$

It has been suggested that respiration is a linear function of μ, and thus also of g (Shuter 1979, reviewed by Geider 1992):

$$r = a\mu + b$$

The empirical constants, a and b, have approximate values of 0.15 and 0.03 d^{-1}, respectively. Unfortunately, the data presented here does not include information on specific rates of respiration. In the discussion below, r is neglected and only g is considered. Since the lowest values for μ in the *Skeletonema* database is about 0.1, over three times greater than the expected value for r, this assumption does not appear to greatly bias our analysis. In my analysis I also assume that values for carbon and chlorophyll within the continuous culture do not vary significantly over the day.

The rate of light absorption by a cell suspension obviously varies with the numerical concentration of cells, the cellular concentration and composition of pigments, and the spectral distribution of the irradiance. It is often convenient to represent the rate of light absorption by the cell suspension as the product of the concentration of chlorophyll, Chl, the chlorophyll-specific absorption cross section, $a_{chl}(\lambda)$, and scalar irradiance, $E_o(\lambda)$,

$$F_a = Chl \int_{700}^{400} \dot{a}_{chl}(\lambda) E_o(\lambda) d\lambda$$

Because spectral information and light levels are commonly reported in units of photosynthetically available radiation (PAR),

$$E_o = \int_{700}^{400} E_o(\lambda) d\lambda$$

I will define a chlorophyll-specific absorption coefficient that is weighted by the spectral distribution of the growth illumination, a_{chl},

$$a_{chl} = \int_{700}^{400} a_{chl}(\lambda) E_o(\lambda) d\lambda \ (E_o)^{-1}$$

Now,

$$F_a = a_{chl} E_o \mathrm{Chl}$$

In the following discussion I will examine variations in growth rate, quantum yield, and cellular concentrations of chlorophyll in continuous cultures of the marine diatom, *Skeletonema costatum*. Finally, a mathematical model of this analysis will be presented.

Light absorption and growth in continuous culture

The relationship between cellular light absorption and photosynthesis is well illustrated by studies of the steady-state growth of the marine diatom *Skeletonema costatum*. One study by Yoder (1979) examined the growth rate and cellular concentration of chlorophyll *a* of cells limited by light intensity and temperature. In this study, *Skeletonema* was grown in a turbidostat at five temperatures (0, 5, 10, 16, 22 °C) and at five light intensities. Cultures were also grown under a number of different photoperiods; however, values for the ratio of cellular carbon-to-chlorophyll *a* were reported only for a 12:12 light:dark cycle.

Another study by Sakshaug et al. (1989) examined the growth rate and cellular concentration of chlorophyll *a* of *S. costatum* grown at constant temperature (15 °C) and limited by either light intensity, photoperiod, or rate of nutrient supply. Cultures were grown continuously by daily dilution; the medium was of an elemental composition that ensured nitrogen limitation. The cultures were grown under varying combinations of light intensity (12 to 1200 μmol m^{-2} d^{-1}), photoperiod (6:18, 14:10, 24:0 light:dark), and dilution rate (0.1 to 1.4 d^{-1}). The matrix of light intensities, dilution rates, and photoperiods was not complete.

We will explore the results of these two studies by examining the terms in the following phenomenological equation:

$$g = \Phi_c \, a_{chl} \, E_o \, D \, (\theta)^{-1} \qquad (1)$$

where θ is the ratio of cellular carbon-to-cellular chlorophyll *a* (gm-at carbon (gm chlorophyll *a*)$^{-1}$).

Both of these studies consisted of measurements of E_o, D, θ, and μ. To calculate the quantum yield from eq. 1, I assumed that a_{chl} was a constant value (16 m^2 (gm Chl *a*)$^{-1}$) and that the specific respiration rate was an insignificant fraction of the carbon-specific photosynthetic rate (i.e. $\mu = g$). Both assumptions can be questioned to a limited extent, particularly for cultures grown at extremely low light levels or very short photoperiods.

Acclimation to temperature

The acclimative response of *S. costatum* to variations in growth temperature is summarized in figure 1. For these measurements, the photoperiod was 12 hours and nutrients were supplied at a rate sufficient to saturate growth rate. Figure 1A indicates that at 0 °C the nutrient-saturated growth rate is about 0.3 d^{-1}, and the maximum μ increases with increasing temperature until maximum growth rate is achieved. We also note from this figure that the irradiance

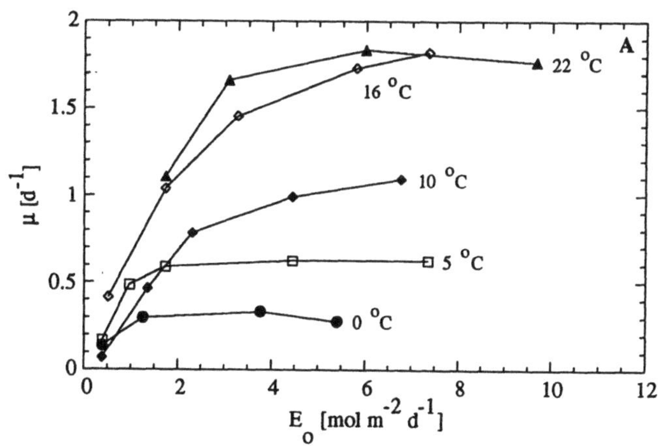

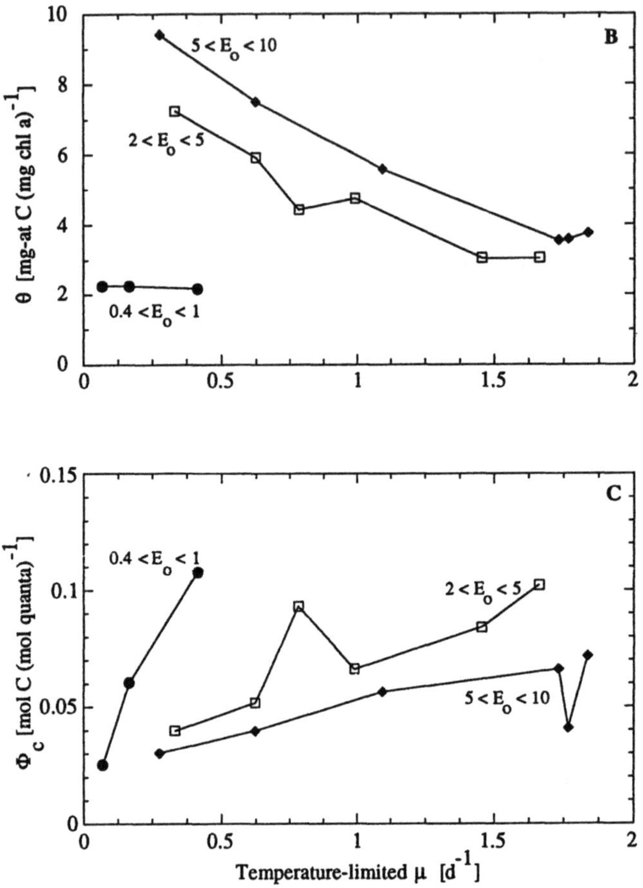

Fig. 1. (A) The specific growth rate of *Skeletonema costatum* as a function of growth irradiance for cells grown at different temperatures. (B) The carbon-to-chlorophyll *a* ratio and the quantum yield of photosynthesis (C) as a function of temperature-limited growth rate for cells grown at various levels of growth irradiance.

sufficient to saturate growth decreases with temperature. At 0 °C an irradiance of 1.5 mol m^{-2} d^{-1} is sufficient to saturate growth, at higher temperatures an irradiance of about 6 mol m^{-2} d^{-1} is required to reach maximum growth rate.

In figure 1B the carbon-to-chlorophyll *a* ratio, θ, for a given irradiance is shown as a function of the temperature-limited specific growth rate, μ(T). Because the photoperiod is constant for all the data, the growth rate at a given irradiance is uniquely determined by the temperature. For example, the coordinates for μ(T) and θ for the line connecting light intensities between 5 and 10 moles m^{-2} d^{-1} represent values for increases with temperature as one looks from left to right. Two features of this figure are noteworthy. First, increases in irradiance correspond with increases in the values for θ. Second, the response to changes in temperature differ with irradiance. At low values of E_o ($0.4 < E_o <1$), θ changes little with temperature-limited growth rate; however, for larger values of E_o (e.g. $5< E_o <10$) θ decreases with increases in temperature-limited growth rate.

In figure 1C, the quantum yield calculated from eq. 1 at a given irradiance is shown as a function of μ(T). We note that Φ_c decreases with increases in E_o. In addition we note that the response to temperature-limitation differs with irradiance. At low values of E_o, Φ_c increases sharply with increases in μ(T) (line $0.4 < E_o < 1$), but at high values of E_o (e.g. line $5 < E_o < 10$), Φ_c has only a small dependence on μ(T).

A comparison of figures 1B and 1C indicates a dichotomy in the acclimation of *S. costatum*. At high light levels, temperature limitation leads to increases in the cellular ratio of carbon-to-chlorophyll *a* but little change in the quantum yield. At low light levels, temperature limitation causes little change in the cellular ratio of carbon-to-chlorophyll *a* but sharp decreases in the quantum yield. In the next section, I show that this dichotomy also exists for nutrient-limited growth.

Acclimation to the rate of nutrient supply

Figures 2 and 3 summarize the response of *S. costatum* to variations in nutrient supply at differing light levels (figures 2A and 3A) and at differing photoperiods (figures 2B and 3B).

In figure 2A the carbon-to-chlorophyll *a* ratio, θ, for a given irradiance and a photoperiod of 24 hours is plotted as a function of the nutrient-limited specific growth rate, μ(N). Again, two features of this figure are noteworthy. First, θ increases with increasing E_o. Second, the response to changes in nutrient supply differs with irradiance. At low values of E_o, θ changes little with nutrient-limited growth rate, but at high values of irradiance θ increases sharply with decreases in

μ(N). A similar pattern is observed in figure 3A, where θ at a constant irradiance is shown as a function of nutrient-limited growth rate for different photoperiods. The values for θ are larger with longer photoperiods. In addition, the response to changes in nutrient supply differ with photoperiod. At low values of D, θ varies little with μ(N), but for longer photoperiods θ increases rapidly with decreases in μ(N).

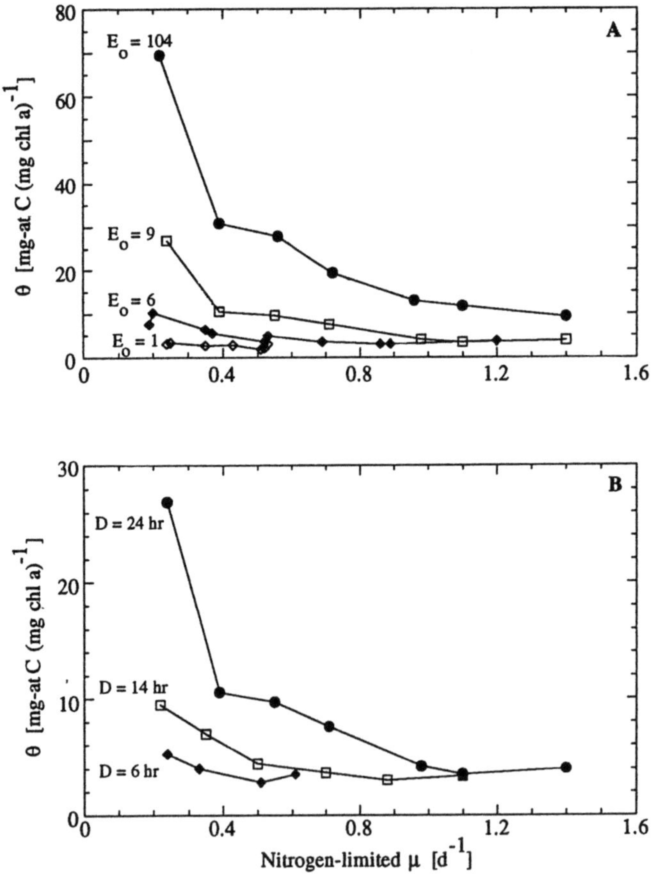

Fig. 2. The carbon-to-chlorophyll *a* ratio of *Skeletonema costatum* as a function of specific growth rate for cells growing at various light intensities (A) and photoperiods (B).

In figure 3A, the quantum yield at a given irradiance and photoperiod is plotted as a function of the nutrient-limited specific growth rate. We note that the values for Φ_C are larger the smaller the values of E_O, and that the response to nutrient-limitation differs with irradiance. At low light levels, Φ_C decreases sharply with decreases in $\mu(N)$, but at high light levels Φ_C shows only a small increase with $\mu(N)$.

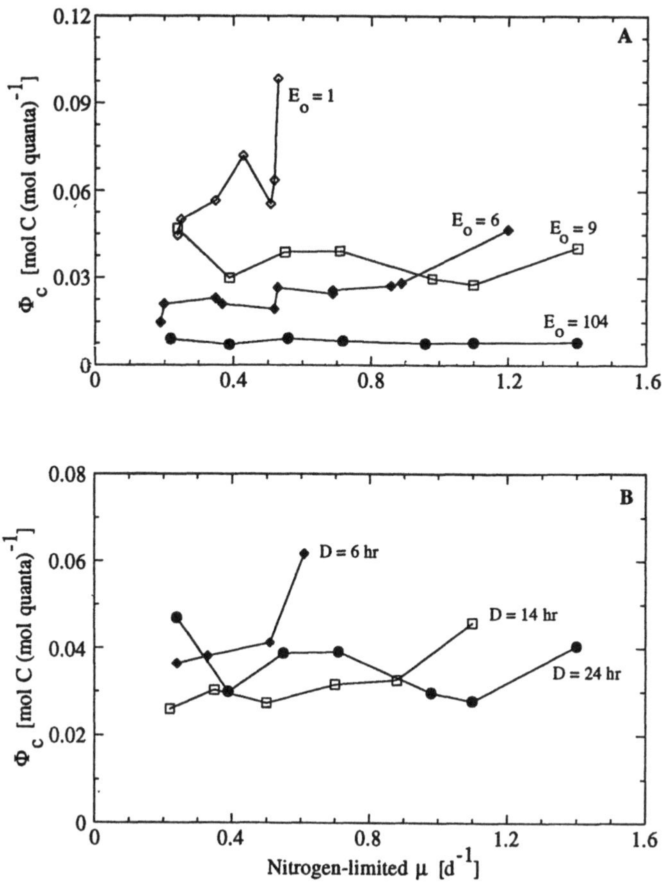

Fig. 3. The quantum yield of photosynthesis for *Skeletonema costatum* as a function of specific growth rate for cells growing at various light intensities (A) and photoperiods (B).

In figure 3B, we note that differences in photoperiod have relatively little effect upon quantum yield. A comparison of figures 3A and 3B therefore indicates that in terms of quantum yield, acclimation to photoperiod is distinctly different from acclimation to light level. Similar to the case for light level, however, changes in Φ_C caused by changes in nutrient supply differ with photoperiod. With long photoperiods Φ_C changes little with $\mu(N)$, but with shorter photoperiods Φ_C appears to decrease with decreases in $\mu(N)$.

The dichotomy mentioned earlier thus appears to hold for nutrient limitation as well as for temperature limitation. When light levels are sufficiently low and photoperiods sufficiently short to limit the growth rate, θ is near the minimum value and varies little with nutrient supply; on the other hand, Φ_C decreases with decreases in nutrient supply. When light levels are high and photoperiods sufficiently long to maintain high growth rates, θ increases and Φ_C remains constant with decreases in the rate of nutrient supply.

Acclimation to photoperiod

In figure 4A the specific growth rate of *S. costatum* is plotted as a function of irradiance for three different photoperiods and a constant temperature. Because the data was obtained by determining the maximum specific growth rates at the given irradiance and photoperiod, the response is considered to be that of nutrient-saturated growth. We see from this figure that for a photoperiod of 24 hours, an irradiance of roughly 0.4 mol m^{-2} h^{-1} is sufficient to saturate growth.

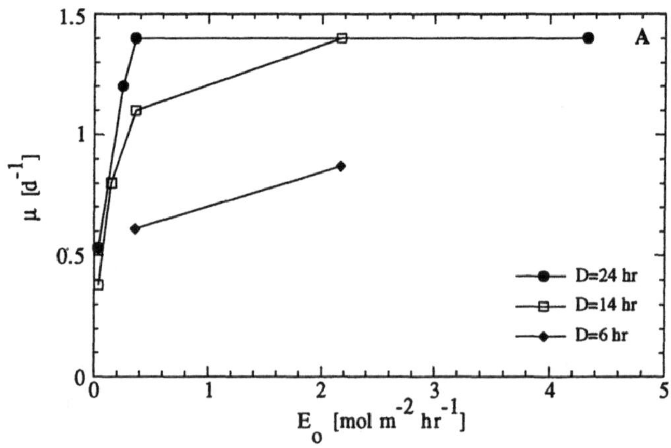

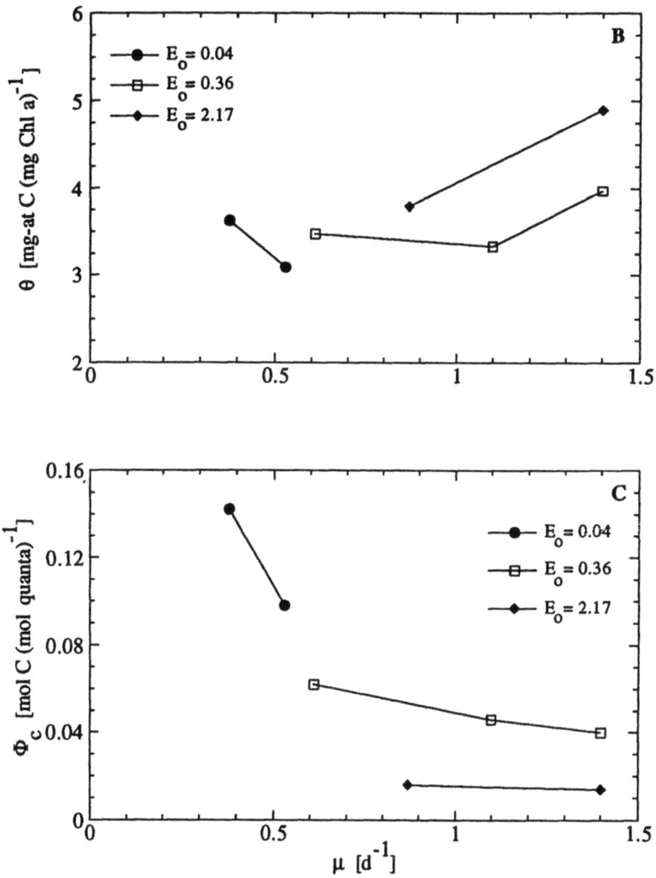

Fig 4. (A) Specific growth rate of *Skeletonema costatum* as a function of irradiance for cells growing at various photoperiods. (B) Changes in the carbon-to-chlorophyll *a* ratio and the quantum yield of photosynthesis (C) of *Skeletonema costatum* as a function of photoperiod-limited growth rate for cells growing at various light intensities.

With shorter photoperiods, a higher irradiance is required to reach maximum growth rate. At the shortest photoperiod (6 hour light period), growth rate is far from saturated at an irradiance of 2.2 mol m^{-2} h^{-1}.

In figure 4B the carbon-to-chlorophyll *a* ratio at a given irradiance is plotted as a function of the photoperiod-limited specific growth rate, $\mu(D)$. Similarly, in figure 4C the quantum yield for a given irradiance is plotted as a function $\mu(D)$. Unfortunately, because of the limitations in number and range of measurements it is difficult to draw definitive conclusions. Figure 4B indicates that at these three irradiances θ appears unchanged by changes in photoperiod. Figure 4C indicates once again that the value of Φ_C is smaller at higher light levels.

Model

Conceptual model

Although the photosynthetic response of phytoplankton to varying light intensity has been examined in considerable detail (e.g. Ryther and Yentsch 1957, Webb et al. 1974, Jassby & Platt 1976, Geider 1990), these models do not attempt to describe the process of acclimation (Cullen 1990), rather they describe changes in photosynthesis-light intensity relationship (P vs I) with acclimation can be well described by obtaining best-fit values for several P vs I parameters. The data for *Skeletonema costatum* presented above describes the growth rate and cellular concentration of chlorophyll for fully acclimated cells, and thus a more detailed description of photosynthetic growth is required than that offered by the instantaneous photosynthetic response function. Numerous models of acclimation by phytoplankton have been published (Bannister 1979, Kiefer and Enns 1976, Bannister and Laws 1980, Shuter 1979, Kiefer and Mitchell 1983; Sakshaug et al. 1989; Smith 1980; Laws and Bannister 1980, Fasham & Platt 1983, Geider 1987, Baumert 1988, Chalup and Laws 1990), and this work has aided in the development of the model presented below.

The model I propose here attempts to mathematically describe the scheme shown in figure 5. Photosynthetic growth is described in terms of discrete photosynthetic units supplying electrons to a pool of reductant that drives the reduction of carbon dioxide and the synthesis of cellular material. I have chosen to explicitly consider only the properties of Photosystem II. In this scheme the pool of electron transport compounds is symbolized by A^{ox} / A^{re}, the oxidized and reduced forms of the carrier pool. Electrons are supplied to this pool by the activity of Photosystem II, symbolized by Chl_{II} / Chl_{II}^*, the ground and excited states of chlorophyll in the photosystem. Although not depicted in this figure, each photosystem is considered to consist of an antenna and a reaction center. Photosystem I is not shown but is assumed to be under metabolic control (Foyer et al.1990) and does not restrict electron flow. While rates of electron supply will depend directly upon the size and cellular concentration of the photosystems and upon light intensity, rates will also be regulated by photoperiod, temperature, and nutrient concentration. Although evidence is limited, it appears that the cellular concentration of the electron transport pool is proportional to the concentration of Photosystem II (Sukenik et al. 1987). Electrons are lost from the electron transport pool by the assimilation of inorganic compounds such as carbon dioxide and nutrients. While the rates of electron assimilation will depend directly upon the cellular concentration of the enzymes of the dark reactions and temperature, rates will also be regulated by photoperiod, nutrient concentration and light intensity.

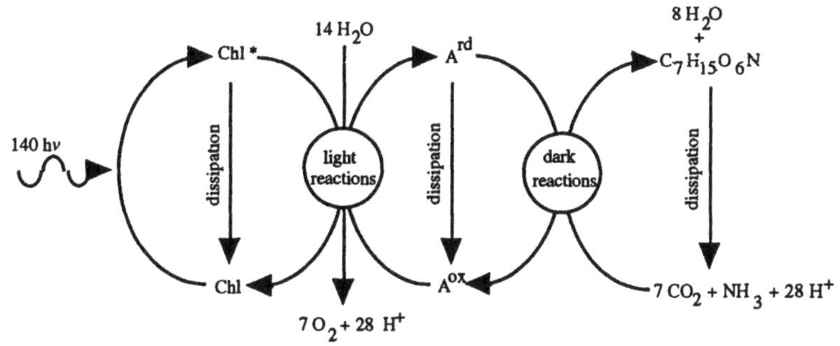

Fig. 5. A conceptual model of growth and adaptation in microalgal cells. Photoautotrophic growth is described as a system of discrete photosynthetic units supplying electrons from a reductant pool, which subsequently reduces carbon dioxide for the synthesis of new cellular material.

Transformations of the system include the dissipation of free energy in Photosystem II by fluorescence and heat production, the dissipation of heat from the electron transport system by short circuiting of electrochemical gradients, and the dissipation of heat and excreted compounds by the "photorespiratory" pathways of the dark reactions. Of course, such dissipation causes reductions in the quantum yield of carbon assimilation. It is assumed that the process of metabolic regulation involves adjustments to maintain growth rate and to minimize the costs that are associated with dissipation. Such regulation can generally be achieved by adjusting the rate of supply of electrons from the light reactions with rates of utilization by the dark reactions. Specifically, this is achieved by varying the cellular concentrations of all three components shown in figure 5: Photosystem II, the electron transport pool, and the enzymes of the dark reactions.

Queing theory

The mathematical description of the scheme shown in figure 5 was derived by combining the concepts of queuing theory (Evans and Fasham 1993, this volume) with the definitions of bio-optical properties described in the introduction. According to this approach, the carbon-specific rate of photosynthesis is determined by the general properties of discrete photosynthetic units. Specifically, the carbon-specific rate of photosynthesis is, under steady-state conditions, a function of the cellular concentration of photosynthetic units, η, the mean rate at which photons will be supplied to the reaction center of the photosynthetic unit, X, and the mean time required for the photosynthetic unit can transform the photon into an electron equivalent of cellular material, τ. I define η as the number of photosynthetic units per unit cellular carbon atom (photosynthetic units (carbon atom)$^{-1}$), X as the rate of photon supply to the reaction center of a photosynthetic unit (photons s^{-1}), and τ as the time required for photochemical transformation (one photon to one electron) by the reaction center (s).

Following the derivation of Asknes (1991), we note that during the time interval t1 N photons will be supplied to a single reaction center,

$$N = X\, t1$$

The time required by the reaction center to process this number of photons, t2, is:

$$t2 = N\, \tau$$

The mean rate at which a single photosynthetic unit transforms photons into electrons, ν, is determined by the Poisson distribution.

$$\nu = \frac{N}{t1 + t2} = \frac{X}{1 + Xt}$$

The rate of electron flow per unit cellular carbon is simply the product $\nu\eta$, which I represent as j_e (mol electrons (mol carbon)$^{-1}$)

$$j_e = \frac{\eta X}{1 + X\tau} \tag{2}$$

The rate of light absorption by the photochemical units is the product ηX, thus the quantum yield of photosynthetic electron flow can be expressed as:

$$\Phi_e = \frac{j_e}{\eta X} = \frac{1}{1 + X\tau} \tag{3}$$

One sees that the quantum yield is solely a function of the dynamics of each photochemical unit. Substitution of the quantum yield term into eq. 2 yields a formulation that for a given steady-state, the specific rate of photosynthetic electron flow increases linearly with increases in η and $1/\tau$, and (superficially contradictorily) with decreases in the quantum yield:

$$j_e = \frac{\eta(1 - \Phi_e)}{\tau} \tag{4}$$

Given this foundation, I now define η, X, and τ in terms of bio-optical parameters. The carbon-specific rate of photosynthesis, g, is the product of the carbon-specific rate of photosynthetic electron flow, the steady-state stoichiometry of carbon fixation to electron flow, j_{ce}, and the photoperiod, D:

$$g = j_e j_{ce} D = \frac{j_{ce} \eta X D}{1 + X\tau} \tag{5}$$

The photochemical unit is characterized by the photochemical cross section of $\sigma(\lambda)$, which itself is equal to the product of the absorption cross section of the unit, $\sigma_a(\lambda)$, and the probability of transfer to the reaction center, Ψ (e.g. Butler 1978). Thus, the rate of supply of photons to the reaction center is:

$$X = \psi\sigma_a E_o \tag{6}$$

I now substitute eq. 6 into 5 to yield a general description of acclimation:

$$g = \frac{j_{ce}\psi\eta\sigma_a E_o D}{1 + \psi\tau\sigma_a E_o} \tag{7}$$

Phytoplankton may acclimate to differences in growth conditions by varying all the dependent variables in this equation. Although information is limited, biophysical studies of acclimation indicate that η and τ are more variable than $\sigma a, \psi$, and j_{ce} (Falkowski et al. 1985). In the case of acclimation to irradiance, decreases in light intensity cause increases in the η and τ. In addition, σ_a increases but to a lesser extent. In the case of acclimation to nutrient supply, decreases in nutrient concentration cause decreases in η and small increases in τ. Interestingly, σ also increases. Although there are few biophysical studies of acclimation to temperature, the limited studies available indicate that decreases in temperature appear to cause decreases in η, and decreases in γ. By analogy with nutrient limitation, I suggest that $\sigma_a\psi$ also increases but to a lesser extent. Finally, the data of *Skeletonema* suggests that decreases in photoperiod cause increases in η, and that τ, σ_a, and ψ change little.

Relationship between biophysical (queing) and physiological parameters

Because the *Skeletonema* database does not include any direct measurements of the dependent variables in eq. 7, one is forced to seek additional relationships that link these biophysical parameters to other physiological parameters that have been repeatedly measured and are better understood.

We now define the photochemical unit as Photosystem II, the unit which oxidizes water and produces oxygen. In the equations above η_{II} is defined as the molar ratio of the cellular concentration of Photosystem II to carbon, τ_{II} is the time required for an exciton that reaches the reaction center of Photosystem II to be converted to a stable, reduced carbon compound, σ_{aII} is the absorption cross section of Photosystem II, and ψ_{II} is the probability that a photon absorbed by the antenna of Photosystem II reaches the reaction center rather than lost as heat or fluorescence.

The carbon-specific absorption cross section contributed by Photosystem II, a_{II} is equal to the product $\eta_{II}\ \sigma_{aII}$. The carbon-specific absorption coefficient of the cell suspension is the sum of contributions from the pigments of Photosystem I, a_I, Photosystem II, a_{II}, and nonphotosynthetic pigments, a_{np}. The fraction of cellular absorption contributed by Photosystem II, f_{II}, is defined by the ratio:

$$f_{II} = a_{II}\ (a_{II} + a_I + a_{np})^{-1}$$

Thus, the carbon-to-chlorophyll ratio of the cell is related to the concentration and absorption cross section of Photosystem II:

$$\theta = a_{chl} f_{II}\ (\eta_{II}\sigma_{aII})^{-1} \tag{8}$$

and the maximum quantum yield of photosynthesis, Φ_c^m, is related to the stoichiometric coefficient and the efficiency with which a photon that has been absorbed by the photosystem is transferred to the reaction center:

$$\Phi_c^m = j_{ce} f_{II} \Psi_{II} \tag{9}$$

In addition, the maximum instantaneous carbon-specific rate of photosynthesis, P^m, can be related to the concentration and turnover time of Photosystem II and to the stoichiometric coefficient:

$$P^m = j_{ce}\eta_{II}\ (\tau_{II})^{-1} \tag{10}$$

Substitution of eq. 8, 9, and 10, into eq. 7 yields the following formulation for the daily carbon-specific rate of photosynthesis:

$$g = \frac{\Phi_c^m a_{chl} E_0 D P^m}{\Phi_c^m a_{chl} E_0 + P^m \theta} \tag{11a}$$

The instantaneous, carbon-specific rate of photosynthesis, P, is:

$$P = \frac{\Phi_c^m a_{chl} E_o P^m}{\Phi_c^m a_{chl} E_o + P^m \theta} \tag{11b}$$

and the quantum yield is:

$$\Phi_c = \frac{\Phi_c^m P^m \theta}{\Phi_c^m a_{chl} E_o + P^m \theta} \tag{12}$$

In eq. 11a I chose to characterize the relationship as that involving two independent variables, Eo and D; two constants, Φ_c^m and a_{chl}; and three dependent variables, P^m, θ, and g. There are also two implicit independent variables, temperature and nutrient supply, which like E_o and D effect the values of all three dependent variables. The key assumption of this model is the constancy of the product Φ_c^m a_{chl}. This assumption is necessitated by the fact that neither of the two parameters were measured in the studies of *Skeletonema*. Φ_c^m and a_{chl} can be easily measured in the laboratory and much less easily in the field. Referring to eq. 8 and 9, one sees that:

$$\Phi_c^m a_{chl} = j_{ce} \psi_{II} \eta_{II} \sigma_{aII} \theta$$

This relationship indicates that this assumption requires that at different acclimated states the stoichiometry of carbon fixation and electron flow remain constant, that the efficiency of energy transfer from the antenna of Photosystem II to the reaction center remains constant, and that the total carbon-specific absorption cross section of Photosystem II units, η σ_{aII}, be inversely proportional to the carbon-to-chlorophyll ratio. This third requirement would be violated if there are significant variations in the distribution of chlorophyll *a* between Photosystems I and II.

In order to obtain a unique mathematical description of acclimation by *Skeletonema* , we will search for two additional formulations; one of the dependence of P^m upon the four environmental variables, and the other of the dependence of θ upon these same variables.

Formulation of P^m

The maximum, *instantaneous*, carbon-specific photosynthetic rate, P^m, is a measure of the cell's maximum or light-saturated capacity to fix carbon (units of h^{-1}). Thus, it is closely related to the maximum or light-saturated, *daily*, carbon-specific growth rate of the cells, g_{sat}. In fact, if for a given temperature and rate of nutrient supply the photoperiod is sufficiently long and the irradiance is sufficiently large to achieve a maximal growth rate, then it is reasonable to expect that:

$$P^m = \frac{g_{sat}(N,T)}{D} \tag{13}$$

In other words, the *instantaneous* carbon-specific photosynthetic rate of cultures growing under light saturation at g_{sat} is expected to be inversely proportional to the photoperiod. This simple relationship may not hold when the cultures have daily carbon-specific photosynthetic rates that are less than g_{sat}. For example, at low irradiance or short photoperiods when g is less than g_{sat}, one could argue that the cells would increase P^m in order to increase growth rate. On the other hand, one might also argue that under such conditions of low energy supply the metabolic costs of maintaining large values of P^m would be too high and the cell would decrease P^m. In addition, if the photoperiod is extremely short, the value of P^m as calculated from eq. 13 may exceed the maximum, instantaneous rate that the cells can photosynthesize.

In order to examine the behavior of P^m in the *Skeletonema* database I have introduced measured values of E_o, D, g (μ=g), and θ into eq. 11a and solved for P^m. I have assumed that the values of a_{chl} and Φ_c^m are 16 m^2 (gm Chl)$^{-1}$ and 0.1 gm-at carbon (mol photon)$^{-1}$, respectively. Invoking eq. 13, I examined the ratio P^m D / $\mu_{sat}(N,T)$ as a function of E_o for the range of temperatures and rates of nutrient supply found in the database. The results indicate that at high irradiance the value of this ratio is close to unity. At lower irradiance this ratio tends to increase, suggesting that the cells respond to limitations in irradiance by increasing P^m. This response is weak, a two to three-fold variation for a change in irradiance that ranges over two orders of magnitude, and the scatter is large. For the reasons mentioned above, eq. 13 will not be valid for cells growing at extremely short photoperiods when P^m would exceed some maximal value.

Unfortunately, the *Skeletonema* database is not complete enough to establish this maximal value for P^m.

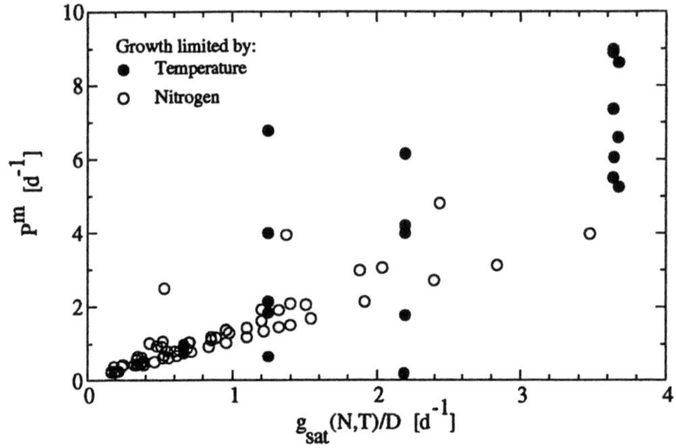

Fig. 6. The maximum, carbon-specific rate of photosynthesis as a function of the predicted value calculated from eq. 19.

In figure 6, the value for P^m calculated from the database is described as a function of the value predicted from eq. 13. The fit is judged satisfactory; thus it appears that the maximum instantaneous carbon-specific photosynthetic rate is a function of nutrient concentration, temperature, and photoperiod but not irradiance.

Formulation of θ

The search for a formulation of the relationship between the ratio of carbon-to-chlorophyll *a*, θ, was based upon two general ideas. First, θ was considered to vary principally with the cellular concentration of Photosystem II units. When the daily supply of photons, E_0D, was small relative to the nutrient and temperature-limited growth rate, $\mu_{sat}(N,T)$, then acclimation would require a high cellular concentration of photosynthetic units and a high value for θ. Likewise, when the daily supply of photons was large relative to the nutrient and temperature-limited growth rate then acclimation would require a low cellular concentration of photosynthetic units and a low value for θ. Second, I speculated that the maintenance of a high cellular concentration of photosynthetic units involves a metabolic cost and that there exists a maximum concentration of units at which growth stops. As the daily supply of photons decreases either because of decreases in irradiance or photoperiod, the signal for the induction of increases in the cellular concentration of photosynthetic units is modified by a negative feedback signal that increases with cellular concentration of photosynthetic units (see Raven 1984).

In order to incorporate the first idea into the formulation of θ, I have included in the equation the nutrient-limited and temperature-limited photosynthetic rate, g(N,T), as well as the irradiance and photoperiod. This index is the ratio of material limitation to energy limitation of growth:

$$\frac{E_0 D}{g_{sat}(N,T)}$$

The second idea was incorporated into the formulation of θ by searching for a function that asymptotically approaches a minimum value of θ^{min} at small and decreasing values for the ratio of radiant energy to nutrient supply, and has a slope equal to Φ_c for large values of this ratio. As illustrated in figure 7, a hyperbola describes well this relationship:

$$\theta = \sqrt{\theta^{min^2} + \left(\frac{\Phi_c^m \, a_{chl} \, E_o \, D}{b \, g_{sat}(N,T)} \right)^2} \tag{14}$$

The values of θ^{min} and b for the best-fit curve are 3 and 40, respectively.

Equations 11a, 11b, 13, and 14 are a bio-optical model of light absorption and growth by *Skeletonema* in a rectified light regime. The inputs to this model include photoperiod, irradiance, and the growth rate at light saturation when either temperature or nutrient concentration are limiting. Although the model was derived from studies of the growth in a rectified photoperiod, I suggest that the model will also apply to growth in a natural light regime in which light levels increase from sunrise to midday and decrease from midday to sunset. Such an assertion is supported by the fact that the two formulations used to calculate the value of the acclimation parameters, P^m and θ, do not require irradiance alone as an input: P^m depends upon the photoperiod (D in eq. 13) and θ depends upon the daily dose of photons (E_oD in eq. 14). In the case of a natural light field, when E_o varies with time, eq. 11b would replace eq. 11a, and the daily carbon-specific rate of photosynthesis would be calculated by integrating this equation from sunrise to sunset. The time scale for changes in P^m and θ will be roughly a day.

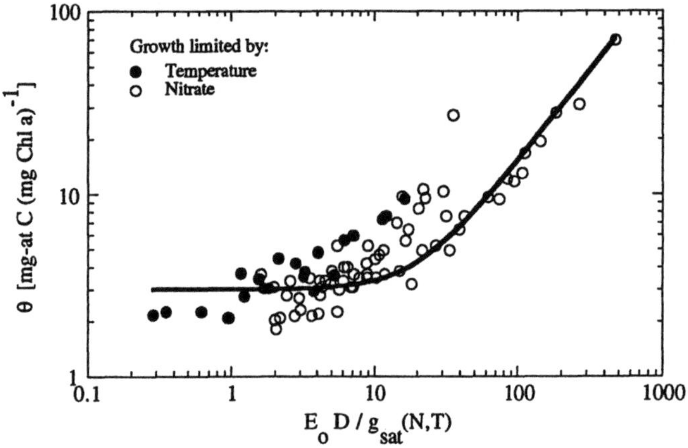

Fig. 7. The carbon-to-chlorophyll a ratio as a function of normalized light level for growth limited by temperature or nitrogen supply. This normalization compares the ratio of material limitation to energy limitation of growth. The line represents a fitted hyperbolic function (eq. 14).

A still more general model would include a formulation for temperature-limited growth, g(T), such as Eppley's formulation:

$$g(T) = a_T\, e^{b_T T}$$

and a formulation of nutrient limited-growth, g(N), such as the Michaelis-Menten function:

$$g(N) = \frac{g^m N}{k_N + N}$$

I invoke Leibig's Law of the most limiting variable: g(N,T) is the smaller of the two values for g(N) and g(T). We now have a description of acclimation in

Skeletonema that is based upon the four environmental variables; temperature, nutrients, photoperiod, and irradiance.

In order to examine this model, I compared growth rates predicted by three models of increasing complexity with rates measured in the *Skeletonema* database. All three models are based upon the phenomenological eq. 1 in which the carbon-specific rate of photosynthesis, g, is formulated as the product of the carbon-specific rate of light absorption, $a_{chl}E_o/\theta$, and the quantum yield, Φ. In the simplest model, figure 8A, Φ is constant. In this figure measured values of the specific growth rate, μ, are plotted as a function of E_oD/θ, a factor that is proportional to the daily, carbon-specific rate of light absorption. One sees that some variability in the variance of μ is associated with variability in the rate of light absorption, but a large amount still remains.

In the second model, figure 8B, Φ is formulated as a function of light level alone and independent of the other three environmental variables. In this model values of g are calculated by introducing values for Eo, D, Φ, and θ into eq. 1.Values for Eo, D, and θ come from the database, while the values for Φ were calculated from eq. 12 assuming a constant "best-fit" value for the product $P^m\,\theta$. Such a model has been presented by Kiefer and Mitchell (1983) and has in a limited number of tests in the field (Marra et al. 1992) provided good estimates of photosynthetic rate. By comparing figures 8A and 8B, we see that the addition of a formulation for light dependent variations in quantum yield increases the accuracy of prediction.

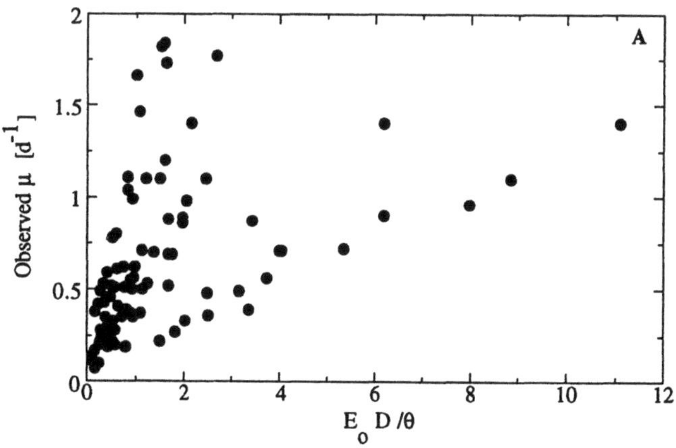

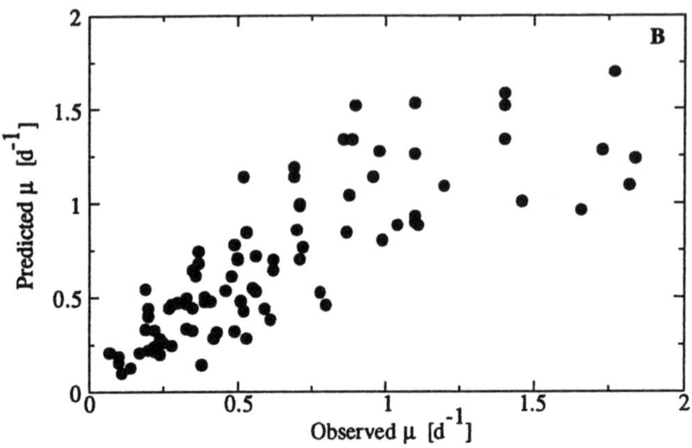

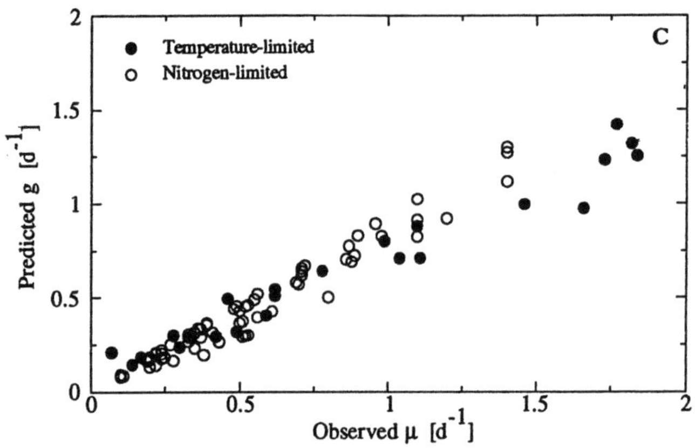

Fig. 8. A comparison of various growth models with measured values for Skeletonema costatum. (A) Observed growth rate as a function of the daily rate of light absorption. (B) Comparison of observed growth rates with a model which includes light dependent variation in the quantum yield of photosynthesis. (C) Comparison of observed growth rates with a detailed model which includes adaptation to temperature, photoperiod, irradiance and nitrogen supply.

In the third model, figure 8C, Φ is formulated as a function of all four environmental parameters, light level, daylength, temperature, and nutrient availability. In this model values of g are calculated from eq. 1. Values for E_0,

D, and θ come from the database. Values for Φ were calculated by introducing measured values for E_o,and calculated values for θ (eq. 14) and P^m (eq. 13) into eq. 12. The values for $g_{sat}(N,T)$ used in the calculation of P^m and θ were culled from the database: in the case of the nutrient-limited chemostat, I simply set $g_{sat}(N,T)=\mu$ for each steady state, and in the case of the temperature-limited turbidostat, I assigned $g_{sat}(N,T)$ a value equal to μ for the brightest light level at a given temperature. A comparison of figures 8A, 8B, and 8C clearly indicates that the most complex model provides the best accuracy.

Conclusions

Examination of the steady-state growth rate and cellular concentration of chlorophyll *a* in *Skeletonema* indicates several basic features of acclimation to differing light intensities, photoperiods, temperatures, and nutrient concentrations. First, the response of *Skeletonema* to variations in these four environmental variables is at least in part characterized by changes in photosynthetic quantum yield and the cellular ratio of carbon-to-chlorophyll *a*. At high irradiance, limitations to growth rate caused by decreases in temperature, nutrient supply, or photoperiod cause large increases in the ratio of carbon-to-chlorophyll *a* and very small decreases in quantum yield. On the other hand, at low irradiance limitations to growth rate caused by decreases in temperature, nutrient supply, or photoperiod cause large decreases in quantum yield and very small increases in the ratio of carbon-to-chlorophyll *a*.

Second, the quantitative relationship between the quantum yield and the rate of carbon fixation appears to conform to the principles of queuing theory. The growth rate of the cells increases with increases in the rate at which photosynthetic units absorb photons, the rate at which the units can carry out photochemical transformations, and the cellular concentration of the units. On the other hand, while the quantum yield of photosynthesis increases with the rate at which the units can carry out photochemical transformations, it decreases with the rate at which photosynthetic units absorb photons, and is independent of the cellular concentration of photosynthetic units. The validity of queuing theory is supported by the observation that when light intensity is most limiting to growth rate, the quantum yield of photosynthesis always decreases with increases in growth rate (eq. 4). This is true at any temperature, nutrient concentration, and photoperiod. Stated in another way, under any condition for growth the photosynthetic quantum yield at a given irradiance is uniquely determined by the photosynthetic response curve.

Third, although the *Skeletonema* database does not provide information on the cellular concentrations, absorption cross sections, or the processing rates of the photosynthetic units, the ratio of the cellular concentration of carbon-to-chlorophyll *a*, the maximum instantaneous carbon-specific rate of photosynthesis, and the light intensity provides information that is closely related to the hit rate, processing time, and concentration of the photosynthetic units. Specifically, the product $P^m \theta$, which is the maximum chlorophyll-specific rate of photosynthesis, and E_o uniquely determines the photosynthetic quantum yield, and thus reflects the hit rate and processing rate of the photosynthetic units. In addition P^m, θ, and E_o determine the daily carbon-specific rate of photosynthesis and thus reflect the hit rate, processing rate, and cellular concentration of the photosynthetic units.

Fourth, differences in the ratio of cellular carbon-to-chlorophyll *a*, and the maximum instantaneous carbon-specific rate of photosynthesis caused by differences in temperature, nutrient concentration, photoperiod and irradiance are quite distinct. P^m increases with increases in temperature and nutrient supply and with decreases in photoperiod; furthermore, it appears to be insensitive to changes in irradiance. More specifically, P^m appears to be well approximated by the ratio of the light-saturated daily rate of photosynthesis (which is determined by temperature and nutrient concentration), to the photoperiod (eq. 13). θ decreases with increases in temperature and nutrient supply and decreases in irradiance and photoperiod. The response of θ to changes in the four environmental parameters (eq. 14) is indexed by ratio of the daily dose of photons to the maximum capacity of the cells to fix carbon. The response is complicated by the fact that there is a minimum ratio of carbon-to-chlorophyll *a* that is approached asymptotically at decreasing irradiance and photoperiods.

Acknowledgments

I thank Rick Reynolds for assistance in development of the model and preparation of this manuscript. The continued support of the Office of Naval Research and the National Aeronautics and Space Administration is duly acknowledged.

References

Asknes, and Egge. 1991. A theoretical model of nutrient uptake in phytoplankton. Mar. Ecol. Prog. Ser. 70:65-72.

Bannister, T.T. 1979. A general theory of steady state phytoplankton growth in a nutrient saturated mixed layer. Limnol. Oceanogr. 14: 386-391.

Bannister, T.T. and E.A. Laws. 1980. Modeling phytoplankton carbon metabolism. in Primary Productivity in the Sea (P.G. Falkowski, ed). Plenum Press, New York. pp. 243- 258.

Baumert, H. 1988. Beitrag zur Physik und numerischen Simulation von Oberflachengewassern unter Berucksichtigung der Wasserbeschaffenheit (Chapter 5). Dissertation, Dr. sc. nat. Technischen Universitat Dresden. 191 pp.

Butler, W.L. 1978. Energy Distribution in the Photochemical Apparatus of Photosynthesis. Ann. Rev. Plant Physiol. 29:345-378.

Chalup, M.S. and E.A. Laws. 1990. A test of the assumptions and predictions of recent microalgal growth models with the marine phytoplankter *Pavlova lutheri*. Limnol. Oceanogr. 35: 583-596.

Cullen, J.J. 1990. On models of growth and photosynthesis in phytoplankton. Deep-Sea Res. 37: 667-683.

Dubinsky Z, P.G. Falkowski, K. Wyman. 1986. Light Harvesting and Utilization by Phytoplankton. Plant Cell Physiol. 27(7):1335-1349.

Falkowski, P.G., Z. Dubinsky, and K. Wyman. 1985. Growth-irradiance relationships in phytoplankton. Plant Cell Physiol. 27: 1335-1349.

Fasham, M.J.R. and T. Platt. 1983. Photosynthetic response of phytoplankton to light: a physiological model. Proc. R. Soc. Lond. 219: 355-370.

Evans, G. and M.J.R. Fasham. this volume.

Foyer, C., R. Furbank, J. Harbinson and P. Horton. 1990. The mechanisms contributing to photosynthetic control of electron transport of carbon assimilation in leaves. Photosynthesis Res. 25: 83-100.

Geider, R.J. 1987. Light and temperature dependence of the carbon to chlorophyll *a* ratio in microalgae and cyanobacteria: implications for physiology and growth of phytoplankton. New Phytologist 106: 1-34.

Geider, R.J. 1990. The relationship between steady state phytoplankton growth and photosynthesis. Limnol. Oceanogr. 35: 971.

Geider, R.J. 1992. Respiration: Taxation without representation? in Primary Productivity and Biogeochemical Cycles in the Sea (P.G. Falkowski and A.D. Woodhead, ed). Plenum Press, New York. pp. 333-360.

Jassby, A.D. and T. Platt. 1976. Mathematical formulation of the relationship between photosynthesis and light for phytoplankton. Limnol. Oceanogr. 21: 540-547.

Kiefer, D.A. and T. Enns. 1976. A steady state model of light, temperature, and carbon limited growth of phytoplankton. in Modeling of Biochemical Processes in Aquatic Systems (R.P. Canale, ed). Ann Arbor Science, Ann Arbor, MI. pp. 319-336.

Kiefer, D.A. and B.G. Mitchell. 1983. A simple, steady state description of phytoplankton growth based on absorption cross section and quantum efficiency. Limnol. Oceanogr. 28: 770-776.

Kolber, Z., J. Zehr, and P.G. Falkowski. 1988. Effects of growth irradiance and nitrogen limitation on photosynthetic energy conversion in Photosystem II Plant Physiol.88: 923.

Laws,E.A. and T.T. Bannister. 1980. Nutrient- and light-limited growth of *Thalassiosira fluviatilis* in continuous culture with implications for phytoplankton growth in the ocean. Limnol. Oceanogr. 25: 457-473.

Marra, J. et al. 1992. Estimation of seasonal primary production from moored optical sensors in the Sargasso Sea. J. Geophys. Res. 97: 7399-7412.

Raven, J. 1984. A cost-benefit analysis of photon absorption by photosynthetic unicells. New Phytologist 98: 593-625.

Ryther, J.H. and C.S. Yentsch. 1957. The estimation of phytoplankton production in the ocean from chlorophyll and light data. Limnol. Oceanogr. 2: 281-286.

Sakshaug, E., D.A. Kiefer and K. Andresen. 1989. A steady state description of growth and light absorption in the marine planktonic diatom *Skeletonema costatum* . Limnol Oceanogr. 34: 198-205.

Shuter, B. 1979. A model of physiological acclimation in unicellular algae. J. Theor. Biol. 78: 519-552.

Smith, R.A. 1980. The theoretical basis for estimating phytoplankton production and specific growth rate from chlorophyll, light, and temperature data. Ecol. Modelling 10: 243-264.

Sukenik, A., J. Bennett and P.G. Falkowski. 1987. Light-saturated photosynthesis: limitation by electron transport or carbon fixation? Biochim. Biophys. Acta: 891-905.

Webb, W.L., M. Newton and D. Starr. 1974. Carbon dioxide exchange of *Alnus rubra* : A mathematical model. Oecologica (Berlin) 17: 281-291.

Yoder, J.A. 1979. Effect of temperature on light-limited growth and chemical composition of *Skeletonema costatum* (Bacillariophyceae). J. Phycol. 15: 362-370.

CARBON: A PHYCOCENTRIC VIEW

John A Raven
Department of Biological Sciences
University of Dundee
Dundee DD1 4HN
UK

Introduction

This paper discusses quantitative aspects of the uptake and assimilation of dissolved inorganic carbon (DIC) by marine phytoplankton that are relevant to the global ocean C cycle. It draws on recent review articles by Raven (1991a,b) and Raven and Johnston (1991, 1992), but some parts are entirely new, for example concerning the influence of algal metabolism on the alkalinity of seawater and the release of dissolved organic C (DOC).

Background: C and RUBISCO

Compared to other resources, DIC varies little in the sea. Air-equilibrium seawater at 20°C contains ~12 mmol m^{-3} CO_2, and even the large departures from air equilibrium observed by Watson *et al* (1991) during a phytoplankton bloom in the North Atlantic reduced this value by less than 20%. Seawater contains more that 2000 mmol m^{-3} of total DIC, and this number varies by an even smaller percentage. Compare this with the concentration of nutrients such as NO_3^-, or photon flux, which vary by orders of magnitude.

Ribulose bisphosphate carboxylase-oxygenase (RUBISCO) is the central and key carboxylase enzyme in the photosynthesis of all O_2-evolving organisms, and its properties are fundamental to most of the ideas in this paper. RUBISCO is a large

NATO ASI Series, Vol. I 10
Towards a Model of Ocean
Biogeochemical Processes
Edited by G.T Evans and M.J R Fasham

molecule ($M_r \approx 550,000$), and its specific reaction rate (< 50 mol CO_2 mol^{-1} enzyme s^{-1} at 25°C) is lower than that of other enzymes of C assimilation or, indeed, of most enzymes, even when all cofactors and substrates, including CO_2, are present at saturating concentrations (Raven 1984, 1991a, b; Raven and Johnston 1991, 1992). This means that, even at CO_2 saturation, a larger fraction of cell protein must be devoted to RUBISCO than to most other enzymes to support the steady-state C flux through metabolic pathways. The situation is exacerbated by the relatively low affinity of the enzyme for CO_2, and by the use of O_2, competitively with CO_2, as an alternative substrate for the enzyme. In an air-equilibrated solution at 20-25°C, the RUBISCO of eukaryotic microalgae is less than half saturated with CO_2 and a very significant fraction of the C fixed is diverted, *via* the oxygenase activity, into phosphoglycolate (see later). Any restriction on CO_2 diffusion to the enzyme would make the situation even worse.

RUBISCO in prokaryotic marine phytoplankton (cyanobacteria, chloroxybacteria) is present in the cytosol; much of it is aggregated into carboxysomes. In eukaryotic marine phytoplankton RUBISCO occurs in the chloroplast stroma, and in most cases some of it is present in pyrenoids, structures possibly analogous to carboxysomes. These considerations are important in terms of the occurrence and identification of CO_2 concentrating mechanisms (discussed below).

CO_2 is the true substrate for RUBISCO, and it generally diffuses readily through biological membranes. These properties make it the obvious exogenous DIC source for photosynthesis. However, two considerations suggest that diffusive CO_2 entry with no catalysis by the cells is not the only mechanism of DIC uptake by phytoplankton cells. The considerations are the extent to which the biomass, and the growth rate, of marine phytoplankton *in situ* are limited by DIC.

Is inorganic C a limiting resource for growth of marine phytoplankton?

We address first the possibility that DIC limits phytoplankton biomass. If biomass contains 6.7 C atoms for every N atom (Redfield, 1958), an extracellular combined N supply of 1.8 mmol m^{-3} is needed to use the 12 mmol m^{-3} of CO_2 in air-equilibrium solution in constructing phytoplankton. There are many times and places, even in well-lit parts of the ocean, where the NO_3^- concentration is higher than 1.8 mmol m^{-3}, and so rapid primary production might in principle exhaust CO_2 before NO_3^-.

In practice, although it is quite common to observe seawater in which combined dissolved N is undetectable and CO_2 is in abundant supply, the reverse has never been observed. This means that, relative to the requirements for phytoplankton biomass production, combined N regeneration does not keep pace with that of CO_2, *via* invasion from the atmosphere and uncatalysed or catalysed (by phytoplankton) use of HCO_3^- as well as by respiration. Thus, C does not limit biomass (but see Raven, 1993, and Riebesell, Wolf-Gladrow and Smetacek, 1993, for a consideration of the possibility of C limitation of the rate of production).

We saw that CO_2 concentrations in seawater at 20-25°C are at, or below, the CO_2 half saturation value ($K_{½(CO2)}$) for marine phytoplankton RUBISCO, and that there are constraints on the extent to which low affinity for CO_2 can be offset by an increased quantity of RUBISCO per unit biomass. Thus, if CO_2 supply to RUBISCO were purely by diffusion with no catalysis by the cells, CO_2 may limit growth rate. However, enrichment studies show that the concentration of DIC in the sea is generally adequate to permit the maximum growth rate attainable under the prevailing concentrations of nutrient elements and flux of photons (Caperon and Smith, 1978). How does this happen?

How do phytoplankton acquire DIC?

Suppose for a moment that a phytoplankton cell contained a known amount of RUBISCO, behaving as it does *in vitro*, surrounded by membrane(s) and diffusion boundary layers of known permeability to CO_2 (P_{CO2}; m s^{-1}). Further suppose that RUBISCO limits the light- and DIC-saturated rate of photosynthesis. If the cell is photosynthesizing at half its maximal rate, then the internal concentration (C_c; mol m^{-3}) of CO_2 must be the known $K_{½}$ (mol m^{-3}) of RUBISCO. If CO_2 entry is by diffusion and the total DIC uptake by the cell is at rate J (mol DIC m^{-2} surface area s^{-1}) then, at steady state, assimilation balances diffusive input:

$$J = P_{CO2} (C_b - C_c) \qquad (1)$$

where C_b is the bulk phase concentration of CO_2 (mol m^{-3}). Solving for C_b, we have

$$C_b = C_c + \frac{J}{P_{CO2}} \qquad (2)$$

In fact, in many of the marine phytoplankton species examined (see Table 1, Column B) C_b is not only less than $C_c + J/P_{CO2}$, but much less than C_c. This is true even when the $K_{½(CO2)}$ of RUBISCO is adjusted to allow for a RUBISCO amount per cell in excess of that needed to account for *in vivo* light and DIC-saturated photosynthesis. Equation (1) accordingly does not apply, and it is generally agreed that the observations are best explained by the occurrence of a CO_2 concentrating mechanism (CCM). This involves an increase in the CO_2 concentration at the site of RUBISCO to values in excess of C_b during steady-state photosynthesis.

Further indirect evidence consistent with the occurrence of a CCM comes from columns C and D in Table 1. Column C related to the CO_2 compensation concentration, which is the ultimate C_b value achieved in a closed system containing photosynthetic cells in the light. Diffusive CO_2 exchange between the bulk medium and RUBISCO should lead to a CO_2 compensation concentration that is predictable from the *in vitro* properties of RUBISCO and the known pathways of dealing with the products of oxygenase activity. For algal RUBISCO and metabolic pathways at air-equilibrium O_2 concentrations in solution, C_b values at the compensation point should exceed 2-3 mmol CO_2 m^{-3}. In all of the cases tested (column C of Table 1) the observed values are well below these predicted values, and are most readily explained by invoking a CCM. Column D of Table 1 shows that almost all of the marine microalgae tested can use HCO_3^-, as tested by the ability to photosynthesise faster than uncatalysed HCO_3^- conversion to CO_2 in the medium can supply CO_2. Almost all organisms that are expressing a CCM can use HCO_3^-, but so can quite a number that are not; so column D provides only a consistency check. Since a CCM causes suppression of oxygenase activity it also prevents the inhibition of photosynthesis by air-equilibrium O_2 concentrations at normal seawater DIC levels which would occur with diffuse CO_2 entry. Data on this point are not common (Raven, 1984) but when available also indicate the occurrence of CCMs (not shown in Table 1).

These four comparisons suggest that CCMs occur quite widely in marine microalgae. However, the most direct evidence is the demonstration of an internal CO_2 concentration in excess of C_b during steady-state photosynthesis (column A of Table 1). The data show that, of the relatively few marine phytoplankters tested, the majority have a CCM according to this criterion, although the experimental conditions used might have favoured the expression of the CCM which is generally subject to induction at low CO_2 concentrations for growth and which shows higher accumulation ratios when experiments are conducted at lower DIC concentrations than occur in normal seawater (Rees 1984; Raven, 1991a, b; Raven and Johnston, 1991). However, taken at face value the data suggest that only the chlorophyte *Stichococcus minor* and the prymnesiophyte (coccolithophorid) *Emiliania huxleyi* rely on diffusive CO_2 entry, whereas the remaining strains exhibit a CCM. The more globally significant of the two organisms that appear to lack CCMs is

Table 1

Occurrence of DIC concentrating mechanisms in marine phytoplankton (from data in Table 2 of Raven, 1991b, with additional data on *Emiliania huxleyi* from Merrett, 1991, and Nimer, Dixon and Merrett, 1992, and on the nomenclature of micromonadophyte strain Ω 48-23, now *Pycnococcus provasolii*, from Guillard *et al.*, 1991).

Higher taxon	Genus, species	A	B	C	D
Cyanobacteria	*Coccochloris peniocystis*	+	+	+	+
	Oscillatoria woronichinii		+		
	Synechococcus sp.	+	+		
Chlorophyceae	*Dunaliella salina*	+			
	Dunaliella tertiolecta	+	+	+	+
	Stichococcus bacillaris		+		+
	Stichococcus minor		-		+
Micromonadophyceae	*Pycnococcus provasolii*		+		+
	Mantoniella squamata		+		+
Eustigmatophyceae	*Monallantus* sp.		+		
	Nannochloropsis occulata		+		
Bacillariophyceae	*Phaeodactylum tricornutum*	+	+	+	+
Prymnesiophyceae	*Emiliania huxleyi*	-	-		-
	Isochrysis galbana	+	+	+	
Rhodophyceae	*Porphyridium cruentum*		+	+	
	Porphyridium purpureum	+	+		
Cryptophyceae	*Chroomonas* sp.		+	+	
Dinophyceae	*Amphidinium carteri*		+	+	

In columns A - D, a + indicates that this indication of a CCM has been found; a - that it has been sought but not found; a blank that it has not been sought.

A: $[CO_2]$ is greater in the cell than in the bulk phase.
B: $K_{½}$ for CO_2 is too small to be consistent with equation (1), even after the adjustments mentioned in the text.
C: The CO_2 compensation concentration for living cells is lower than that for RUBISCO *in vitro*.
D: The cell has an ability to use HCO_3^- directly.

Emiliania huxleyi, both in its own right and as a representative of the coccolithophorids (Holligan and Groom, 1986; Holligan *et al*, 1983).

One might argue that a high $K_{½(CO2)}$ for photosynthesis need not imply a high $K_{½}$ for growth. However, the cases of substantially lower growth $K_{½(CO2)}$ than uptake/assimilation $K_{½}$ measured in short-term experiments relate to elements with a low and variable cell quota (Morel, 1987; Turpin, 1988), whereas the cell quota for C is high and fixed. Work on the freshwater cyanobacterium *Synechococcus leopolensis* shows that the $K_{½}$ for CO_2 for growth is indeed only slightly greater than that for photosynthesis (Turpin, Miller, Parslow, Elrifi and Canvin, 1985). Accordingly, the growth of *Emiliania huxleyi* and *Stichococcus minor* in air-equilibrium solution with all other nutrients, and light, optimally supplied is probably not C saturated. However, especially for *Emiliania huxleyi* which occurs in N-poor (and Fe-poor, Brand, 1991) waters, other nutrients are rarely optimally supplied.

The mechanism by which inorganic C is concentrated in the neighbourhood of RUBISCO is unresolved (Badger and Price, 1992). The two basic categories of mechanism by which such CO_2 concentration occurs in photosynthetic cells are

---

Footnotes to Table 1

The data used in generating this Table are for organisms grown in seawater-based media and ideally should involve air-equilibration of the medium with respect to free CO_2 and to O_2 throughout the culture period. This is crucial since the characteristics of inorganic C acquisition can vary with relatively small changes in concentration and speciation of inorganic C relative to those found in air-equilibrium seawater (Rees, 1984; Johnston and Raven, 1991, 1992 and unpublished). Some of the growth conditions described for the organisms whose properties are summarised in Table 1 suggest that air-equilibrium conditions were not maintained (Johnston and Raven, 1992 and unpublished) and the interpretation of the data in the Table should be considered with this in mind.

Although data from Nimer, Dixon and Merrett (1992) would suggest a + for *Emiliania huxleyi* in column D, they used cell densities that were too low to provide a rigorous test.

active transport of DIC across a membrane(s) without any intermediate involving a C-C bond to the C of CO_2 or HCO_3^-, and a mechanism analogous to C_4 metabolism of higher plants. This latter kind of mechanism involves fixation of DIC just inside the plasmalemma *via* a $(C_3 + C_1)$ carboxylation and regeneration of CO_2 near RUBISCO by a $(C_4 - C_1)$ decarboxylase. Such a process demands a specific location of these two categories of enzyme, and a means of transporting the intermediary C_4 dicarboxylic acid and C_3 monocarboxylic acid between these sites while preventing excessive CO_2 leakage from around RUBISCO. If these requirements are met, the C_4 mechanism could increase the CO_2 concentration around RUBISCO to a level above C_b, while increasing the concentration difference across the plasmalemma relative to a comparable cell with direct CO_2 diffusion to RUBISCO (see equation (1)), at the expense of metabolic energy input.

The alternative mechanism is the transport of HCO_3^- or CO_2 across the plasmalemma and/or a plastid envelope membrane by an energy dependent process which actively maintains a CO_2 concentration near RUBISCO higher than C_b. This is again dependent on prevention of excessive back leakage of DIC.

Distinguishing between these two categories of mechanism is not as easy as may appear at first sight. $(C_3 + C_1)$ carboxylases are ubiquitous in photolithotrophs, operating in parallel with RUBISCO in supplying intermediates unavailable from products of RUBISCO without further carboxylation (Luinenburg and Coleman, 1990; Raven and Farquhar, 1990). Further, some of the functions of the $(C_3 + C_1)$ carboxylations in this biosynthetic mode involve a subsequent $(C_4 - C_1)$ decarboxylation within seconds or minutes (Smith, Gauthier, Dennis and Turpin, 1992; Theodorou, Elrifi, Turpin and Plaxton, 1991). A combination of this latter circumstance with active DIC influx and a small back-leakage would give a 'false positive' result in the classic pulse and chase test for C_4-type metabolism in photosynthesis. In this test a short-term (seconds) 'pulse' of ^{14}C-DIC is followed by a 'chase' of $^{12}CO_2$. The transfer of ^{14}C from a pulse-labelled dicarboxylic acid into products of RUBISCO activity is a good test for C_4 metabolism when CO_2 supply to the $(C_3 + C_1)$ carboxylase is by diffusion, as in C_4 higher plants. However, in cells with a non-leaky DIC pump, any ^{14}C-DIC transported into the cell and subject to the (C_3 + C_1) carboxylation/(C_4 - C_1) decarboxylation mechanism will inevitably end up mainly as substrate for RUBISCO even when there is no 'formal' C_4 photosynthesis. Despite this bias toward finding C_4-like photosynthesis in cells with a CCM based on active DIC transport across membranes, the (incomplete) available data do not favour C_4-like metabolism as an obligate part of CO_2 supply to RUBISCO in marine microalgae with CCMs (Raven, 1984, 1990a; Johnston and Raven, 1989; Kerby and Raven, 1985).

Faute de mieux we are left with the active influx of inorganic C across a membrane (or membranes) as the basis for the marine phytoplankton CCM or, more correctly, CCMs, since they are probably polyphyletic (Raven, 1991a, b). The mechanism(s) of these active transporters, and even their location in eukaryotic cells, which have plastid envelope membranes as well as the plasmalemma between the external medium and RUBISCO, is unclear. This circumstance is in spite of much high quality experimentation (Coleman, 1991; Badger and Price, 1992). Both CO_2 and HCO_3^- can be actively transported across the plasmalemma and, in eukaryotes, there is some evidence for active DIC uptake into chloroplasts (Badger and Price, 1992). Such a role for the plastid envelope might help to reconcile the presence of a CCM with data on marine microalgae with CCM-like gas exchange characteristics indicating negligible DIC or CO_2 accumulation averaged over the whole cell volume when photosynthesising in air-equilibrated seawater. We note that such an inhomogeneous distribution is also implicit in any C_4 mechanism for CCMs, so that inhomogeneity of inorganic C concentration, with a higher CO_2 concentration around RUBISCO than elsewhere, cannot be used to distinguish between CCMs based on C_4 metabolism and those based on active DIC flux across membranes.

A currently favoured model for the operation of the cyanobacterial CCM involves the release of HCO_3^- from the transport mechanism into the cytosol regardless of the DIC species taken from the medium (Badger and Price, 1992). Much of the RUBISCO is aggragated into crystalline arrays termed carboxysomes which are believed to be the sole location of carbonic anhydrase inside the cells

(Badger and Price, 1992). This allows RUBISCO to be exposed to a high concentration of CO_2 without incurring as much leakage of CO_2 through the plasmalemma as would result from a uniformly high level of CO_2 throughout the cell caused by CO_2 release from the transporter to the cytosol and/or carbonic anhydrase activity throughout the cell. This proposed mechanism requires further testing; the same applies to an even greater extent to an analogous mechanism in eukaryotic phytoplankton using pyrenoids, the RUBISCO-containing structures in many algal chloroplasts, instead of carboxysomes.

Control of the expression of CCMs

Even when a CCM is genetically encoded in an organism, the necessary enzymes, transporters and structures may not be produced in all circumstances. Most relevant work is on freshwater microalgae and cyanobacteria. Aizawa and Miyachi (1986) and Aizawa, Nakamura and Miyachi (1986) show that, at external CO_2 concentrations some 50 times air-equilibrium, expression of the CCM is much reduced, particularly in eukaryotes. Rees (1984) and Johnston and Raven (1991, 1992, and unpublished) studied the effects on the diatom *Phaeodactylum tricornutum* of varying DIC (and free CO_2) for growth in the range 0.2-5.0 times the normal, air-equilibrated seawater value. These workers showed that the DIC and CO_2 affinity (i.e. $1/K_{½}$), and the capacity to maintain an intracellular CO_2 concentration in excess of that in the medium, varies inversely with, and rapidly adjusts to, the external DIC concentration. These data do not immediately address the influence on the CCM of the 20% draw-down of free CO_2 levels reported by Watson *et al* (1991). This rapid (over a period of hours) acclimation to external DIC availability means that experiments on photosynthetic characteristics as a function of experimental DIC concentration must be of short duration, otherwise acclimation to the experimental DIC concentrations will be significant. This constraint also applies to the examination of the responses of CCMs to N and photon availability during growth. Examples of these effects for N are seen for freshwater green microalgae where the CCM is expressed even at high DIC levels when combined N is scarce (Beardall, Griffiths and Raven, 1982; cf. Beardall, Roberts and Millhouse, 1991). As these authors point out, this may be because less of the N-rich RUBISCO is required for a given growth rate when the intracellular CO_2 concentrations are higher. For light limitation, Beardall (1991) showed that the CCM of the freshwater cyanobacterium *Anabaena* is expressed less at low photon flux densities for growth. This finding is related to the energetics of the CCM (see below).

A conceptual problem with the switch from diffusive CO_2 entry to a CCM relates to the low P_{CO2} for the plasmalemma which is required for some models of CCMs in order to reduce DIC back-leakage and the higher P_{CO2} needed for diffusive CO_2 entry. One possibility (Raven, 1991b) is that the high P_{CO2} needed for diffusive CO_2 supply is a function of a lipid-solution contribution at the low end of the range shown in Table 2, and some protein components of the membrane which mediates passive CO_2 transport. Conversion of these transporters into DIC pumps (requiring additional protein components) would reduce P_{CO2} to the 'lipid solution' value. However, it is unclear how DIC would be assimilated during the switch; a further problem is the different extent of switching as a function of external DIC concentration. More satisfying in this regard are the pyrenoid- or carboxysome-based mechanisms (see above) which do not necessarily call for large changes in P_{CO2} at the plasmalemma during the switch.

Use of the natural abundance of stable C isotopes as a relatively non-invasive probe of diffusive CO_2 entry

The natural abundance of stable isotopes of C is widely used in studies of photosynthesis in terrestrial plants with diffusive flux of CO_2 to RUBISCO to determine the extent to which the rate of photosynthesis under a given set of conditions is limited by diffusion of CO_2.

The general context of the definition of limitation in the analysis of metabolic control was outlined by Kacser and Burns (1973) and used in the context of C_3 photosynthesis in land plants by Ball, Woodrow and Berry (1987) and Woodrow and Berry (1988). Control analysis defines the control coefficient of a system component N with respect to flux J (C^J_N) as:

$$C^J_N = \frac{\partial J.N}{\partial N.J} = \frac{\partial \ln J}{\partial \ln N} \qquad (3)$$

The sum of all control coefficients in a system is unity. Thus, for the flux of CO_2 through a number of components in series, the control coefficient for one component is a measure of the fraction of the total limitation which resides in that step. If we consider photosynthesis in a cell with diffusive supply of CO_2 to RUBISCO, we can consider the pathway as consisting of two components, (1) CO_2 diffusion and (2) RUBISCO activity and related biochemical and biophysical reactions. When bulk phase CO_2 concentration limits the rate of photosynthesis, with changes in photosynthetic rate directly proportional to small changes in CO_2 concentration from the normal value, the control strength of the two components can be closely approximated by the CO_2 concentration drops which occur in relation to the two components (Jones, 1973; Farquhar, Ehleringer and Hubick, 1989).

If C_b is the bulk phase CO_2 concentration and C_c is the steady-state CO_2 concentration in the chloroplast, the fractional limitation by diffusion for an aquatic plant is $(C_b - C_c)/C_b$, and that by RUBISCO is C_c/C_b (ignoring a small correction for CO_2 compensation concentration). The significance of methods such as $^{13}C/^{12}C$ natural abundance measurement which give non-invasive (up to the time of sampling, killing and measuring the cells) estimates of C_b/C_c is that they provide an independent estimate of the restriction of phytoplankton photosynthesis by CO_2 transport in the case of CO_2 diffusion. Furthermore, estimation of C_c/C_b by the techniques used in collecting the data in column A of Table 1 becomes increasingly inaccurate as C_c/C_b decreases, especially when $C_b > C_c$. C_c/C_b can be used, in conjunction with measurements of the area-based CO_2 influx in photosynthesis, to compute P_{CO2} (equation (1)) for comparison with estimates from the properties of lipid membranes (Table 2).

Our analysis has involved the coccolithophorid *Emiliania huxleyi*, whose maximum specific growth rates and photosynthetic DIC assimilation characteristics are discussed by Nimer, Dixon and Merrett (1992) and Raven and Johnston (1991), using the $^{13}C/^{12}C$ data for cultures from Sikes and Wilber (1982). We use the equations for computing C_c/C_b from the $^{13}C/^{12}C$ of extracellular CO_2 and of the organic C in the cells, and the required values of discrimination between $^{13}CO_2$ and $^{12}CO_2$ in diffusion through water and in fixation catalysed by RUBISCO, given by Raven and Farquhar (1990) and Maberly, Raven and Johnston (1992) for the analysis of diffusive CO_2 entry and fixation in aquatic plants. The result of these calculations is C_c/C_b not greater than 0.64 for *Emiliania huxleyi*; even lower values can be computed for other coccolithophorids whose $^{13}C/^{12}C$ has been measured in culture (*Hymenomonas carterae*: Sikes and Wilber, 1982; *Coccolithus pelagicus*: Falkowski, 1991). This upper limit on C_c/C_b of 0.64 in cultured coccolithophorids implies a fractional limitation of photosynthesis by CO_2 diffusion of at least 0.36. This is higher than would be expected for the P_{CO2} values widely attributed to biological membranes. Thus, for a specific growth rate of 30×10^{-6} s^{-1} in continuous saturating light at 20°C for *Emiliania huxleyi* (Raven and Johnston, 1991), and a cell radius of 3 μm, the net C influx is 313 nmol m^{-2} s^{-1} (assuming 125 kg organic C per m^3 cell volume). Using equation (1), with J = 313 nmol m^{-2} s^{-1}, C_b = 11.8 mmol m^{-3} (air-equilibrated seawater at 20°C), and C_c = 7.04 mmol m^{-3} (C_c/C_b = 0.64), P_{CO2} is 6.55×10^{-5} m s^{-1}. This is almost two orders of magnitude lower than the commonly accepted P_{CO2} values for biological membranes i.e., ~ 3.5×10^{-3} m s^{-1}, although much lower values of 2×10^{-6} m s^{-1} have also been reported: Raven, 1984, 1991a,b; Table 2). This mismatch is not eliminated by considering further membranes which must be crossed by CO_2 (4 envelope membranes of the coccolithophorid chloroplast) and the diffusion boundary layer around the cells (equation (7) of Raven, 1986a). Considering slower-growing cells would mean even lower computed P_{CO2} values.

Table 2

Permeability coefficient (P; m s^{-1}) for various solutes relevant to inorganic C supply to RUBISCO and to leakage of low M_r solutes. Data relate to the 'lipid solution' mechanism.

Solute	Membrane	P(m s^{-1})	Reference
CO_2	*In vitro* bilayer; plasmalemma	$2\text{-}3500 \times 10^{-6}$	Raven (1984, 1991b); Gimmler & Hartung (1988); Baier *et al.* (1990); Gimmler *et al.* (1990)
HCO_3^- anion	*In vitro* bilayer	$1\text{-}5 \times 10^{-11}$	Raven (1984)
Formic acid	*In vitro* bilayer	1×10^{-4}	Raven (1984)
Acetic acid	*In vitro* bilayer; plasmalemma	$2.4\text{-}66 \times 10^{-6}$	Raven (1984); Gimmler & Hartung (1988)
Propionic acid	*In vitro* bilayer	$2\text{-}6 \times 10^{-4}$	Raven (1984)
Glycolic acid	Method of Stein(1967); (1)	$1\text{-}10 \times 10^{-8}$	Raven (unpublished)
Lactic acid	*In vitro* bilayer	5×10^{-7}	Raven (1984)
Maleic acid	*In vitro* bilayer	4×10^{-7}	Raven (1984)
Ethanol	Plasmalemma	2×10^{-5}	Lieb & Stein (1986)
$Maleate^{2-}$ anion	*In vitro* bilayer	4×10^{-11}	Raven (1984)
Malic acid	Calculated from Collander Plot and EtOH/H_2O distribution	$1\text{-}2 \times 10^{-8}$	Lüttge & Smith (1984)
Citric acid	Plasmalemma	5×10^{-8}	Gimmler & Hartung (1988)
Glycerol	*In vitro* bilayer; plasmalemma	$3\text{-}540 \times 10^{-11}$	Raven (1984); Lieb & Stein (1986); Gimmler & Hartung (1988)
Erythritol	Plasmalemmma	$6\text{-}67 \times 10^{-12}$	Lieb & Stein (1986); Gimmler & Hartung (1988)

Solute	Membrane	P(m s^{-1})	Reference
Mannitol, Sorbitol	*In vitro* bilayer; plasmalemma	$1\text{-}10\times10^{-12}$	Raven (1984); Gimmler & Hartung (1988)
Glucose	*In vitro* bilayer; plasmalemma	$1\text{-}10\times10^{-12}$	Raven (1984); Gimmler & Hartung (1988)
Sucrose	Plasmalemma	4×10^{-11}	Gimmler & Hartung (1988)
Histamine	*In vitro* bilayer	3×5.10^{-7}	Raven (1984)
Urea	*In vitro* bilayer; plasmalemma	$6\text{-}40\times10^{-9}$	Raven (1984); Lieb & Stein(1986)
Formamide	*In vitro* bilayer	$7\text{-}8\times10^{-7}$	Raven (1984)
Acetamide	*In vitro* bilayer	$2\text{-}4\times10^{-7}$	Raven (1984)
2-Amino-isobutyric acid	Plasmalemma	3×10^{-10}	Gimmler & Hartung (1988)
Glycine	Method of Stein(1967); (2)	10^{-12}	Raven (unpublished)

The solutes for which permeability coefficient data are available are not necessarily those present in the highest concentrations inside marine phytoplankton cells. The compatible solutes, of which only one or two molecular species are found in a given phytoplankton genotype, are present at the highest concentration, but are only represented in the Table by glycerol, mannitol, sorbitol and (?) sucrose; proline, glycine betaine (and other betaines) and β-dimethylsulphonium propionate are not represented. The zwitterionic character of these N- and S-containing compatible solutes suggests low P values (similar to mannitol or sorbitol). Maleic acid is included as the only dicarboxylic acid whose P value has been directly determined. Compounds such as histamine, maleic acid, propionic acid and ethanol are likely to be present only at very low concentrations.

Footnotes

(1) Taking into account observations on the homologue lactic acid.

(2) Assuming 11 potential H-bonds per molecule (Ginsburg & Stein, 1987).

Before accepting this low computed P_{CO2} we must point out that there is the possibility that the $^{13}C/^{12}C$ ratios on which the analysis was based did not originate from completely air-equilibrated cultures; such an absence of air-equilibrium would tend to decrease the predicted P_{CO2} (see Johnston and Raven, 1992). Further work is needed to establish if the P_{CO2} was under-estimated as a result of this technical problem. It is clear that the relatively low discrimination against ^{13}C in coccolithophorids is not a function of unusually high contributions to CO_2 fixation from carboxylases other than RUBISCO which show low $^{13}C/^{12}C$ discrimination. This conclusion is based on data from Paasche (1964), Descolas-Gros and Fontugne (1985, 1990) and Arnelle and O'Leary (1992) showing that coccolithophorids do not have high rates of (C_3 + C_1) carboxylation and that the enzyme responsible (phosphoenolpyruvate carboxykinase) has a similar $^{13}C/^{12}C$ discrimination to RUBISCO (cf. Dauby, 1989; Reiskind and Bowes, 1991).

Resolution of the extent of $^{13}C/^{12}C$ discrimination in air-equilibrated cultures of coccolithophorids will help the quantitative interpretation of factors limiting DIC use by coccolithophorids, and will improve the interpretation of latitudinal trends in the $^{13}C/^{12}C$ of natural phytoplankton assemblages (Rau, Takahashi and Des Marais, 1989), at least to the extent that the component species rely on CO_2 diffusion. $^{13}C/^{12}C$ measurements are, as will be seen below, also useful in interpreting the behaviour of cells using CCMs.

Comparison of the costs of CO_2 fixation in terms of other resources when a CCM is operative and when CO_2 enters by diffusion

Raven (1988, 1990a,b, 1991a,b), Raven and Johnston (1991, 1992) and Raven and Lucas (1985) consider observations and theoretical effects of the mechanism of inorganic C acquisition by microalgae on the resource (photons, N, Fe, Mn, Mo) costs of CO_2 fixation from extant seawater inorganic C concentrations and pH. Photons are, of course, consumed in photosynthesis, so their cost is measured as mol photon consumed per mol C assimilated. The chemical nutrients act catalytically in C assimilation, or are required to produce catalysts of C assimilation, and their cost is expressed as mol nutrient in the cell required to permit fixation of one mol C per second.

Organisms that concentrate CO_2 incur a photon cost for active DIC transport; organisms depending on CO_2 entry by diffusion incur a photon cost related to RUBISCO oxygenase activity. Which of the costs is less depends on the fraction of the pumped DIC which leaks out of the cells in the case of cells operating a CCM, and on the fate of the phosphoglycolate generated by RUBISCO oxygenase in the cells lacking a CCM (Raven, 1984, 1990a, 1991a,b; Raven and Johnston, 1991, 1992; Raven and Lucas, 1985). Such experimental data as are available support the view that phytoplankton cells using a CCM can have lower photon costs of CO_2 assimilation than those using diffusive entry of CO_2. This in turn implies a very low fractional leakage of pumped DIC, and values at the high (and most widely accepted) end of the range of P_{CO2} values (Table 2) are very difficult to accommodate. We note that the high fractional leakage which can be deduced from $^{13}C/^{12}C$ data, using the approach of Sharkey and Berry (1985), is susceptible to an interpretation which involves less energy input. This possibility, as yet not rigorously tested, involves the DIC pump operating to exchange DIC across the plasmalemma at no net energy cost as well as catalysing net, energy-requiring DIC influx (Raven, 1990a; Raven and Johnston, 1991). Further, the possible role for carboxysomes and, perhaps, pyrenoids which was mentioned earlier (Badger and Price, 1992) could decrease leakage of CO_2 to values consistent with the observed photon yields and even with plasmalemma P_{CO2} values at the high end of the range in Table 2, and still be consistent with the DIC exchange explanation of the stable isotope data.

Cells that use a CCM can afford to produce smaller quantities of N-rich compounds like RUBISCO and enzymes of phosphoglycolate metabolism. These savings can more than outweigh the N costs of DIC pumping. Such predictions (Raven, 1984, 1991a, 1991b; Raven and Johnston, 1991, 1992) are borne out by such data as are available (Beardall, Griffiths and Raven, 1982; Raven, 1984, 1991a,b). This is especially the case if the cell N content in the numerator of the definition of N cost is corrected for the subsistence N content, i.e. the minimum cell N content for survival (Ågren, 1985; Turpin, 1988). Such results are predicted for Fe, Mn and Mo, though few of the required experiments have been performed (Raven, 1988, 1990b, 1991a,b; Sunda, Swift and Huntsman, 1991).

Alkalinity of surface waters in relation to microalgal processes

The effects of photosynthesis and calcification on alkalinity are well understood (Stumm and Morgan, 1981). Mineralisation of organic N to yield NH_4^+ increases alkalinity, while NH_4^+ assimilation decreases alkalinity (Brewer and Goldman, 1976): coccolith formation decreases soluble alkalinity, while coccolith dissolution increases soluble alkalinity. Recycled productivity, involving the cycles of C and N, does not alter alkalinity in the steady state.

New productivity (*sensu* Eppley, 1990) involves net production of organic matter based on exogenous inputs (mainly by net inflow of nutrient-rich water) into the marine euphotic zone. Upwellings generally inject combined N as NO_3^-, whose assimilation generates alkalinity. A cell with the Redfield ratio (Redfield, 1958) of organic C to organic N generates up to 0.151 mol OH^- (excreted) per mol organic C produced when NO_3^- is the N source (Brewer and Goldman, 1976; Raven, 1980, 1984). This increases alkalinity in upwelling regions, and reduces the tendency for the high CO_2 concentration (from respiration in deep water) in upwelled water to escape to the atmosphere. However, the whole cycle, including processes (ammonification, nitrification) in the dark depths of the ocean produce NO_3^- again, leads to no net change in alkalinity.

There is one route by which phytoplankton metabolism can export acidity from the ocean. Dimethylsulphonium propionate (DMSP; $(CH_3)_2S^+CH_2CH_2COO^-$, i.e. $C_5H_{10}SO_2$) is produced by some halophilic flowering plants, some rhodophyte and ulvophyte (green) benthic marine algae, and by prymnesiophyte (including coccolithophorid), dinophyte and, possibly, chrysophyte planktonic marine microalgae (Keller, Bellows and Guillard, 1989; Malin, Turner and Liss, 1992; Raven, 1986b). It can be present in high concentrations; the prymnesiophyte *Chrysochromulina polylepis* has almost 400 mol m^{-3} DMSP (Keller, Bellows and Guillard, 1989), or, assuming 125 kg C per m^3 fresh weight, 0.0384 mol DMSP per mol organic C, 0.192 of cell organic C is in DMSP. The high concentration is needed because of its role as a 'compatible solute', generating osmolarity without damaging enzymes.

Synthesis of DMSP from CO_2, H_2O and SO_4 yields 2 mol OH^- per mol DMSP produced:

$$5CO_2 + 6H_2O + SO_4^{2-} \rightarrow C_5H_{10}SO_2 + 8O_2 + 2OH^- \quad (4)$$

Breakdown of DMSP, catalysed by a base or enzymically, yields volatile dimethylsulphide (DMS) and acrylic acid:

$$C_5H_{10}SO_2 \rightarrow (CH_3)_2S + CH_2{=}CHCOO^- + H^+ \quad (5)$$

and acrylic acid is broken down to CO_2 and H_2O:

$$CH_2{=}CHCOO^- + H^+ + 3O_2 \rightarrow 3CO_2 + 2H_2O \quad (6)$$

Adding equations (4-6)

$$2CO_2 + 4H_2O + SO_4^{2-} \rightarrow (CH_3)_2S + 5O_2 + 2OH^- \quad (7)$$

This DMS-related acidity transfer from the ocean surface to the atmosphere, unlike sedimentation of $CaCO_3$ or of organic C generated with NO_3^- as N source which respectively decrease and increase surface alkalinity, can occur without the need for 'new' production. DMSP can be synthesised and the resulting DMS volatalized from non-limiting resources (CO_2, H_2O, SO_4^{2-}, photons) at the ocean surface in 'recycled' (with respect to N) productivity. With a global production of 0.6-1.8 Tmol volatalized DMS from the oceans each year (Andreae and Jaeschke, 1992;

Keller, Bellows and Guillard, 1989), 1.2-3.6 Tmol OH^- are generated globally in surface ocean waters each year. For comparison, global oceanic primary productivity is some 2,500 Tmol C per year.

The volatalized DMS has a short residence time in the atmosphere (less than one day: Newman, Krouse and Grinenko, 1991; Malin, Turner and Liss, 1992) and ultimately produces H_2SO_4

$$(CH_3)_2S + 5O_2 \rightarrow 2H_2O + 2CO_2 + 2H^+ + SO_4^{2-} \quad (8)$$

Addition of equation (7), which is the sum of equations (4) - (6) and reflects the net synthesis of DMS in the surface ocean, to equation (8), which reflects the breakdown of DMS in the atmosphere, yields

$$2H_2O \rightarrow 2OH^- + 2H^+ \quad (9)$$

where the H_2O and OH^- are in the ocean surface and the H^+ is in the atmosphere.

Since not all of this H^+ is precipitated over the oceans, in pre-industrial days it would have reflected a net transfer of acidity to the continents. At the moment, the reverse transfer of some of the acidity generated by oxidation of reduced S in the fossil fuels on land is likely to more than offset the 'natural' acidity transfer (Newman, Krouse and Grinenko, 1991; Howarth and Stewart, 1992).

Dissolved organic C efflux: mechanisms and functions

Although much of the loss of organic compounds is from damaged cells ('sloppy' feeding by zooplankton; lysis by viruses) it is of interest to consider the rate of loss from healthy cells (Fogg, 1983). Bjornsen (1988) makes a strong case that the loss is due to non-catalysed passive (by diffusion through the lipid phase) efflux, based on movement across the plasmalemma and through the diffusion boundary layer in response to differences in concentration and, for electrolytes, in electrical potential difference which is negative-inside across the plasmalemma (see Raven, 1984, 1986a). If we assume a spherical cell and combine equation (5) of Bjornsen (1988) with explicit spherical geometry we obtain

$$E = \frac{3PF}{r} \tag{10}$$

where

E = Efflux of organic C in low M_r solutes as a fraction of the total cell organic C (s^{-1})

P = Mean permeability of the plasmalemma to low M_r organic solutes ($m\ s^{-1}$)

F = Fraction of the total cell organic C which is present as low M_r solutes

r = cell radius

This analysis assumes that there is no organic C in the bulk phase. If there is 125 kg C per m^{-3} cell volume, 0.45 kg C per kg dry weight and a density of 1 Mg m^{-3}, there is 14423 mol total C m^{-3}. With Bjornsen's assumption of F = 0.1 we obtain a concentration of 1442 mol C in low M_r solutes per m^3 cell water. If there is an average of five C atoms per molecule in organic acids, amino acids, sugars, sugar phosphates, nucleotides, etc., then the concentration of these low M_r solutes is 288 mol m^{-3}. This is a plausible value for marine phytoplankton (but would clearly be impossibly high for many freshwater phytoplankton: see Raven, 1984). Again with Bjornsen's values of $r = 5\times10^{-6}$ m and $P = 10^{-11}\ m\ s^{-1}$, then $E = 6\times10^{-7}\ s^{-1}$ for a cell with a doubling time of 24h, and hence a specific growth rate of $8\times10^{-6}\ s^{-1}$. Thus, 7.5% of the cell C is lost per day.

The value of P used by Bjornsen (1988), i.e. $10^{-11}\ m\ s^{-1}$, is reasonable (Raven, 1984, Table 2). The cellular constituents in Table 2 with the higher P values are present in low concentrations in the cells. For the weak electrolytes (organic acids, amines) this is exacerbated by having pK_a values well removed from cytosol pH so that most of such a solute is present as the dissociated (charged) forms with much lower P values. The rate of loss as a fraction of biomass is inversely proportional to r. Invariance of loss as a fraction of growth with cell size requires that the growth rate is also inversely proportional to r; however, specific growth rate declines less rapidly with increasing r, so that we would predict that lipid solution loss of low M_r organic solutes should be a lower fraction of C accumulation in growth in larger cells (Raven, 1986a). Reabsorption of some of the effluxed C would reduce the net loss.

Some reliable estimates of organic C efflux by phytoplankton exceed what is expected of lipid solution efflux (Bjornsen, 1988; Fogg, 1983). This implies that there are mediated efflux mechanisms, perhaps because efflux has some advantages for the cell. Suggested advantages for the cell of the presence just outside the plasmalemma of the effluxed solutes include a UV-B screen (Raven, 1991c); a means of intercepting damaging hydroxyl radicals generated by UV-B outside the cells (Palenik, Price and Morel, 1991); and involvement in photoreduction of ferric iron to the more usable ferrous form (Finden, Tipping, Jaworski and Reynolds, 1984; Waite and Morel, 1984).

Another possibility is that the 'deliberate' efflux of organic C may help to limit the extent of photoinhibition (Bjornsen, 1988; Fogg, 1975, 1983). When growth is limited by the supply of nutrients rather than of light, the photons absorbed by the photosynthetic pigments are less readily used in constructive ways (i.e. the production of organic matter incorporated into cell material) and may thus be used destructively (photoinhibition). While the energy cost of repair of photoinhibition, and the 'opportunity cost' of a reduced photosynthetic performance (Raven, 1989) are not very important liabilities under high light, low nutrient conditions, photoinhibition can constrain the ability of cells to respond quickly when it moves to lower light and availability of higher nutrients deeper in the water column. The half-time of recovery from photoinhibition upon transfer to lower photon flux densities can be two hours or more, a substantial fraction of the resource-saturated cell generation time, and larger than the time needed for vertical motion entrained in water movements (Raven, 1989; Kyle, Osmond and Arntzen, 1987). Since the cell quota for C has limited amplitude (Raven, 1984; Turpin, 1988) the use of continued photoproduction of organic C as a 'constructive' energy sink could be permitted by the loss of excess organic C by uncoupled respiration to CO_2 or, as proposed here, by efflux to the medium (Fogg, 1975, 1983). While attractive, this hypothesis needs critical evaluation. Is there a feedback control of photosynthetic activity by the build-up of immediate products of photosynthesis which have not been consumed in growth? If not, then there must be some overflow of organic C, and a regulated (mediated) uncoupled respiration to CO_2 or a mediated organic C efflux is unavoidable (Fogg, 1975, 1983; cf Kazarinoff, 1981; Siekevitz, 1972). The question marks in this paragraph mean that the question 'can enhanced consumption of photoproduced NADPH and ATP *via* reactions catalysed by the carboxylase or oxygenase activities of RUBISCO and the associated downstream reactions alleviate photoinhibitory damage' (Osmond, 1981; Kaplan, 1981; Kyle, Osmond and Arntzen, 1987) must currently be answered as 'not proven'.

Conclusions

The major implications of the discussion in this paper for biogeochemical ocean processes are as follows.

(1) Limitation of growth of marine phytoplankton cells by the inorganic C supply from air-equilibrated seawater is unlikely except for a limited range of taxa (e.g. the coccolithophorids) under the exceptional circumstances of a saturating supply of light and nutrients other than C.

(2) The use of the natural abundance of C isotopes to study the mechanism of inorganic C assimilation by microalgae has substantial, but incompletely fulfilled, promise. The use of $^{13}C/^{12}C$ values for ecological and palaeoecological interpretations of marine C cycling, currently based mainly on diffusive CO_2 entry, should take into account the occurrence of CO_2 concentrating mechanisms (Raven, 1991a).

(3) There is a significant (if minor) contribution to surface seawater alkalinity increase related to the synthesis of DMSP with subsequent loss of DMS to the atmosphere. However, in global terms, any acidity transfer from the oceans to the continents via DMS and its oxidation product H_2SO_4 in pre-industrial times has been swamped by the reverse flux of oxidized sulphur from the burning of fossil fuel.

(4) Much of the excretion of organic C by 'healthy' phytoplankton cells can be accounted for by unavoidable leakage through the lipid part of the plasmalemma. The proposed function(s) of any additional organic C efflux via specific proteinaceous transporters are, as yet, not experimentally validated.

Acknowledgements

Work on the assimilation of inorganic C by marine algae in the author's laboratory is funded by NERC.

Inputs from Dr A M Johnston have, as always, been invaluable.

References

Ågren GI (1985) Theory for growth of plants derived from the nitrogen productivity concept. Physiol. Plant 64:17-28

Aizawa K, Miyachi S (1986) Carbonic anhydrase and CO_2 concentrating mechanisms in microalgae and cyanobacteria. FEMS Microbiology Reviews 39:215-233

Aizawa K, Nakamura V, Miyachi S (1986) Variations of PEPc activity in *Dunaliella* associated with changes in atmospheric CO_2 concentration. Plant Cell Physiol 26:1199-1203

Andreae MO, Jaeschke WA (1992) Exchange of sulphur between biosphere and atmosphere over temperate and tropical regions. In Sulphur Cycling on the Continents (ed. by RW Howarth, JW Stewart, MV Ivanov) pp 27-61. John Wiley Chichester

Arnelle DR, O'Leary MH (1992) Binding of carbon dioxide to phosphoenolpyruvate carboxykinase deduces from carbon kinetic isotope effects. Biochemistry 31:4363-4368

Badger MR, Price D (1992) The CO_2 concentrating mechanism in cyanobacteria and microalgae. Physiol Plant 84:606-615

Baier M, Gimmler H, Hartung W (1990) The permeability of the guard cell plasma membrane and tonoplast. J exp Bot 41:351-358

Ball JT, Woodrow IE, Berry JA (1987) A model predicting stomatal conductance and its contribution to the control of photosynthesis under different environmental conditions. In Progress in Photosynthesis Research Vol IV (ed. by J Biggins) pp 221-224 Martinus Nishoff Publishers Dordrecht

Beardall J (1991) Effects of photon flux density on the 'CO_2-concentrating mechanism' of the cyanobacterium *Anabaena variabilis*. J Plankton Res (Supplement) 13:133-141

Beardall J, Griffiths H, Raven JA (1982) Carbon isotope discrimination and the CO_2 concentrating mechanism in *Chlorella emersonii*. J Exp Bot 33:729-737

Beardall J, Roberts S, Millhouse J (1991) Effect of nitrogen limitation on uptake of inorganic carbon and specific activity of ribulose-1,5-bisphosphate carboxylase/oxygenase in green microalgae. Can J Bot 69:1146-1150

Bjornsen PK (1988) Phytoplankton exudation of organic matter: why do healthy cells do it? Limnol Oceanogr 33:151-154

Brand L E (1991) Minimum iron requirement of marine phytoplankton and the implications for the biogeochemical control of new production. Limnol Oceanogr 36:1756-1771

Brewer PG, Goldman JC (1976) Alkalinity changes generated by phytoplankton growth. Limnol Oceanogr 21:108-117

Caperon J, Smith DF (1978) Photosynthetic rates of marine algae as a function of inorganic carbon concentration. Limnol Oceanogr 23:704-708

Coleman JR (1991) The molecular and biochemical analyses of CO_2-concentrating mechanisms in cyanobacteria and microalgae. Plant Cell Environ 14:861-867

Dauby P (1989) The stable isotope ratios in benthic food webs of the Gulf of Calvi, Corsica. Continent Shelf Res 9:181-195

Descolas-Gros C, Fontugne MR (1985) Carbon fixation in marine phytoplankton: carboxylase activities and stable carbon-isotope ratios; physiological and palaeclimatological aspects. Mar Biol 87:1-6

Descolas-Gros C, Fontugne MR (1990) Stable isotope fractionation by marine phytoplankton during photosynthesis. Plant Cell Environm 13:207-218

Eppley RW (1990) New production: history, methods, problems. In Productivity of the Ocean: Past and Present (ed. by W H Berger, V S Smeterack, G Wefer) pp 85-97. John Wiley Chichester.

Falkowski PG (1991) Species variability in the fractionation of ^{13}C and ^{12}C by marine phytoplankton. J Plankton Res (Supplement) 13:21-28

Farquhar GD, Ehleringer JR, Hubick KT (1989) Carbon isotope discrimination and photosynthesis. Ann Rev Plant Physiol Plant Mol Biol 40:503-537

Finden DAS, Tipping E, Jaworski GHM, Reynolds CS (1984) Light-induced reduction of natural Fe(III) oxide and its relevance to phytoplankton. Nature 309:783-784

Fogg GE (1975) Biochemical pathways in unicellular plants. In Photosynthesis and Productivity in Different Environments (ed. by JP Cooper) pp 437-457. Cambridge University Press

Fogg GE (1983) The ecological significance of extracellular products of phytoplankton photosynthesis. Bot Mar 26:3-14

Gimmler H, Hartung W (1988) Low permeability of the plasma membrane of *Dunaliella parva* for solutes. J Plant Physiol 133:165-172

Gimmler H, Weiss C, Baier M, Hartung W (1990) The conductance of the plasmalemma for CO_2. J exp Bot 41:785-794

Ginsburg, H, Stein, WD (1987) Biophysical analysis of novel transport pathways induced in red blood cell membranes. J Membrane Biol 96:1010

Guillard RLL, Keller MD, O'Kelly CJ, Floyd GL (1991) *Pycnococcus provasolii* gen. et sp. nov., a coccoid prasinoxanthin-containing phytoplankter from the western North Atlantic and Gulf of Mexico. J Phycol 27:39-47

Holligan PM, Violl M, Harbour DS, Camus P, Champagne-Phillipe M (1983) Satellite and ship studies of coccolithophore production along a continental shelf edge. Nature 304:339-342

Holligan PM, Groom SB (1986) Phytoplankton distributions along the shelf edge. Proc Roy Soc Edinburgh 88B:239-263

Howarth RW, Stewart JWB (1992) The interactions of sulphur with other element cycles in ecosystems. In Sulphur Cycling on the Continents (ed. by RW Howarth, JW Stewart, MV Ivanov) pp 67-84. John Wiley Chichester

Johnston AM, Raven JA (1989) Extraction, partial purification and characterization of phosphoenolpyruvate carboxykinase from *Ascophyllum nodosum*. J Phycol 25:568-576.

Johnston AM, Raven JA (1991) The acquisition of inorganic carbon by *Phaeodactylum tricornutum*. Br Phycol J 26:95

Johnston AM, Raven JA (1992) The acquisition of DIC by the marine diatom *Phaeodactylum tricornutum*. Br Phycol J 27:94

Jones HG (1973) Limiting factors in photosynthesis. New Phytol 72:1089-1094

Kacser H, Burns JA (1973) The control of flux. Symp Soc Exp Biol 27:65-104

Kaplan A (1981) Photoinhibition in *Spirulina platenis*: response of photosynthesis and HCO_3^- uptake capability to CO_2-depleted conditions. J Exp Bot 32:669-677.

Kazarinoff MN (1981) There is an energy cost for catalytic turnover which arises due to enzyme degradation. Arch Biochem Biophys 208:131-141

Keller MD, Bellows WK, Guillard RLL (1989) Dimethylsulfide production and marine phytoplankton: an additional impact of unusual blooms. In Novel Phytoplankton Blooms. Causes and Impacts of Recurrent Brown Tides and Other Unusual Blooms (ed by EM Cosper, VM Bricelj, EJ Carpenter), pp 101-115. Coastal and Estaurine Studies Volume 35 Springer-Verlag Berlin

Kerby NW, Raven JA (1985) Transport and fixation of inorganic carbon by marine algae. Adv in Bot Res 11:71-113

Kyle DJ, Osmond CB, Arntzen CJ (eds) (1987) Photoinhibition. Eslevier Amsterdam

Lieb WR, Stein WD (1986) Non-Stokesian nature of transverse diffusion within human red cell membranes. J membrane Biol 92:111-119

Luinenburg I, Coleman JR (1990) A requirement for phosphoenolpyruvate carboxylase in the cyanobacterium *Synechococcus* PCC 7942. Arch Microbiol 154:471-474

Lüttge E, Smith JAC (1984) Mechanism of passive malic-acid efflux from vacuoles of the CAM plant *Kalanchoe daigremontiana*. J Membr Biol 81:149-158

Maberly SC, Raven JA, Johnston AM (1992) Discrimination between ^{12}C and ^{13}C in marine plants. Oecologia 91:481-492

Malin G, Turner SM, Liss PS (1992) Sulfur: the plankton/climate connection. J Phycol 28:590-597

Merrett MJ (1991) Inorganic carbon transport in some marine microalgal species. Can J Bot 69:1032-1039

Morel FMM (1987) Kinetics of nutrient uptake and growth in phytoplankton. J Phycol 23:137-150

Newman L, Krouse HR, Grinenko VA (1991) Sulphur isotope variations in the atmosphere. In Stable Isotopes in the Assessment of Natural and Anthropogenic Sulphur in the Environment (ed. by HR Krouse and VA Grinenko) pp. 133-176. John Wiley Chichester

Nimer NA, Dixon GK, Merrett MJ (1992) Utilization of inorganic carbon by the coccolithophorid *Emiliania huxleyi* (Lohman) Kamptner. New Phytol 120:153-158

Osmond CN (1981) Photorespiration and photoinhibition. Some implications for the energetics of photosynthesis. Biochim Biophys Acta 639:77-98

Paasche E (1964) A tracer study of the inorganic C uptake during coccolith formation and photosynthesis in the coccolithophorid *Coccolithus huxleyi*. Physiol Plant Suppl III:1-82

Palenik B, Price NM, Morel FMM (1991) Potential effects of UV-B on the chemical environment of marine organisms: a review. Environm Pollution 70:117-130

Rau GA, Takahashi T, Des Marais DJ (1989) Latitudinal variations in plankton $\delta^{13}C$ implications for CO_2 and productivity in past oceans. Nature 341:516-518

Raven JA (1980) Nutrient transport in microalgae. Adv Microb Physiol 21:47-226

Raven JA (1984) Energetics and Transport in Aquatic Plants. AR Liss. New York

Raven JA (1986a) Physiological consequences of extremely small size for autotrophic organisms in the sea. Photosynthetic Picoplankton (ed. by T Platt, WKW Li) pp 1-70 Can Bull Fisheries Aquatic Sciences 214

Raven JA (1986b) Biochemical disposal of excess H^+ in growing plants? New Phytol 104:75-206

Raven JA (1988) The iron and molybdenum use efficiencies of plant growth with different energy, carbon and nitrogen sources. New Phytol 109:279-287

Raven JA (1989) Fight or flight: the economics of repair and avoidance of photoinhibition of photosynthesis. Functional Ecol 3:5-19

Raven JA (1990a) Use of isotopes in estimating respiration and photorespiration in microalgae. Mar Microbial Food Webs 4:59-86

Raven JA (1990b) Predictions of Mn and Fe use efficiencies of phototrophic growth as a function of light availability for growth and C assimilation pathway. New Phytol 116:1-18

Raven JA (1991a) Implications of inorganic C utilization:ecology, evolution and geochemistry. Can J Bot 69:908-924

Raven JA (1991b) Physiology of inorganic C acquisition and implications for resource use efficiency by marine phytoplankton: relation to increased CO_2 and temperature. Plant Cell Environm 14:779-794

Raven JA (1991c) Responses of aquatic photosynthetic organisms to increased solar UV-B. J Photochem Photobiol 9:239-244

Raven JA (1993) Limits to growth rates. Nature 361:209-210

Raven JA, Farquhar GD (1990) The influence of N metabolism and organic acid synthesis on the natural abundance of C isotopes in plants. New Phytol 116:505-529

Raven JA, Johnston AM (1991) Inorganic carbon acquisition mechanisms in marine phytoplankton and their implications for the use of other resources. Limnol Oceanogr 36:1701-1714

Raven JA, Johnston AM (1992) Algal DIC pumps and atmospheric CO_2. In

Photosynthetic Carbon Metabolism and Regulation of Atmospheric CO_2 and O_2 (ed. by NE Tolbert, J Preiss) in press. Oxford University Press

Raven JA, Lucas WJ (1985) Energy costs of carbon acquisition. In Inorganic Carbon Uptake by Aquatic Photosynthetic Organisms (ed. by WJ Lucas, JA Berry) pp. 305-324. American Society of Plant Physiology Rockville Maryland USA

Redfield AC (1958) The biological control of chemical factors in the environment. Amer Sci 46:204-221

Rees TAV (1984) Sodium-dependent photosynthetic oxygen metabolism in a marine diatom. J Exp Bot 35:332-337

Reiskind JB, Bowes G (1991) The role of phosphoenolpyruvate carboxykinase in a marine macroalga with C_4-like photosynthetic characteristics. Proc Nat Acad Sci Wash 88:2883-2887

Riebesell U, Wolf-Gladrow DA, Smetacek V (1993). Carbon dioxide limitation of marine phytoplankton growth rates. Nature 361:249-251

Sharkey TD, Berry JA (1985) Carbon isotope fractionation of algae as influenced by an inducible CO_2 concentrating mechanism. In Inorganic Carbon Uptake by Aquatic Photosynthetic Organisms (ed. by WJ Lucas, JA Berry) pp 389-401. American Society for Plant Physiology Rockville Maryland USA

Siekevitz P (1972) The turnover of proteins and the usage of information. J Theoret Biol 37:321-324

Sikes CS, Wilber KM (1982) Functions of coccolith formation. Limnol Oceanogr 27:18-26

Smith RG, Gauthier DA, Dennis DT, Turpin DH (1992) Malate- and pyruvate-dependent fatty acid synthesis in leucoplasts from developing castor bean endosperm. Plant Physiol 99:1233-1238

Stein WD (1967) The Movement of Molecules across Cell Membranes. Academic Press New York

Stumm W, Morgan JJ (1981) Aquatic Chemistry. An Introduction Emphasizing Chemical Equilibria in Natural Waters 2nd Edn. Wiley Interscience New York

Sunda WG, Swift DG, Huntsman SA (1991) Low iron requirement for growth in oceanic phytoplankton. Nature 351:55-57

Theodorou ME, Elrifi IR, Turpin DH, Plaxton WC (1991) Effects of phosphorus limitation on respiratory metabolism in the green alga *Selenastrum minutum*. Plant Physiol 95:1089-1095

Turpin DH (1988) Physiological mechanisms in phytoplankton resource competition. In Growth and Reproductive Strategies of Freshwater Phytoplankton (ed. by CD Sandgren) pp 316-368. Cambridge University Press

Turpin DH, Miller AG, Parslow JS, Elrifi IR, Canvin DT (1985) Predicting the kinetics of DIC-limited growth from the short-term kinetics of photosynthesis in *Synechococcus leopolensis* (cyanophyta) J Phycol 21:409-418

Waite DT, Morel FMM (1984) Photoreductive dissolution of colloidal iron oxides in natural waters. Environm Sci Technol 18:860-868

Watson AJ, Robinson C, Robinson JE, Williams PJB, Fasham MJR (1991) Spatial variability in the sink for atmospheric carbon dioxide in the North Atlantic. Nature 350:50-53

Woodrow IE, Berry JA (1988) Enzymatic regulation of photosynthetic CO_2 fixation in C_3 plants. Ann Rev Plant Physiol Plant Mol Biol 39:533-594

TOWARDS A GENERAL DESCRIPTION OF PHYTOPLANKTON GROWTH FOR BIOGEOCHEMICAL MODELS

John J. Cullen
Dalhousie University
Department of Oceanography
Halifax, Nova Scotia, B3H 4J1
Canada

R.J. Geider[1], J. Ishizaka[2], D.A. Kiefer[3], J. Marra[4], E. Sakshaug[5], J.A. Raven[6]

1. Introduction

The growth of phytoplankton is fundamentally important to biogeochemical cycling in the sea, and models of this process are essential to describing the fluxes of carbon, nitrogen, and many other elements in the ocean. We address here nitrogen-based models that predict the photosynthesis and growth of phytoplankton for use in basin-scale simulations of marine biogeochemical processes. These models describe light absorption by photosynthetic pigments and biological transformations of carbon and nitrogen, so they must specify, implicitly or explicitly, the cellular chemical composition of phytoplankton (i.e., chlorophyll *a*, C and N) and photosynthesis per unit chlorophyll *a* as a function of irradiance (P^B vs. E).

It is important to recognize that neither the relative proportions of cellular constituents nor the P^B vs. E relationship are constant (Fig. 1): rather, for

[1] College of Marine Studies, University of Delaware, Lewes, Delaware 19958-1298, USA
[2] National Institute for Resources and the Environment, 16-3 Onogawa, Tsukuba, Ibaraki, 305 Japan
[3] Department of Biological Sciences, University of Southern California, Los Angeles, California 90089-6911 USA
[4] Lamont-Doherty Geological Observatory of Columbia University, Palisades, New York 10964 USA
[5] Trondhjem Biological Station, University of Trondheim, The Museum, Bynesveien 46, N-7018 Trondheim, Norway
[6] Department of Biological Sciences, University of Dundee, Dundee, DD1 4HN, United Kingdom

NATO ASI Series, Vol I 10
Towards a Model of Ocean Biogeochemical Processes
Edited by G T Evans and M J R Fasham

individual species in culture they vary with the growth rate of phytoplankton (μ; d^{-1}) as influenced by daylength (D), irradiance (E), limiting nutrient (here we discuss dissolved inorganic nitrogen, N), and temperature (T). Our task is to describe how this variation can be predicted and how these predictions can be incorporated into numerical simulations of biogeochemical dynamics in the ocean. Construction and validation of a general model is well beyond the scope of this report. Likewise, a comprehensive review of growth models and supporting experimental studies, though appropriate to the task, is not possible in this context. Instead, we present a general analytical framework and some quantitative relationships that can be examined for consistency with experimental data. It would seem reasonable, and perhaps essential, to compare our predictions explicitly with those of others who have performed thorough reviews and analyses of some aspects of microalgal growth (e.g. Shuter, 1979; Laws & Bannister, 1980; Zevenboom & Mur, 1984; Falkowski et al., 1985; Geider et al., 1986a; Geider, 1987; Langdon, 1988; Laws & Chalup, 1990; Kiefer, this volume), but this is an extremely demanding task that, unfortunately, could not be attempted.

Some members of this working group are uncomfortable even suggesting that a general, physiologically-based model of phytoplankton growth can be used to describe primary production in the sea. First, as we will show, experimental data are insufficient to resolve critical questions. The second complication is taxonomic diversity: there are important, but incompletely resolved, biochemical and physiological differences between groups that are surely important in the sea (Geider, 1992a), but we are a long way from predicting ecological success of different taxa in general models of ocean processes (see Margalef, 1978). Finally, we acknowledge the difficulties in extrapolating results of laboratory studies to the ocean. Not only is it technically demanding to simulate important characteristics of the marine environment (e.g., vertical mixing, intermittent nutrient supply, and interspecies interactions), but it is extremely difficult to validate model predictions because growth rates and chemical composition of natural phytoplankton cannot be measured directly (Eppley, 1980). Clearly, realism must be sacrificed in our modeling efforts.

Acknowledging misgivings expressed by some members of this working group, we suggest how phytoplankton growth might be described in a numerical model of biogeochemical processes in the sea. The foundation is an analytical description of adapted growth and chemical composition as a function of D, E,

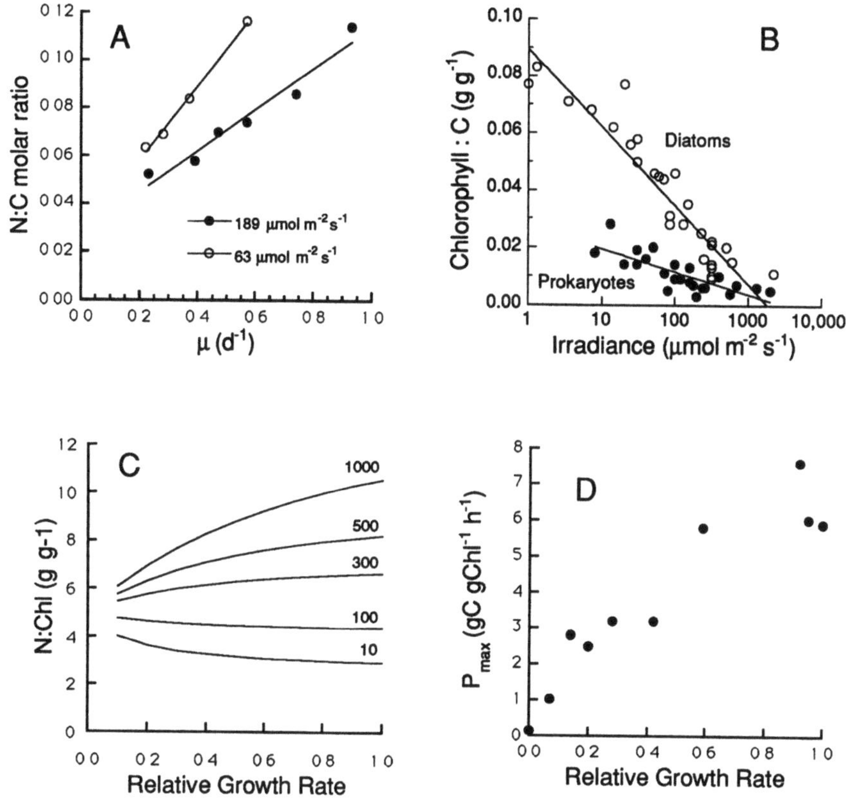

Fig. 1. Chemical and physiological characteristics of phytoplankton in culture. A. The molar N:C ratio of *Pavlova lutheri* as a function of nutrient-limited growth rate (d^{-1}) for two different irradiances, 63 and 189 μmol m^{-2} s^{-1} (from Chalup & Laws, 1990). B. Chl:C (g:g) of cultures as a function of growth irradiance for diatoms and prokaryotes, illustrating substantial taxonomic differences (Geider, 1992a). C. Predictions of N:Chl (g:g) as a function of N-limited growth rate for different growth irradiances in μmol m^{-2} s^{-1} (Laws & Chalup, 1990). Relative growth rate is μ normalized to the nutrient-saturated growth rate at that irradiance and temperature. D. Maximal rate of normalized photosynthesis, P_{max} (gC $gChl^{-1}$ h^{-1}) for the diatom *Chaetoceros gracilis* as a function of N-limited growth rate in continuous culture at 160 μmol m^{-2} s^{-1} (data from Thomas & Dodson, 1972). Relative growth rate as in C. Some subsequent studies have found a similar pattern, and others have found no relationship between P_{max} and N-limited growth rate (see Cullen *et al.*, 1992).

N, and T. This steady-state model should be developed as a guide for a numerical model that is appropriate for dynamic simulations where steady-state is not achieved. Although our presentation provides few answers, it gives a framework for integrating information on microalgal growth, in order to improve predictions of growth rates and chemical composition for global models. Also, our laboured struggle to find generalizations highlights the pressing need for further research to resolve unanswered questions central to biogeochemical modeling.

2. Analytical Model of Adapted Growth

Trué to a well-established tradition (cf. Geider *et al.*, 1986 and references therein), the nascent *DENT* (daylength, irradiance, nutrient, temperature) model is an energy budget, based on C. Such models describe the relationships between specific growth rate, the ratio of cellular C to chlorophyll *a*, and the efficiency of photosynthesis. Because biogeochemical models for the ocean are often based on N, the cellular N:C ratio should also be predicted. This steady-state model would serve as a guide for a dynamic model, which, given enough time under constant conditions, would converge upon the same growth rate and chemical composition as the steady-state model.

The steady-state *DENT* model is based on a simple equality:

$$\mu + r = \frac{P^C}{C}. \tag{1}$$

That is, the gross growth rate (net growth rate, μ, plus respiration, r: d^{-1}) equals the gross photosynthetic rate (P^C, gC $cell^{-1}$ d^{-1}) normalized to cellular C (gC $cell^{-1}$). Excretion is implicit in the respiration term. Equation 1 is made meaningful by describing P^C as a function of light absorption and photosynthetic efficiency, in one of many possible representations:

$$\mu + r = D \cdot E \cdot a_{chl} \cdot \phi \cdot \frac{Chl}{C} \tag{2a}$$

where D is dimensionless daylength (h/24h), E is mean irradiance during the photoperiod (μmol m^{-2} s^{-1}), a_{chl} is the Chl-specific absorption coefficient (m^2 gChl^{-1}), ϕ is photosynthetic quantum yield (mol C (mol photons)$^{-1}$), and *Chl* is cellular chlorophyll *a* (gChl cell^{-1}). The product, $D \cdot E \cdot a_{chl} \cdot \phi \cdot Chl$ is gross cellular photosynthesis, P^C. Evidence for the applicability of the energy balance comes primarily from an examination of the intraspecific responses of μ, Chl and C (Kiefer & Mitchell, 1983; Sakshaug et al., 1989; Nielsen, 1992), with few independent, direct measurements of all of the variables in eq. 2a.

The energy balance (eq. 2a) does not predict growth rate or photosynthesis as functions of D, E, N and T. It is essentially an accountant's ledger, constraining the mathematical relationships among parameters according to a budget for photons and carbon atoms, thereby providing a unified approach for examining genetic and phenotypic adaptation in phytoplankton. A predictive *DENT* model must specify rate processes (i.e., μ and r), chemical composition (i.e., $C{:}Chl$), and the photon (ϕ) and light-harvesting (a_{chl}) efficiencies as functions of temperature, irradiance, and nutrient concentrations within the constraint that an energy balance is achieved. The key to a *DENT* model is to describe the functional dependencies between these parameters as functions of D, E, N, and T. Cellular N:C should also be modeled. Many studies have dealt with subsets of this problem (e.g., Bannister, 1979; Shuter, 1979; Dubinsky et al., 1986; Raven & Geider, 1988; Laws & Chalup, 1990; Langdon, 1988), and, in principle, these results can be integrated into a comprehensive description for a species in culture (Kiefer & Cullen, 1992; Kiefer, this volume). Here we suggest a general form for an analytical representation, emphasizing the possible relationships between parameters.

Parameterizing respiration. In order to further constrain the energy balance, it is desirable to specify the relation between respiration and growth. Following Shuter (1979), respiration can be assigned to maintenance respiration (r_o with units of d^{-1}) which is independent of growth rate, and biosynthesis which is proportional to growth rate:

$$r = r_o + \beta\mu. \tag{2b}$$

The cost of biosynthesis, β, is dimensionless. The revised energy balance is:

$$\mu(1+\beta)+r_{o}=D\cdot E\cdot a_{chl}\cdot\phi\cdot\frac{Chl}{C}. \tag{2c}$$

Hence, by specifying growth rate, we specify the respiration rate as well. We assume that r_0 is a physiological "constant" and that the dependence of respiration on *D, T,* and *N* can be treated in terms of their effects on μ. The cost of biosynthesis will depend primarily on the rate of protein synthesis, and it may be more appropriate to normalize respiration to protein synthesis rather than growth (this may be particularly important under N- or P- limiting conditions in which phytoplankton typically accumulate carbohydrate and lipid energy reserves). Unfortunately, there is very little evidence which can be employed to determine the values of r_0 and β, and it is possible that these parameters depend on taxon and growth condition (for reviews see Geider, 1987; Langdon, 1992).

Although eq. 2b provides an attractive approach to dealing with the problem of parameterizing phytoplankton respiration, it is an approach that is based on inference rather than hard data (see Geider, 1992b). Finally, in parameterizing respiration, we do not explicitly consider the roles of photorespiration (i.e., the reactions initiated by oxygenation of RuBP by the photosynthetic enzyme RUBISCO) and the Mehler reaction (i.e., O_2 photoreduction by components of the photosynthetic electron transport chain) in energy dissipation by phytoplankton. Rather, we assume that these processes can be treated implicitly through their effect on the quantum efficiency of photosynthesis (ϕ) as discussed below.

The temperature function. Analytical representations of *DENT* interactions would be simplified if environmental factors could be nondimensionalized. A temperature function (Fig. 2A) describes the effect of suboptimal temperature on biosynthetic processes:

$$g_T^* = f(T) \tag{3}$$

where g_T^* is a dimensionless factor that scales maximum rate at a given temperature to the maximum rate at optimum temperature. The function could be used to describe relative growth rate directly, but we will apply it here to temperature-dependent processes that determine growth rate rather than to growth rate *per se*. Nonetheless, we are compelled to use nutrient- and light-saturated growth rate as the best indication of temperature dependence. No analytical representation for $f(T)$ is presented because we are discussing approaches to the modeling effort rather than implementation. Eppley (1972) presented an analytical representation of $f(T)$ for μ that is still being used to predict maximum growth rates of phytoplankton. That equation provides an upper bound for growth rate based on the fastest growing species (i.e., diatoms and chlorophytes) at their optimum temperature (Raven & Geider, 1988). It does not describe the response of individual species to temperature, which is often more pronounced than the Eppley (1972) curve (Li & Morris, 1982), nor is it applicable to all species. In particular, dinoflagellates and large diatoms are characterized by lower growth rates (Raven & Geider, 1988).

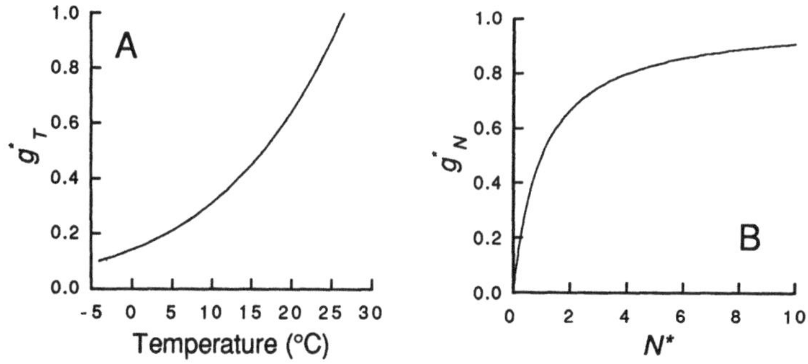

Fig. 2. General features of *DENT* functional relationships. A. The dimensionless temperature function, $g^{\bullet}_{T}$, shown here as an arbitrary Arrhenius function of the form $g^{\bullet}_{T}$ = A·exp(-B/(T-273)), where A and B are constants and T is temperature (°C). B. The dimensionless nutrient function, $g^{\bullet}_{N}$ (eq. 4), with the concentration of limiting nutrient (N^*, dimensionless) as described in eq. 5.

The nutrient function. As with the temperature function, it is hoped that the nutrient function will describe the effects of nutrient supply on the processes that determine growth rate, yet the first approximation is an equation relating gross growth rate to nutrient supply. Nutrient-limitation of gross specific growth rate (the dimensionless factor g_N^*) is described with the Monod equation:

$$g_N^* = \frac{N^*}{1+N^*}, \tag{4}$$

where N^* is the effective nutrient concentration, defined as $[N]/K_m$, with [N], the nutrient concentration, and K_m, the half-saturation constant for growth, in units of μM. Equation 4 is adequate to describe the nutrient dependence of growth rate when uptake of an non-substitutable limiting nutrient is catalyzed by a single transport mechanism (Morel, 1987). This appears to be the case for P- and Si- limitation of growth. It also applies when growth is limited by either nitrate or ammonium. Tentatively, effective nutrient concentration during growth on both nitrate and ammonium is modeled additively:

$$N^* = \frac{[NH_4]}{K_{m(NH_4)}} + \frac{[NO_3]}{K_{m(NO_3)}}, \tag{5}$$

with the recognition that nutrient interactions (such as ammonium-inhibition of nitrate uptake), once defined, can be accommodated by treating the saturation constants as variables influenced by the presence of other nutrients.

Unfortunately, the appropriate treatment of growth when nitrate and ammonium are simultaneously present at suboptimal concentrations cannot be specified at present because considerable uncertainty surrounds the interaction of these different nitrogen forms on phytoplankton nitrogen assimilation and growth (Dortch, 1990). Progress in resolving this uncertainty has been hindered by the fact that, until recently, the detection limits (approximately 50 nM) of our techniques for measuring nitrate and ammonium have exceeded the concentrations at which these nutrients limit phytoplankton growth (K_m<50 nM). Although this technical limitation has recently been overcome, the new techniques for measuring nitrate and ammonium at nM concentrations have rarely been employed in ecophysiological investigations (Garside & Glover, 1991). The situation is further complicated by recognition that phytoplankton can assimilate urea and amino acids (Antia et al., 1991). Finally, the role of trace elements as limiting factors for phytoplankton growth, although the subject of recent investigations, remains uncertain (Bruland et al., 1991; Martin et al., 1991; Morel et al., 1991). In the absence of a mechanistic understanding of the interaction of nitrate, ammonium and other nutrients, we cannot endorse any particular description of nitrate-ammonium interactions. The data required for reaching an informed opinion on this matter are simply not available.

The nutrient function (eq. 4) is assumed to act multiplicatively with the temperature function (eq. 3) with respect to growth rate. Thus, consistent with at least one study (Zevenboom et al., 1980), at any nutrient concentration, the absolute uptake rate (per unit cellular N) scales directly with g_T^* so that the half-saturation constant is independent of temperature.

Quantum yield of photosynthesis. For a model of balanced phytoplankton growth under constant irradiance during the light period, the quantum yield for gross photosynthesis at the growth irradiance is required (designated ϕ(E)). This is not the same thing as an instantaneous measure of photosynthetic quantum yield versus E for any one growth condition, and few data are available for making the appropriate generalizations. Determination of ϕ(E) requires simultaneous measurement of the rates of light absorption, respiration (in the light!) and net photosynthesis at the growth irradiance. Most of our inferences on the light dependence of ϕ(E) are in fact based on measurements of the quantum efficiency of growth. The quantum efficiency of growth (not gross photosynthesis at the growth irradiance) has been measured under nutrient-replete (Kok, 1952; Myers, 1980; Pirt et al., 1980; Morel et al., 1987; Osborne & Geider, 1987) and nutrient-limited (Chalup & Laws, 1990) conditions. However, these measurements are fraught with technical difficulties and considerable controversy surrounds the absolute values (Osborne & Geider, 1987).

It is well recognized, nonetheless, that the quantum yield of photosynthesis (ϕ, mol C mol photons^{-1}) declines with increasing irradiance as rate of absorption of photons exceeds the rate at which they can be processed by the photosynthetic apparatus. A relatively simple cumulative one-hit model (Dubinsky et al., 1986; Peterson et al., 1987; Eilers & Peeters, 1988; Falkowski, 1992) describes photosynthetic quantum yield as a function of E during exposures to pulses of light, but that exponential model is not necessarily appropriate for describing steady-state ϕ during acclimated growth. Thus, for nutrient-saturated growth at one temperature, we specify a general function,

$$\phi = \phi_{max} \cdot f(E/E_k). \tag{6}$$

The maximum quantum yield in low light for that temperature and nutrient concentration is ϕ_{max} and E is scaled to a saturation irradiance, E_k so that photosynthesis (absorbed irradiance $\cdot$ ϕ) is a saturating function of irradiance. Because it has been shown that that the quantum yield of photosynthesis during growth is a function of instantaneous irradiance rather than cumulative exposure during the day (Sakshaug et al., 1989), daylength is not included in eq. 6. The observations of Chan (1978) for diatoms and dinoflagellates, in which the growth rate was linearly related to the Chl:protein ratio, suggest that light utilization efficiency, hence ϕ(E), varies little between species.

In the simplest view, temperature is expected to affect the rate at which photons can be processed (Li & Morris, 1982; Raven & Geider, 1988), but not their rate of absorption, so E_k should be a function of g_T^*. Thus, to account for T, we replace E_k in eq. 6 with $E_k \cdot g_T^*$. At this stage of the modeling effort, we defer responding to indications that ϕ_{max} is depressed at very low temperatures (Tilzer et al., 1986; Raven & Geider, 1988) or other suggestions that photosynthetic efficiency is enhanced at lower temperatures (Davison, 1991). Clearly, these types of responses would further complicate efforts at generalization.

The effect of nutrient limitation on ϕ(E) is much more difficult to describe because data are scarce: measurements of photosynthetic characteristics have been made during nutrient-limited growth, but ϕ_{max} and E_k for growth, as in eq. 6, are not the same as ϕ_{max} and E_k estimated from short-term measurements of photosynthesis (Myers, 1970; Cullen, 1990) or by fluorescence-based probes of photosynthesis (Falkowski, 1992). Even short-term measurements present a confusing picture. Published data for the maximum normalized rate of photosynthesis during nutrient-limited growth are not consistent (Cullen et al., 1992).

The weight of available evidence suggests that ϕ_{max} is a nonlinear function of nutrient-limited growth rate in continuous culture (Herzig & Falkowski, 1989; Chalup & Laws, 1990; Falkowski, 1992), so ϕ_{max} should be a function of g_N^* in an equation describing variability of ϕ:

$$\phi_{max} = \phi_{opt} \cdot f_1(g_N^*), \tag{7a}$$

where $f_1(g_N^*)$ approaches 1.0 (i.e., ϕ_{max} approaches the theoretical maximum, ϕ_{opt}) as g_N^* approaches 1.

It is not at all clear how nutrient limitation influences E_k, the saturation parameter for photosynthesis at growth irradiance, so an unspecified function of nutrient limitation is included in an extremely general equation for variation of quantum yield as a function of E, N, and T:

$$\phi = \phi_{opt} \cdot f_1(g_N^*) \cdot f\left(\frac{E \cdot f_2(g_N^*)}{E_K \cdot g_T^*}\right). \tag{7b}$$

Here, f_1 and f_2 are potentially, but not necessarily, different functions of g_N^*. It is possible (and it would be very convenient) if both $f_1(g_N^*)$ and $f_2(g_N^*)$ were the

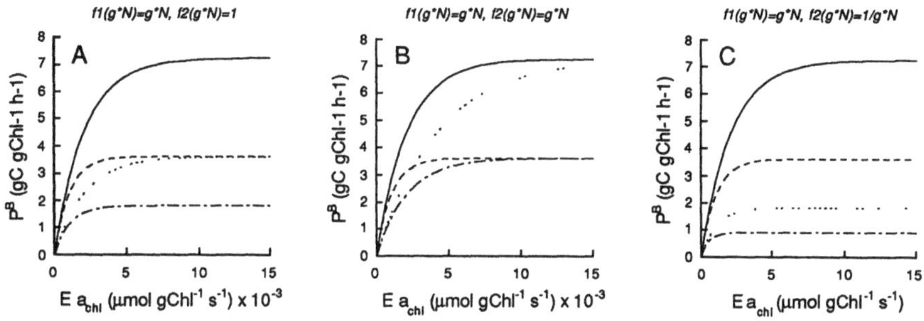

Fig. 3. The relationship between normalized photosynthesis at growth irradiance, P^B (gC $gChl^{-1}$ h^{-1}) and absorbed irradiance, $E \cdot a_{chl}$ (μmol photons mg Chl^{-1} s^{-1}) as predicted by different variants of eq. 7b. (—) Nutrient-saturated growth at optimal temperature: $g_N^* = g_T^* = 1.0$; (···) nutrient-limited growth at optimal temperature: $g_N^* = 0.5$, $g_T^* = 1.0$; (---) nutrient-saturated growth at suboptimal temperature: $g_N^* = 1.0$, $g_T^* = 0.5$; (- · -) nutrient-limited growth at suboptimal temperature: $g_N^* = 0.5$, $g_T^* = 0.5$. A. Results from assuming that nutrient-limitation has no effect on the scaling of E_k (i.e., both light-saturated and light-limited ϕ(E) are affected the same). B. Nutrient-limitation increases E_k as if the efficiency of electron transport were reduced but capacity per unit chlorophyll *a* is unaffected by nutrition. C. Nutrient-limitation decreases E_k as if capacity were more sensitive to nutrient limitation than efficiency. These different predictions can be tested with measurements of μ, C:Chl, and a_{chl} on continuous cultures.

same, but that has yet to be resolved. Contrasting predictions of eq. 7b can be examined by plotting normalized photosynthesis, P^B (gC $gChl^{-1}$ h^{-1}) vs. absorbed irradiance using different nutrient-limitation functions, f_1 and f_2 (Fig. 3). These predictions can be compared to results form laboratory experiments to resolve which functional relationships apply. The variation of the specific absorption coefficient for chlorophyll *a* (a_{chl}), which also influences P^B, is discussed below.

Carbon:chlorophyll ratio. The C:Chl of phytoplankton, θ, is a sensitive indicator of physiological state (Geider, 1987). The cellular content of Chl, relative to C, decreases when nutrients limit growth, thereby reducing the absorption of light, and bringing the rate of light absorption more-or-less into balance with the rate of energy demand. At a given irradiance, θ is inversely proportional to temperature, reflecting cellular regulation that allows more light to be absorbed when higher temperatures permit higher rates of synthesis per unit protein and a greater demand for energy. Chlorophyll content increases when light limits growth, partially compensating for lower photon flux with higher cellular absorption. A fundamental limitation to this compensating cellular regulation is θ_{min}, the minimum C:Chl (i.e., maximum Chl content; Geider, 1987; Langdon, 1988). Along the lines of Kiefer (this volume), the interactions of *D*, *E*, *N*, and *T* in determining θ are tentatively described as:

$$\theta = f(\theta_{min}, \frac{D \cdot E \cdot a_{chl}}{g_T^*}) \cdot \frac{1}{g_N^*}. \tag{8}$$

That is, with respect to temperature, θ is a function of absorbed irradiance relative to g_T^*, constrained by a maximum Chl relative to C (Fig. 4A). As will be shown below, a_{chl} covaries with θ, so eq. 8 might be simplified by altering the function to include the effect of varying a_{chl} implicitly. Nutrient limitation is modeled as having a direct linear relationship with Chl relative to C (e.g. Laws & Bannister, 1980). That is, nutrient limitation is expressed as an increase of θ until the C-specific growth rate matches the growth rate allowed by the nutrient supply (Bannister & Laws, 1980). In eq. 8, this representation violates an energy balance to the extent that ϕ(E) changes with g_N^* (see eq. 7b).

A value for θ_{min} can be estimated in three different ways. First, we can draw on investigations of the macromolecular composition of the photosynthetic apparatus (Geider, 1987). Second, we can rearrange the energy balance (eq. 2) to calculate θ from E and a_{chl} at the compensation irradiance where μ=0 and r=r_0. Third, we can empirically measure θ in cells grown at high temperature and low irradiance (Geider, 1987). Minimum values for θ of 7 to 10 gC $gChl^{-1}$ have been observed in chlorophytes and diatoms at temperatures near 25°C (see also Langdon, 1988). However, dinoflagellates and cyanobacteria are characterized by higher values of θ (Chan, 1978; Geider, 1992a). In the dinoflagellates, high θ is associated with low growth rate, whereas in the cyanobacteria high θ arises

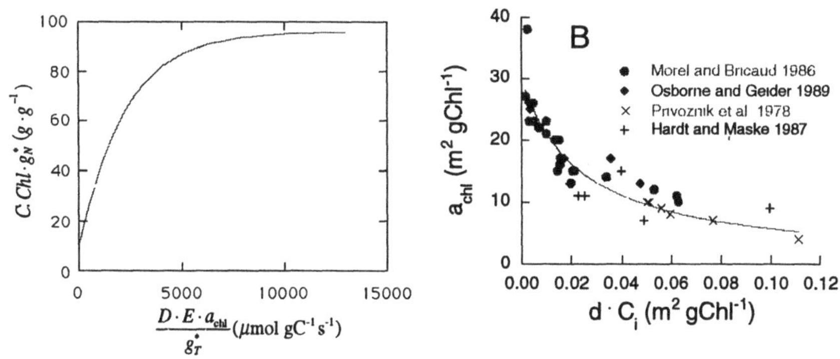

Fig. 4. Variation of θ and a_{chl}. A. The product of C:Chl (g:g) and the nutrient-limitation function $g_N^{\bullet}$, as a function of absorbed irradiance over the day, scaled by the temperature function. In this illustration, the θ vs. E relationship has the same form as the photosynthesis (ϕ(E) $\cdot$ E) vs. E relationship so that during nutrient-sufficient growth, light-limitation of gross growth rate is compensated by increased Chl relative to C. The constraint is that maximum Chl content is limited by θ_{min}. The predicted relationships between θ , μ, and E are roughly consistent with less general functions presented by Geider (1987) and Langdon (1988). Here, nutrient limitation ($g_N^{\bullet}$) is related directly to the amount of Chl relative to C rather than to the efficiency of light utilization. This assumption should be modified when the effects of nutrition on ϕ(E) are understood (see text). B. The specific absorption coefficient for Chl, a_{chl} (m^2 $gChl^{-1}$), as a function of $d{\cdot}C_i$, the product of cellular diameter (d, m) and intracellular chlorophyll concentration (C_i, gChl m^{-3}). Symbols represent results from different studies, compiled by Geider (1992a), the source for this comparison. The line is the best fit to the equation, $a_{chl} = a_{sol}{\cdot}(1/(k{\cdot}d{\cdot}C_i)+1)$: a_{sol}= 30.3 m^2 $gChl^{-1}$, k (an arbitrary coefficient) = 43.4 gChl m^{-2}, R^2 = 0.84. More rigorous analytical descriptions of this relationship are presented by Morel and Bricaud (1981). It should be possible to describe a_{chl} as a function of d and θ by relating C_i to θ with an empirical function.

because light harvesting is primarily by phycobiliproteins rather than by chlorophyll-carotenoid protein complexes. There is insufficient data available for characterizing θ_{min} and the irradiance dependence of θ for other phytoplankton taxa such as the coccolithophorids.

Absorption coefficient for chlorophyll. Normalized photosynthesis, P^B, is frequently measured in the field and predicted in models of biogeochemical cycling. For most studies of phytoplankton growth, P^B can be estimated from measurements of C, Chl and μ (Laws & Bannister, 1980). However, for a general model of phytoplankton growth, it is prudent to describe P^B as a function of light absorption and quantum yield,

$$P^B = \phi \cdot E \cdot a_{chl}, \tag{9}$$

where a_{chl} (m^2 $gChl^{-1}$) is the specific absorption coefficient for chlorophyll *a*. Differences between patterns of P^B and ϕ are attributable to changes in a_{chl}. Unfortunately, a_{chl} has not been measured as frequently as P^B, so validation of quantum-yield based *DENT* models is difficult although some success has been reported (Marra et al., 1992). It is possible to estimate a_{chl} from optical principles or from empirical relationships, however. A recent example (Geider, 1992) provides us with the basis for proposing that

$$a_{chl} = f(d, \theta), \tag{10}$$

where d is cell diameter (Fig. 4B). A maximum value for a_{chl} of about 30 m^2 $gChl^{-1}$ is approached for small cells and high θ. Basically, self-shading decreases a_{chl} as the internal Chl concentration increases, with the effect being greater for larger cells (Morel & Bricaud, 1981; Kirk, 1983). The influence of variable accessory pigmentation (Bidigare et al., 1992) is not explicitly described here, but, as mentioned above, it can be important, particularly for cyanobacteria.

The *DENT* equation. The details of the functional relationships between parameters (eqs. 3-8, 10) must be examined carefully for consistency with available data from cultures. As anyone who has tried it can attest, comparing measurements from different growth conditions can be extremely difficult. We feel that the exercise can be facilitated by scaling the independent variables with the dimensionless parameters g_T^* and g_N^*. For example, the similarities between acclimation to lower temperature and to brighter light (Davison, 1991) are qualitatively reconciled when irradiance is scaled to g_T^*, as in eq. 7b. This analytical simplification is based on the concept of relative growth rate as described by Goldman (1980). It is conceptually awkward to use relative growth rate to scale parameters that predict relative growth rate (a problem of circularity that plagues this pursuit), but we have to start somewhere.

Our efforts are directed toward a predictive representation of eq. 2:

$$\mu = \frac{E \cdot D \cdot Chl \cdot a_{chl} \cdot \phi}{C} - r = f(D, E, N, T), \tag{11}$$

as well as prediction of P^B through substitution in eq. 9. Although several models have described various aspects of the functional relationships, sometimes in ample detail, the ultimate goal has not yet been achieved, and some of us don't think that we are very close. The principal problems are the apparent complexity of the interactions of environmental factors, and the scarcity of data (sets of growth experiments during which μ, Chl, C and a_{chl} were measured). A great number of growth studies have been performed, however, and indications are that the observations can be incorporated into general descriptions if the appropriate scaling factors are applied. The important thing, we think, is the structure for expressing the relationships:

1) maximum rates are a function of temperature (g_T^*);
2) nutrient-limitation of growth g_N^* can be expressed as a function of ammonium and nitrate concentrations, scaled to their half-saturation constants;
3) C:Chl is a function of absorbed irradiance scaled to g_T^*, and also g_N^*;
4) a_{chl} is a function of Chl and cell diameter; and
5) quantum yield (i.e., ϕ (E)) is a function of g_T^* and probably g_N^*.

Whatever the real patterns are, we hope to see them using this analytical context.

Practical advice. Equations 3 to 8 and 10 can be substituted into a predictive *DENT* model, but the result is horrendously complex and not particularly instructive, even if the unspecified functions are replaced with analytical representations. Nonetheless, even a complex equation could be used to make predictions for a biogeochemical model of phytoplankton growth. Experience has shown, however, that simplicity is the key for interdisciplinary acceptance of a phytoplankton model. How can the *DENT* model be simplified for use in numerical simulations?

One way to simplify the energy balance is to combine terms. For example, even though both ϕ_{max} and a_{chl} are functions of nutrient-limited growth rate at a given irradiance and temperature, it is possible that their product (see eq. 9) is acceptably constant (Kiefer & Mitchell, 1983), allowing us to express the energy balance as a function of the P^B vs. E relationship and the C:Chl ratio (see Geider, 1990). One form for such a model could be

$$\mu + r = \frac{Chl}{C} \cdot D \cdot P^B_{max}(1 - e^{-E/E_k}), \tag{12}$$

(Cullen, 1990) where both P^B_{max} (gC gChl^{-1} h^{-1}; the maximum adapted rate of normalized photosynthesis) and E_k are functions of g^*_T and possibly g^*_N. Given that the left-hand side of the equation can be specified by g^*_T and g^*_N, C:Chl is algebraically constrained. Equation 12 is practical because it has three essential parameters for biogeochemical modeling: growth rate, chemical composition and P^B vs. E. However, the simplification of modeling P^B rather than $\phi \cdot a_{chl}$ compromises its utility as a physiological description of growth and chemical composition of phytoplankton, leaving us with a problem of describing empirically how P^B vs. E and μ vs. E are influenced by D, E, N, and T. That may be the best we can do.

Equation 12 may fail to hold, even as an approximation. For example, Laws and Chalup (1990) concluded that the product of the quantum yield coefficient (ϕ) and specific absorption coefficient (a_{chl}) was hyperbolically related to nutrient limited growth rate. Thus, detailed growth models can uncover relationships that might be used to modify the oversimplified relationship in eq. 12.

Predicting the ratio of N:C. Cellular nitrogen can be included explicitly in growth models (Shuter, 1979; Laws & Chalup, 1990), and its variation can be interpreted mechanistically in terms of the C and N required for biosynthetic proteins and the C that is stored in carbohydrates and lipids. The energy-budget approach does not consider N, an important component of biogeochemical models, so the ratio of cellular N:C must be described as a function of what is predicted by the *DENT* model. Variation of N:C is described by Laws and Chalup (1990) as a simple linear function of nutrient-limited growth rate:

$$N{:}C = f(g_N^*). \quad (13)$$

Data are insufficient to resolve completely the variability of N:C as a function of *D*, *E*, *N*, and *T*.

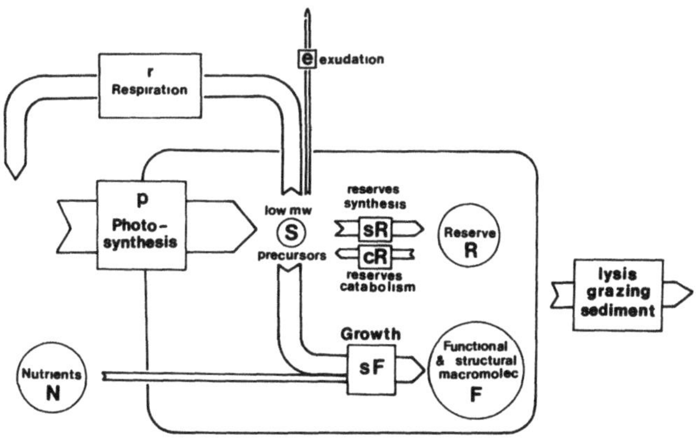

Fig. 5. Diagrammatic representation of a numerical simulation of phytoplankton photosynthesis and growth in a variable environment, from Lancelot et al. (1991b), where the mathematical formulation can be found. Structural and functional macromolecules, *F*, are composed of about 85% protein, a major part of the nutrients taken up by the cell; the reserve pool, *R*, consists of lipids and polysaccharide, storage for carbon. We suggest that this model can form the basis of a dynamic simulation of phytoplankton growth that, under steady conditions, would converge on the predictions of a *DENT* model.

Limitations of a steady state model. The result of a *DENT* steady-state model is a set of predictions that should be very useful for choosing conversion factors (C:Chl and C:N) and for specifying P^B vs. *E* in different environments as a function of assumed nutrient limitation. This would be useful in the biogeochemical models presently in use. However, a steady-state model cannot describe the photosynthesis, growth, and chemical composition of phytoplankton in variable environments because environmental conditions change too rapidly for steady-state to be assumed. In theory, this type of prediction can be accomplished with a dynamic model of photosynthesis, nutrient uptake, and allocation of cellular carbon on the cellular level (Lancelot et al., 1991a; Lancelot et al., 1991b; Fig. 5). The dynamic model also obviates the problem of assuming constant E during the photoperiod in the steady-state model. We hoped to describe the dynamic model in some detail and to suggest modifications, but this was not possible in the time available. For example, future models could incorporate recently-developed descriptions of P^B vs. E under variable irradiance (Baumert & Uhlmann, 1983; Denman & Marra, 1986; Pahl-Wostl & Imboden, 1990; Janowitz & Kamykowski, 1991). It is clear to us that further development of dynamic models is warranted, with one objective being agreement between the dynamic and steady-state *DENT* model for constant conditions.

4. Other considerations

Our working group considered many aspects of modeling phytoplankton growth for biogeochemical models, but we concentrated on a *DENT* model, trying to tell numerical modelers how one might be constructed. This goal proved to be elusive. Even the most general approaches were hampered by lack of information and insufficient time to assimilate and apply the huge amounts of experimental data that are available. Other considerations, which are important to any modeling effort, further complicate our work, but should be kept in mind as models are developed:

1) The size of phytoplankton has important physiological, optical, trophodynamic, and biogeochemical consequences. Phytoplankton should be modeled as at least two classes (big and small), with different functional responses. For example, it could be assumed that diatoms dominate the large size class. That is, only the big cells require Si, and sinking is important only in the large size class — faster when nutrients are depleted (Steele & Yentsch, 1960; see also Waite et al., 1992).

2) Limitation of phytoplankton growth by CO_2 is a special case (Raven & Johnston, 1991), which we consider to be unimportant at this stage of model development.

3) We cannot suggest *DENT* models for limitation of phytoplankton growth by iron or other trace elements (Morel et al., 1991). It would probably be best to omit iron from models and identify its possible influence from aberrant model behavior.

4) It should be recognized that, under certain conditions, respiration can be an extremely important component of energy flow (Geider, 1992b) and that simplified treatments of r can lead to poor predictions under those circumstances.

Although numerous other recommendations can be listed, we stop here because the situation as presented is sufficiently intimidating.

5. Caveats

Our generalized descriptions of phytoplankton growth are quite idealistic, yet still vague and incomplete. If appropriate *DENT* models are developed and validated, they are likely to apply initially to only a few species of phytoplankton in culture. It should be clearly recognized that other factors (e.g., taxonomy and environmental variability) will have an influence on predictions. We have no quantitative approach for predicting species composition for natural phytoplankton, thus we expect problems with general models. With time, taxon-specific predictions can be developed and compared with nature. Unfortunately, validation of predictions (i.e. μ, C:Chl and C:N) is complicated by the contribution of detritus to measurements of particulate C and N: we cannot measure directly the growth rate or chemical composition of phytoplankton (Eppley, 1980).

6. Conclusions

Models of global biogeochemical processes in the sea require information on the effects of daylength, irradiance, temperature and nutrients on the biochemical and physiological properties of phytoplankton. Although much has been achieved in describing important interactions, construction of a general descriptive model has proven very difficult. This is a task that requires time, a great deal of study, additional lab work, clear focus, and collaboration. Progress may accelerate if large research programmes recognize explicitly the need to develop models of phytoplankton growth, channeling resources accordingly. The obstacles to formulating good, general models of phytoplankton are daunting indeed, but the need for progress cannot be ignored. The alternative is to use inadequate approximations of C:Chl and P^B vs. E in basin-scale biogeochemical models.

Acknowledgments. Partially supported by grants from NASA, the Office of Naval Research, and NSERC Canada to JJC. Thanks are extended to W.K.W. Li for information on temperature limitation.

References

Antia NJ, Harrison PJ, Oliveira L (1991) The role of dissolved organic nitrogen in phytoplankton nutrition, cell biology and ecology. Phycologia. 30:1-89

Bannister TT (1979) Quantitative description of steady state, nutrient-saturated algal growth, including adaptation. Limnol Oceanogr. 24:76-96

Baumert H, Uhlmann D (1983) Theory of the upper limit to phytoplankton production per unit area in natural waters. Int Revue ges Hydrobiol. 68:753-783

Bidigare RR, Prézelin BB, Smith RC (1992) Bio-optical models and the problems of scaling. In: Falkowski PG, Woodhead A, (eds) Primary Productivity and Biogeochemical Cycling in the Sea, Plenum, New York, p 175-212

Bruland KW, Donat JR, Hutchins DA (1991) Interactive influences of bioactive trace metals on biological production in oceanic waters. Limnol Oceanogr. 36:1555-1577

Chalup MS, Laws EA (1990) A test of the assumptions and predictions of recent microalgal growth models with the marine phytoplankter *Pavlova lutheri*. Limnol Oceanogr. 35:583-596

Chan AT (1978) Comparative physiological study of marine diatoms and dinoflagellates in relation to irradiance and cell size. I. Growth under continuous light. J Phycol. 14:396-402
Cullen JJ (1990) On models of growth and photosynthesis in phytoplankton. Deep-Sea Res. 37:667-683
Cullen JJ, Yang X, MacIntyre HL (1992) Nutrient limitation of marine photosynthesis. In: Falkowski PG, Woodhead A, (eds) Primary Productivity and Biogeochemical Cycles in the Sea, Plenum, New York, p 69-88
Davison IR (1991) Environmental effects on algal photosynthesis: temperature. J Phycol. 27:2-8
Denman KL, Marra J (1986) Modelling the time dependent photoadaption of phytoplankton in fluctuating light. In: Nihoul JCJ, (ed) Marine interfaces ecohydrodynamics, Elsevier, Amsterdam, p 341-360
Dortch Q (1990) The interaction between ammonium and nitrate uptake in phytoplankton. Mar Ecol Prog Ser. 61:183-201
Dubinsky Z, Falkowski PG, Wyman K (1986) Light harvesting and utilization by phytoplankton. Plant Cell Physiol. 27:1335–1349
Eilers PHC, Peeters JCH (1988) A model for the relationship between light intensity and the rate of photosynthesis in phytoplankton. Ecol Modelling. 42:199-215
Eppley RW (1972) Temperature and phytoplankton growth in the sea. Fish Bull. 70:1063-1085
Eppley RW (1980) Estimating phytoplankton growth rates in the central oligotrophic oceans. In: Falkowski PG, (ed) Primary Productivity in the Sea, Plenum, New York, p 231–242
Falkowski PG (1992) Molecular ecology of phytoplankton photosynthesis. In: Falkowski PG, Woodhead A, (eds) Primary Productivity and Biogeochemical Cycles in the Sea, Plenum, New York, p 47-67
Falkowski PG, Dubinsky Z, Wyman K (1985) Growth-irradiance relationship in phytoplankton. Limnol Oceanogr. 30:311-321
Garside C, Glover HE (1991) Chemiluminescent measurements of nitrate kinetics: I. *Thalassiosira pseudonana* (clone 3H) and neritic assemblages. J Plankton Res. 13 Suppl.:5-19
Geider RJ (1987) Light and temperature dependence of the carbon to chlorophyll a ratio in microalgae and cyanobacteria: implications for physiology and growth of phytoplankton. New Phytol. 106:1-34
Geider RJ (1990) The relationship between steady state phytoplankton growth and photosynthesis. Limnol Oceanogr. 35:971-972
Geider RJ (1992a) Quantitative phytoplankton ecophysiology: implications for primary production and phytoplankton growth. ICES mar Sci Symp. 194: (in press)

Geider RJ (1992b) Respiration: Taxation without representation. In: Falkowski PG, Woodhead A, (eds) Primary Productivity and Biogeochemical Cycles in the Sea, Plenum, New York, p 333-360

Geider RJ, Osborne BA, Raven JA (1986a) Growth, photosynthesis and maintenance metabolic cost in the diatom *Phaeodactylum tricornutum* at very low light levels. J Phycol. 22:39-48

Geider RJ, Platt T, Raven JA (1986b) Size dependence of growth and photosynthesis in diatoms: a synthesis. Mar Ecol Prog Ser. 30:93-104

Goldman JC (1980) Physiological processes, nutrient availability, and concept of relative growth rate in marine phytoplankton ecology. In: Falkowski PG, (ed) Primary Productivity in the Sea, Plenum, New York, p 179-194

Herzig R, Falkowski PG (1989) Nitrogen limitation in *Isochrysis galbana* (Haptophyceae). I. Photosynthetic energy conversion and growth efficiencies. J Phycol. 25:462-471

Janowitz GS, Kamykowski D (1991) An eulerian model of phytoplankton photosynthetic response in the upper mixed layer. J Plankton Res. 13:983-1002

Kiefer DA, Cullen JJ (1991) Phytoplankton growth and light absorption as regulated by light, temperature, and nutrients. Polar Res. 10:163-172

Kiefer DA, Mitchell DG (1983) A simple, steady state description of phytoplankton growth based on absorption cross section and quantum efficiency. Limnol Oceanogr. 28:770–776

Kirk JTO (1983) Light and Photosynthesis in Aquatic Ecosystems, Cambridge University Press, Cambridge

Kok B (1952) On the efficiency of chlorella growth. Acta Botanica Neerlandica. 1:445-476

Lancelot C, Billen G, Veth, C, Becquevort S, Mathot S (1991a) Modelling carbon cycling through phytoplankton and microbes in the Scotia-Weddell Sea area during sea ice retreat. Mar Chem. 35:305-324

Lancelot C, Veth C, Mathot S (1991b) Modelling ice-edge phytoplankton bloom in the Scotia-Weddell sea sector of the Southern Ocean during spring 1988. J Mar Syst. 2:333-346

Langdon C (1988) On the causes of interspecific differences in the growth-irradiance relationship for phytoplankton. II. A general review. J Plankton Res. 10:1291-1312

Langdon C (1992) The significance of respiration in phytoplankton production. ICES mar Sci Symp. 194:(in press)

Laws EA, Bannister TT (1980) Nutrient- and light-limited growth of *Thalassiosira fluviatilis* in continuous culture, with implications for phytoplankton growth in the ocean. Limnol Oceanogr. 25:457–473

Laws EA, Chalup MS (1990) A microalgal growth model. Limnol Oceanogr. 35:597-608

Li WKW, Morris I (1982) Temperature adaptation in *Phaeodactylum tricornutum* Bohlin: photosynthetic rate compensation and capacity. J Exp Mar Biol Ecol. 58:135-150

Margalef R (1978) Life forms of phytoplankton as survival alternatives in an unstable environment. Oceanol Acta. 1:493-509

Marra J *et al.* (1992) Estimation of seasonal primary production from moored optical sensors in the Sargasso Sea. J Geophys Res. 97:7399-7412

Martin JH, Gordon RM, Fitzwater SE (1991) The case for iron. Limnol Oceanogr. 36:19

Morel A, Bricaud A (1981) Theoretical results concerning light absorption in a discrete medium, and application to specific absorption of phytoplankton. Deep-Sea Res. 28:1375-1393

Morel A, Lazzara L, Gostan J (1987) Growth rate and quantum yield time response for a diatom to changing irradiances (energy and color). Limnol Oceanogr. 32:1066-1084

Morel FMM (1987) Kinetics of nutrient uptake and growth in phytoplankton. J Phycol. 22:1037-150

Morel FMM, Hudson RJM, Price NM (1991) Limitation of productivity by trace metals in the sea. Limnol Oceanogr. 36:1742-1755

Myers J (1970) Genetic and adaptive physiological characteristics observed in the Chlorellas. In: Malek I, (ed) Prediction and measurement of photosynthetic productivity, Center for Agricultural Publishing and Documentation, Wagenigen, p 447–454

Myers J (1980) On the algae: thoughts about physiology and measurements of efficiency. In: Falkowski PG, (ed) Primary productivity in the sea, Plenum Press, New York, p 1-16

Nielsen MV (1992) Irradiance and daylength effects on growth and chemical composition of *Gyrodinium aureolum*. J Plankton Res. 14:811-820

Osborne BA, Geider RJ (1987) Photon requirement for growth of the diatom *Phaeodactylum tricornutum* (Bacillariophyceae). Plant Cell Environ. 10:17

Pahl-Wostl C, Imboden DM (1990) DYPHORA - A dynamic model for the rate of photosynthesis of algae. J Plankton Res. 12:1207-1221

Peterson DH, Perry MJ, Bencala KE, Talbot MC (1987) Phytoplankton productivity in relation to light intensity: A simple equation. Est Coastal Shelf Sci. 24:813-832

Pirt SJ, Lee YK, Richmond A, Watts-Pirt M (1980) The photosynthetic efficiency of chlorella biomass growth with reference to solar energy utilization. J Chem Tech Biotech. 30:25-34

Raven JA, Geider RJ (1988) Temperature and algal growth. New Phytol. 110:441-461

Raven JA, Johnston AM (1991) Mechanisms of inorganic-carbon acquisition in marine phytoplankton and their implications for use of other resources. Limnol Oceanogr. 36:1701-1714

Sakshaug E, Kiefer DA, Andresen K (1989) A steady state description of growth and light absorption in the marine planktonic diatom *Skeletonema costatum*. Limnol Oceanogr. 34:198-205

Shuter B (1979) A model of physiological adaptation in unicellular algae. J Theor Biol. 78:519–552

Steele JH, Yentsch CS (1960) The vertical distribution of chlorophyll. J Mar Biol Ass UK. 39:217-226

Thomas WH, Dodson AN (1972) On nitrogen deficiency in tropical Pacific oceanic phytoplankton. II. Photosynthetic and cellular characteristics of a chemostat-grown diatom. Limnol Oceanogr. 17:515-523

Tilzer MM, Elbrächter M, Gieskes WW, Beese B (1986) Light-temperature interactions in the control of photosynthesis in Antarctic phytoplankton. Polar Biol. 5:105-112

Waite AM, Thompson PA, Harrison PJ (1992) Does energy control the sinking rate of marine diatoms? Limnol Oceanogr. 37:468-477

Zevenboom W, Mur LR (1984) Growth and photosynthetic response of the cyanobacterium *Microcystis aeruginosa* in relation to photoperiodicity and irradiance. Arch Microbiol. 139:232-239

Zevenboom W, van Donk E, Mur LC (1980) Effect of temperature on nitrate-limited *Oscillatoria agardhii*: a preliminary study. In: Zevenboom W, *Oscillatoria agardhii*: a comparative investigation of continuous cultures and natural populations of a cyanobacterium, Ph.D. Thesis, Universiteit van Amsterdam, Amsterdam

MODELLING ZOOPLANKTON

Thomas R. Anderson
James Rennell Centre for Ocean Circulation,
Gamma House,
Chilworth Research Centre,
Chilworth, Southampton,
SO1 7NS, United Kingdom.

V. Andersen[1], H.G. Fransz[2], B.W. Frost[3], O. Klepper[4], F. Rassoulzadegan[1], F. Wulff[5]

1. Introduction

Marine zooplankton are a diverse group of organisms. Different species have different foraging strategies, life cycles, patterns of diel vertical migration, etc. Reducing such complexity to a minimal number of compartments in large-scale ecosystem models therefore poses a significant problem. The working group concentrated their discussions on the following topics, each of which need to be considered when tackling this problem:

(i) The functional roles of zooplankton in the marine food web. If different types of zooplankton are responsible for different system processes, then they would probably have to be modelled as separate entities.

(ii) Model structure. A suitable structure must be designed to incorporate different functional groups and trophic levels. Factors influencing whether different groups of zooplankton should be modelled

[1] Station Zoologique, Observatoire Océanologique, BP 28, F-06230, Villefrance-sur-Mer, France
[2] Netherlands Institute for Sea Research, P.O. Box 59, 1790 AB Den Burg, Texel, The Netherlands
[3] School of Oceanography, WB-10, University of Washington, Seattle, Washington 98195, USA
[4] Natl. Inst. Publ. Health Env. Protection, P.O. Box 1, 3720 BA Bilthoven, The Netherlands
[5] Institute of Marine Ecology, University of Stockholm, Box 6801, S-113 86 Stockholm, Sweden

NATO ASI Series, Vol I 10
Towards a Model of Ocean
Biogeochemical Processes
Edited by G T Evans and M J R Fasham

explicitly (as state variables) or implicitly (in the model equations) are discussed.

(iii) Complexity within functional groups. A major difficulty when constructing large-scale ecosystem models is the derivation of universal parameters for zooplankton, because of the enormous variety of species that occurs in the ocean. This variety is investigated in some detail, so as to provide insight into the modelling difficulties involved. Recommendations can then be made on how future modelling research can proceed.

2. Functional roles of zooplankton

The group identified three major roles for zooplankton in the marine system: grazing (control of biomass stocks), nutrient regeneration, and transport of material to depth, either via sinking material or by direct transport via diel vertical migration. The group envisaged the simple food web shown in Figure 1 as representing how zooplankton fulfil the above roles. Here, it is assumed that the smallest phytoplankton are mostly

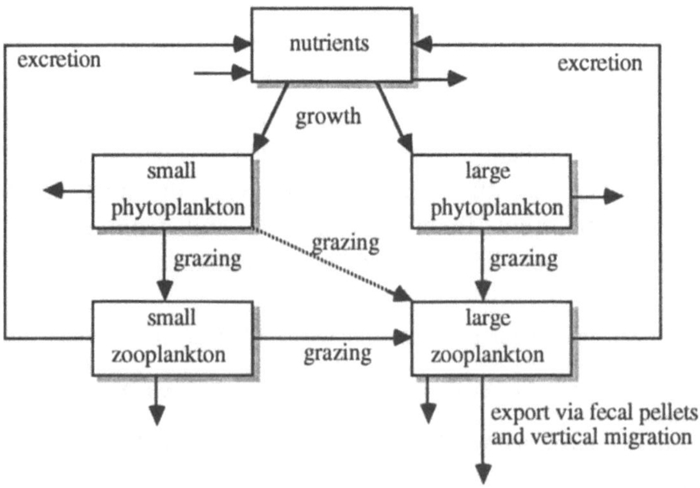

Fig. 1. Hypothetical food web illustrating roles of zooplankton in the marine system. Unlabelled arrows are unspecified flows. The dotted arrow represents a relatively small flux.

consumed by small zooplankton (microzooplankton), and that large phytoplankton are consumed by larger zooplankton (meso- and macrozooplankton). The sinking speed of fecal pellets is positively related to their size (Small *et al.*, 1979); the smallest are produced by protozoans, intermediate sizes by copepods and the largest by euphausiids, salps and fish (Fowler & Knauer, 1986). A significant export is then only effected by the larger types (Tseytlin, 1982). Additionally, because the relative energy expenditure for swimming increases with size (Klyashtorin, 1978) as well as swimming speed, direct transport of material by vertical migration is generally only carried out by the larger zooplankton (Angel, 1985). Many of the world's oceans, and in particular the tropical oligotrophic ocean and the stratified waters of the temperate summer, are dominated by small phytoplankton (Cushing, 1989). In such areas the grazing and nitrogen excretion will be dominated by small zooplankton, but the particulate export flux would nevertheless be effected by large zooplankton. A model with a single zooplankton compartment might therefore grossly overestimate the export flux.

RECOMMENDATION: On the basis of functional types, it was agreed that microzooplankton and larger zooplankton should be separated as distinct groups in ecosystem models. The larger zooplankton could possibly also be usefully separated into meso- and macrozooplankton, because the sinking rates of their fecal pellets differ significantly.

3. Model structure

Having identified two or three groups of zooplankton that should be treated as separate entities in models, how should the model be structured to incorporate these? In particular, what determines whether a zooplankton functional type or trophic level should be explicitly represented as a state variable, or implicitly included in the model equations? Several factors were identified that were of potential importance:

(i) *Coupling with the rest of the system.* If a zooplankton group is very tightly coupled with the rest of the system, then an explicit representation might be unnecessary. This might, for example, be the case with microzooplankton, which have very high maximum grazing rates and may in a quickly responding feedback loop be very tightly coupled to phytoplankton biomass. A single state variable could then replace separate compartments for phytoplankton and microzooplankton. However, one should remember that even only a short time lag between the population response of microzooplankton to a phytoplankton bloom could have a very significant impact on the simulated bloom dynamics. It is therefore important to consider the relative time scales of the dynamics of the grazer and its food before implementing such an approach.

(ii) *Ability to parameterise explicit model compartments.* There may be considerable variability between different types of organism within a trophic level or functional group. When constructing large-scale models, the best results may be achieved by using simple implicit representations for the more complex groups (typically higher trophic levels), because of an inability to derive adequate parameterisations for explicit representations. A poor parameterisation of an explicitly modelled zooplankton group may rapidly cause erroneous feedbacks in the remainder of the system. The issue of variability within functional groups is discussed in the next main section.

(iii) *Stability requirements of the system.* If model compartments are linked in series, then the number of explicitly modelled trophic levels will largely determine the dynamics of phytoplankton and nutrients in a modelled ecosystem, assuming steady state (Fransz *et al.*, 1991; Frost & Franzen, 1992). Consider the two food webs shown in Figure 2. An even number

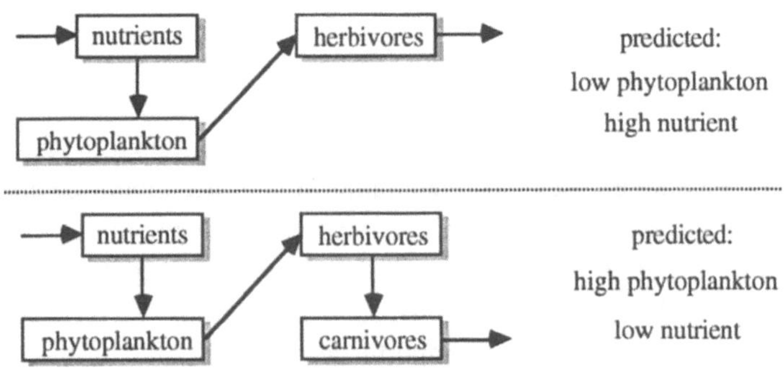

Fig. 2. Theoretical steady-state predictions for two simple ecosystem models.

of trophic levels results in a simulated high phytoplankton, low nutrient, regime, with the opposite occurring when an odd number of trophic levels is applied. This problem is less severe if the compartments are not linked solely in series (e.g. consider the dotted arrow in Figure 1), and if modelled systems depart from steady state. Nevertheless, modellers must be aware of it.

An implicit formulation is always used to close the highest explicit trophic level. Steele and Henderson (1992) have demonstrated the importance of taking care when selecting closure terms in plankton models. A simple linear mortality term for closure (constant specific rate) could be justified as the simplest and most economical way to close the system. A density-dependent closure term could be justified if it is assumed that some undefined predator population has a biomass proportional to its prey (the highest trophic level). More complex functions could be used to take account of feeding responses of predators.

(iv) *Long-term trends.* It may be impossible to adequately simulate these if simple implicit formulations are used. For example, one might envisage a change in the relative dominance of different species in response to climate change; an explicit treatment of the various species, in order to correctly couple them with the changing environment, might be the only way to model this.

4. Complexity within functional groups

Each zooplankton functional group may represent a diverse range of organisms. For example large herbivorous grazers include copepods, salps and euphausiids; these organisms may differ markedly in the way in which they interact with the rest of the food web. Such diversity is most apparent in the large zooplankton: they may have long and complicated life histories, and exhibit much variation in their ability to exploit different resources. Much of the following discussion will therefore concentrate on this group. Microzooplankton have much shorter life spans, and their population growth is relatively tightly coupled to their prey, and hence modelling problems caused by diversity within the group are not so severe.

The relative dominance of different organisms may vary geographically or seasonally. If the dynamics of different organisms within a single group are sufficiently variable to make a universal parameterisation impossible, then further subdivision of zooplankton into more compartments would be necessary. The following aspects of zooplankton dynamics were identified as being particularly important: grazing, growth dynamics, life cycles, mortality and diel vertical migration. These are now considered in turn.

4.1 Grazing

Zooplankton have a wide range of feeding mechanisms, which influence the size and quality of ingested food. For example, copepods collect particles typically in the size range 5 μm to 100 μm by mechanical sieving (Boyd, 1976). In contrast, salps collect particles in the range 1 μm to 1000 μm by pumping water through a mucous net (Madin, 1974). The grazing parameters associated with dominant grazers of different types may therefore vary considerably.

Diet selection was identified as also being important. For example, depending on stage and species, calanoid copepods may be herbivorous, carnivorous or even cannibalistic, to various extents which are affected by the size, shape, abundance and behaviour of food (Paffenhöfer & Knowles, 1980; Daan *et al.*, 1988). Microzooplankton can also feed selectively by size (Fenchel, 1980) and food quality (Burkhill *et al.*,1987). Chosen grazing functions should therefore take account of switching, multi-resource availability, and preferences for those resources (e.g. Fasham *et al.*, 1990).

4.2 Growth dynamics

Zooplankton use ingested food for growth and reproduction, the remainder being egested, respired or excreted as waste products. The relationships which describe these processes are by no means simple, and often vary between organisms or as functions of environmental variables. A few examples will serve to illustrate the point:

(i) The assimilation efficiency of the copepod *Calanus pacificus* was shown by Landry *et al.* (1984) to be influenced by feeding history. Animals subjected to low food had higher efficiencies. Assimilation efficiency was also shown to vary seasonally in this species (Hasset & Landry, 1990).

(ii) Respiration rates depend on temperature (e.g. Mullin & Brooks,

1970), as well as size and developmental stage of individuals. For example, Paffenhöfer and Gardner (1984) suggested that early copepodids may not be able to reduce their respiration rate in response to low levels of food, unlike later stages which may be able to make such a change (Cowles & Strickler, 1983).

(iii) Copepods subjected to a periodic variation in food supply at high latitudes store much assimilated food as lipid. Hence, a wide range in C:N ratios between copepods of different latitudes is observed (Båmstedt, 1986).

(iv) Egg production rate per copepod female can be a function of food level and temperature (e.g. Fransz *et al.*, 1989), but internally stored lipids can also be used to time spawning periods well in advance of the algal blooming season (e.g. Fransz, 1988).

The group agreed that more experimental work was necessary to verify models of growth-related processes. Great care would be needed when parameterising different types of organism covering extensive geographical ranges.

4.3 Mortality

Zooplankton mortality arises from a number of sources, including predation, starvation, parasitism and senescence (Ohman, 1986). Care must therefore be taken when parameterising this process. The major source of mortality in herbivorous mortality is often predation (e.g. Ohman, 1986). In general, the specific mortality rate due to predation might be assumed to be density-dependent (e.g. Frost, 1987). In reality, however, the situation is often more complicated. For example, predators on copepods will consist of numerous species, each of which will interact with the numbers and size distribution of their prey (Steele & Frost, 1977). Predatory zooplankton and fish have complex life history cycles; these are often coupled to the population dynamics of their prey (e.g. Terazaki & Miller, 1986).

At some point in a model, a mortality (closure) term must be applied to the highest explicitly modelled trophic level. This is discussed in Section 3.

4.4 Diel vertical migration

A wide variety of zooplankton undergo diel vertical migration, commonly feeding in surface waters at night, and migrating to depths by day. This was identified as potentially having an important bearing on system dynamics in two ways (Angel, 1989a,b; Longhurst & Harrison, 1989). Firstly, by feeding in the surface and defecating at depth, migrant organisms could accelerate the fecal pellet flux from the upper ocean (Andersen & Nival, 1991). Secondly, migration may be important in oligotrophic waters because it causes a redistribution of essential nutrients such as nitrogen (Angel, 1989a). Avoidance of visual predators is a commonly hypothesised selective pressure for the evolution of such behaviour (e.g. Frost, 1988). Migration may vary seasonally or interannually because of variations in predator abundance (Frost, 1988). Also, when food is abundant, migration may stop in order to maximise growth (Koslow & Ota, 1981). According to the review of Forward (1988), light is generally agreed to be the most important direct cue for the behaviour. The depth to which zooplankton migrate may be based on vulnerability to detection by predators (Frost, 1988); this might change seasonally or with predator type. Some species undergo reversed migration in order to avoid non-visual predators (Hutchinson, 1967; Frost & Bollens, 1992), whereby they migrate to depth at night after spending the day in surface waters.

Clearly, the parameterisation of diel vertical migration in models will not be easy, because of behavioural differences between species, and variations (geographical and temporal) in environment. The discussion group considered that the impact of this migration on overall system dynamics might be a second order effect, which could be adequately represented using a simple parameterisation, or be neglected (Tseytlin, 1982). If fecal pellets sink fast, the exact position at which they are egested might not have a large impact on overall system dynamics. Microzooplankton, such as ciliates and naupliar stages of copepods, do not migrate. Mesozooplankton such as copepods have a relatively small migration amplitude (e.g. Huntley & Brooks, 1982), compared to that of macrozooplankton and micronekton (Andersen & Sardou, 1992; Andersen *et al.*, 1992). If most grazing (and hence excretion) is done by microzooplankton, then the impact of diel vertical migration on nutrient redistribution might not be particularly significant. Nevertheless, the group considered it essential to test such reasoning properly using small-scale models such as that of Andersen and Nival (1991), which simulated euphausiid diel vertical migration as a function of irradiance and food concentration. This would provide information about the level of detail at which this process should be treated in larger-scale models.

4.5 Life history cycles

The life histories of marine zooplankton are varied and complex. The group discussion centred on identifying important aspects of life cycles, and on how these might be incorporated into ecological models. Much of the discussion revolved around different examples; four such examples are given here for illustrative purposes:

(i) North Atlantic. The characteristic diatom production of this region is directly grazed by calanoid copepods, principally *Calanus finmarchicus* (Parsons & Lalli, 1988). This species shows two main generations during the year (Williams & Conway, 1988). Stage V copepodites overwinter at depth (several hundred metres); during this time individuals may become dormant (Hirche, 1983), reducing metabolism and utilising a fat reserve until the following spring.

(ii) Subarctic Pacific. The dominant copepod of this region is *Neocalanus plumchrus* (Parsons & Lalli, 1988). This has a single generation per year and a long overwintering phase (Miller & Clemons, 1988), spent at depth. A large overwintering population of microzooplankton, which have a much shorter life span, is probably present which serves as an intermediate link between primary producers and large-sized copepods.

(iii) Antarctic. The zooplankton assemblage of this system is dominated by copepods and krill(*Euphausia superba*). Krill adults are very large compared to other pelagic herbivores. This euphausiid has a two to four year life cycle comprising egg, larvae, juvenile and adult (Ikeda, 1985). Unlike copepods which migrate to depth over winter, krill undergoes an extensive horizontal migration to ice cover where they continue feeding (Marschall, 1988).

(iv) Salp blooms. At times, salps form large blooms and totally dominate the zooplankton biomass (e.g. Wiebe *et al.*, 1979). The reproductive cycle is an alternation between an asexually reproducing solitary generation and a sexually reproducing aggregate generation (Alldredge & Madin, 1982; Braconnot *et al.*, 1988).

These examples highlight the vast variation that exists in life history cycles between species and geographical areas. Based on discussion of life histories such as those above, the following problems were recognised:

(i) To what extent is it necessary to recognise variation between different developmental stages in a single life history in models?

(ii) There is enormous variation between the life histories of dominant grazers in different marine systems. How can this be incorporated into global-scale models?

The first problem was addressed by considering how different stages impact on the remainder of the marine system. Three major effects were identified. Firstly, stages which do not involve reproduction put a brake on the rate of population increase under favourable growth conditions. The spring bloom in the North Atlantic may be underexploited by grazers because of this (Colebrook, 1979). Secondly, a period of dormancy over winter conserves stocks under unfavourable conditions, ready to respond to food when it becomes available. The treatment of dormancy is particularly difficult. It is thought that the cue which controls this phenomenon is photoperiod (Miller *et al.*, 1991). At a given photoperiod zooplankton could be made (in the model) to migrate to depth and remain there dormant (essentially doing nothing) until a similar cue in spring causes them to re-emerge at the surface. Alternatively, food concentrations were proposed as a possible cue. However, if this were imposed in a model, highly variable food concentrations (over time) would cause the organisms to "yo-yo" to and from deep water in ridiculous fashion. Thirdly, different developmental stages may have quite different grazing and growth dynamics (discussed above).

The variation in life histories between the dominant grazers of different systems poses a severe modelling problem. For example, the proposed cueing of dormancy with photoperiod does not apply to krill, which migrates horizontally and keeps on feeding through the winter. Different organisms undergo quite different modes of reproduction (e.g. compare salps and copepods); this affects their population response to changes in food. A single type of organism has varying dynamics between different geographic locations (e.g. compare copepods in the subarctic Atlantic and Pacific oceans).

4.6 Modelling strategy

Several general strategies can be used to solve the modelling problems described above, but they are by no means simple:

(i) Different types of zooplankton, e.g. copepods, euphausiids, salps, could each be separately included in large-scale models. Several problems can arise when using this approach:

- Excessive computational demand.

- Geographical and temporal variation in parameters within a single type of organism. It is to be expected that species will achieve local adaptation over broad geographic ranges primarily by adjusting their phenology (Miller *et al.*, 1991). This problem may be partly overcome by relating such variation to known environmental variables. For example, the generation time of copepods could be related to temperature (e.g. McLaren, 1978). If ecological models could use the appropriate environmental cues then such adaptation could be advantageously included.

- Inadequate parameterisations in global-scale models may cause incorrect geographical distributions of organisms to occur.

- Model stability may be hard to achieve if a large number of components are included, or even a small number of components with very detailed dynamics. For example, Parslow (1981) found that ecosystem models which simulated *Neocalanus* spp. were not robust if life history patterns were explicitly simulated in detail.

This approach would therefore require detailed parameterisations of the different organisms in order to correctly simulate their coupling with different marine ecosystems. It would also probably require a detailed model of other system components, e.g. phytoplankton, bacteria, etc.

(ii) Different parameterisations for zooplankton could be used for different geographic areas. While this may be acceptable for regional-scale models, it is generally not desirable for global-scale models (Murphy, this volume). This is particularly the case for models which are to be used to study phenomena such as climate change; a single environmentally-dependent parameter set is then highly desirable.

(iii) The explicit treatment of some aspects of zooplankton dynamics, e.g. life history cycles, could be largely eliminated. Essential features, such as dormancy, could be retained in a simple manner, but others, such as numerous developmental stages and cohorts, could be omitted. Rather than reducing the predictive ability of models, this approach could improve it on large-scale models because of the wide variation that exists in zooplankton and their dynamics on such scales.

RECOMMENDATION: In future, all the approaches outlined above should be adopted to try and assess what the most suitable level of detail to be used in zooplankton modelling is. There is a need for small-scale models to investigate in detail the impact of different types of zooplankton on the functioning of the marine ecosystem. The results of such simulations could provide useful information to aid simplified parameterisations for larger-scale models. Very often there is also a need for improved data on many aspects of zooplankton in order to sufficiently parameterise models.

References

Alldredge, AL, Madin, LP (1982) Pelagic tunicates: unique herbivores in the marine plankton. BioSci 32: 655-663

Andersen, V, Nival, P (1991) A model of the diel vertical migration of zooplankton based on euphausiids. J Mar Res 49: 153-175

Andersen, V, Sardou, J (1992) The diel migrations and vertical distributions of zooplankton and micronekton in the Northwestern Mediterranean Sea. 1. Euphausiids, mysids, decapods and fishes. J Plank Res (in press)

Andersen, V, Sardou, J, Nival, P (1992) The diel migrations and vertical distributions of zooplankton and micronekton in the Northwestern Mediterranean Sea. 2. Siphonophores, hydromedusae and pyrosomids. J Plank Res (in press)

Angel, MV (1985) Vertical migrations in the oceanic realm: possible causes and probable effects. In: Rankin, MA (ed.) Migration: mechanisms and adaptive significance. Contributions in Marine Science, Suppl. to vol. 27, 45-70

Angel, MV (1989a) Vertical profiles of pelagic communities in the vicinity of the Azores Front and their implications to deep ocean ecology. Prog Oceanogr 22: 1-46

Angel, MV (1989b) Does mesopelagic biology affect the vertical flux? In: Berger, WH, Smetacek, VS, Wefer, G (eds.) Productivity of the Ocean: Present and Past. John Wiley & Sons, 155-173

Båmstedt, U (1986) Chemical composition and energy content. In: Corner, EDS, O'Hara, SCM (eds.) The biological chemistry of marine copepods. Oxford Univ Press, 1-58

Boyd, CM (1976) Selection of particle sizes by filter-feeding copepods: A plea for reason. Limnol Oceanogr 21: 175-180

Braconnot, J-C, Choe, S-M, Nival, P (1988) La croissance et le développement de *Salpa fusiformis* Cuvier (Tunicata, Thaliacea). Ann Inst océanogr, Paris 64: 101-114

Burkhill, PH, Mantoura, RFC, Llewellyn, CA, Owens, NJP (1987) Microzooplankton grazing and selectivity of phytoplankton in coastal waters. Mar Biol 93: 581-590

Colebrook, JM (1979) Continuous plankton records: seasonal cycles of phytoplankton and copepods in the North Atlantic Ocean and the North Sea. Mar Biol 51: 23-32

Cowles, TJ, Strickler, JR (1983) Characterisation of feeding activity patterns in the planktonic copepod *Centropages typicus* Kroyer under various feeding conditions. Limnol Oceanogr 28: 106-115

Cushing, DH (1989) A difference in structure between ecosystems in strongly stratified waters and in those that are only weakly stratified. J Plank Res 11: 1-13

Daan, R, Gonzalez, SR, Breteler, WCMK (1988) Cannibalism in omnivorous calanoid copepods. Mar Ecol Prog Ser 47: 45-54

Fasham, MJR, Ducklow, HW, McKelvie, SM (1990) A nitrogen-based model of plankton dynamics in the oceanic mixed layer. J Mar Res 48: 591-639

Fenchel, T (1980) Suspension feeding in ciliated protozoa: functional response and particle size selection. Microb Ecol 6: 1-11

Forward, RBJnr (1988) Diel vertical migration: zooplankton photobiology and behaviour. Oceanogr Mar Biol Ann Rev 26: 363-393

Fowler, SW, Knauer, CA (1986) Role of large particles in the transport of elements and organic compounds through the oceanic water column. Prog Oceanogr 16: 147-194

Fransz, HG (1988) Vernal abundance, structure and development of epipelagic copepod populations of the eastern Weddell Sea (Antarctica). Polar Biol 9: 107-114

Fransz, HG, Gonzalez, SR, Breteler, WCMK (1989) Fecundity as a factor controlling the seasonal population cycle in *Temora longicornis* (Copepoda, Calanoida). In: Ryland, JS & Tyler, PA (eds.) Reproduction, genetics and distributions of marine organisms. Olsen & Olsen, Fredensborg, Denmark, 83-89

Fransz, HG, Mommaerts, JP, Radach, G (1991) Ecological modelling of the North Sea. Neth Jnl Sea Res 28: 67-140

Frost, BW (1987) Grazing control of phytoplankton stock in the open subarctic Pacific: a model assessing the role of mesozooplankton, particularly the large calanoid copepods *Neocalanus* spp. Mar Ecol Prog Ser 39: 49-68

Frost, BW (1988) Variability and possible adaptive significance of diel vertical migration in *Calanus pacificus*, a planktonic marine copepod. Bull Mar Sci 43: 675-694

Frost, BW, Bollens, SM (1992) Variability of diel vertical migration in the marine copepod *Pseudocalanus newmani*. Can J Fish Aquat Sci (in press)

Frost, BW, Franzen, NC (1992) Grazing and iron limitation in the control of phytoplankton stock and nutrient concentration: a chemostat analogue of the Pacific equatorial upwelling zone. Mar Ecol Prog Ser (in press)

Hassett, RP, Landry, MR (1990) Seasonal changes in feeding rate, digestive enzyme activity, and assimilation efficiency of *Calanus pacificus*. Mar Ecol Prog Ser 62: 203-210

Hirche, H-J (1983) Overwintering of *Calanus finmarchicus* and *Calanus helgolandicus*. Mar Ecol Prog Ser 11: 281-290

Huntley, M, Brooks, ER (1982) Effects of age and food availability on dielvertical migration of *Calanus pacificus*. Mar Biol 71: 23-31

Hutchinson, GE (1967) A treatise on limnology. Vol II. John Wiley & Sons, New York

Ikeda, T. (1985) Life history of Antarctic krill *Euphausia superba*: a new look from an experimental approach. Bull Mar Sci 37: 599-608

Klyashtorin, LB (1978) Estimation of energy expenditures for active swimming and vertical migrations in planktonic crustaceans. Oceanology 18: 91-94

Koslow, JA, Ota, A (1981) The ecology of vertical migration in three common zooplankters in the La Jolla Bight, April-August 1967. Biol Oceanogr 1: 107-134

Landry, MR, Hassett, RP, Fagerness, V, Downs, J, Lorenzen, CJ (1984) Effect of food acclimation on assimilation efficiency of *Calanus pacificus*. Limnol Oceanogr 29: 361-364

Longhurst, AR, Harrison, WG (1989) The biological pump: profiles of plankton production and consumption in the upper ocean. Prog Oceanogr 22: 47-123

Madin, LP (1974) Field observations on the feeding behaviour of salps (Tunicata: Thaliacea). Mar Biol 25: 143-147

Marschall, H (1988) The overwintering strategy of Antarctic krill under the pack-ice of the Weddell Sea. Polar Biol 9: 129-135

McLaren, IA (1978) Generation lengths of some temperate marine copepods: estimation, prediction, and implications. J Fish Res Bd Can 35: 1330-1342

Miller, CB, Clemons, M (1988) Revised life history analysis for large grazing copepods in the subarctic Pacific Ocean. Prog Oceanogr 20: 293-313

Miller, CB, Cowles, TJ, Wiebe, PH, Copley, NJ, Grigg, H (1991) Phenology in *Calanus finmarchicus*; hypotheses about control mechanisms. Mar Ecol Prog Ser 72: 79-91

Mullin, MM, Brooks, ER (1970) Growth and metabolism of two planktonic, marine copepods as influenced by temperature and type of food. In: Steele, JH (ed.) Marine food chains. Univ California Press, 74-95

Murphy, E (1993) Global extrapolation (this volume)

Ohman, MD (1986) Predator-limited population growth of the copepod *Pseudocalanus* sp. J Plank Res 8: 673-713

Paffenhöfer, G-A, Gardner, WS (1984) Ammonium release by juveniles and adult females of the subtropical marine copepod *Eucalanus pileatus*. J Plank Res 6: 505-513

Paffenhöfer, G-A, Knowles, SC (1980) Omnivorousness in marine planktonic copepods. J Plank Res 2: 355-365

Parslow, JS (1981) Phytoplankton-zooplankton interactions: data analysis and modelling (with particular reference to Ocean Station P [50°N, 145°W] and controlled ecosystem experiments). PhD thesis, Univ British Columbia. Cited In: Frost, BW (1987) Grazing control of phytoplankton stock in the open subarctic Pacific Ocean: a model assessing the role of mesozooplankton, particularly the large calanoid copepods *Neocalanus* spp. Mar Ecol Prog Ser 39: 49-68

Parsons, TR and Lalli, CM (1988) Comparative oceanic ecology of the plankton communities of the subarctic Atlantic and Pacific Oceans. Oceanogr Mar Biol Ann Rev 26: 317-359

Small, LF, Fowler, SW (1979) Sinking rates of natural copepod fecal pellets. Mar Biol 51: 233-241

Steele, JH, Frost, BW (1977) The structure of plankton communities. Phil Trans Roy Soc Lond B 280: 485-534

Steele, JH, Henderson, EW (1992) The role of predation in plankton models. J Plank Res 14: 157-172

Terazaki, M, Miller, CB (1986) Life history and vertical distribution fo pelagic chaetognaths at Ocean Station P in the subarctic Pacific. Deep-Sea Res 33: 323-337

Tseytlin, VB (1982) Transport of organic matter in daily vertical migrations of pelagic animals in trophic regions of the ocean. Oceanology 22: 614-618

Wiebe, PH, Madin, LP, Haury, LR, Harbison, GR, Philbin, LM (1979) Diel vertical migration by *Salpa aspera* and its potential for large-scale particulate organic matter transport to the deep-sea. Mar Biol 53: 249-255

Williams, R, Conway, DVP (1988) Vertical distribution and seasonal numerical abundance of the Calanidae in oceanic waters to the south-west of the British Isles. Hydrobiologia 167/168: 259-266

MICROBIAL PROCESSES AND THE BIOLOGICAL CARBON PUMP

T.Frede Thingstad
Dept. of Microbiology & Plant Physiology
University of Bergen
Jahnebk.5
N-5007 Bergen
Norway

Introduction

Since the content of carbon in the worlds oceans is about 60 times that of the atmosphere, and since there is a large exchange of carbon dioxide between the oceanic reservoir and the atmosphere, an understanding of the mechanisms regulating storage of carbon in the oceans interior is central to the discussion of the greenhouse effect. The low atmospheric content of CO_2 during the last ice-age (e.g. Barnola et al., 1987) also strongly suggests that there must be mechanisms through which the ocean's storage capacity may change. By producing and degrading organic C and $CaCO_3$, the biological processes of the ocean are part of this geochemical cycle. These biological processes are complex and are governed by the physiology and behavior of organisms whose diameters span 7 orders of magnitude (viruses to whales), with generation times from hours to years, and with complex trophic relationships. Intimately linked to these biological processes are the diffusive and advective transport processes of the ocean, regulating the supply of nutrients to the primary production in the photic zone. Any simulation model would obviously have to be based on extensive simplifications of these coupled processes, simplifications that require an understanding of how carbon storage is coupled to both biological and physical processes. The aim of the following discussion is not to develop a model of the biological C-pump, but to try to identify which microbiological mechanisms that may be important for the behavior of such models. We will only discuss processes in the part of the of the ocean where primary production is nutrient-limited, although this may exclude regions with processes potentially important

NATO ASI Series, Vol. I 10
Towards a Model of Ocean
Biogeochemical Processes
Edited by G. T. Evans and M. J. R. Fasham

to the C-budget of the ocean (Sarmiento & Toggweiler,1984; Siegenthaler & Wenk,1984), and we will only discuss properties related to the 'organic'- or 'soft-tissue' pump, not the biological production and transport of $CaCO_3$. Recent formulations of models related to the effect of biological processes on nutrient and carbon distribution in the ocean may be found in e.g Maier-Reimer et Bacastow (1990) and Najjar et al. (1992), and models of the photic zone food web incorporating microbial processes have recently been presented by Taylor & Joint (1990) and by Ducklow & Fasham (1991).

Key features of the biological mechanisms.

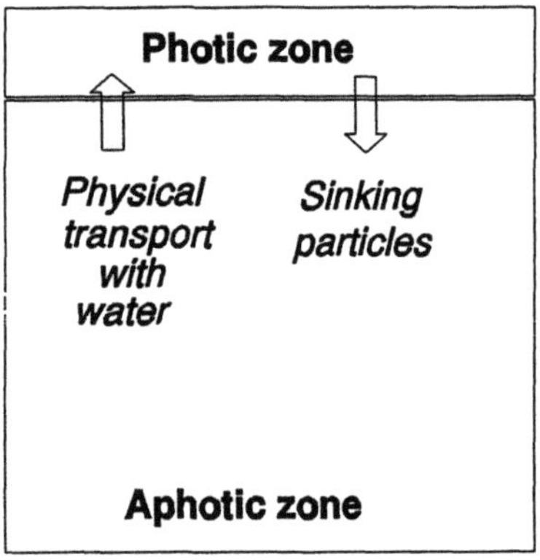

Figure 1: Principle of operation of the biological 'carbon pump' discussed. At equilibrium, the transport of nutrients and carbon into the aphotic layer by sinking particles must be balanced by the processes transporting carbon and nutrient back to the photic zone (storage in sediments neglected). High concentrations of carbon in the aphotic zone (=high storage capacity for carbon) can be obtained if 1)nutrient concentration in the aphotic zone is high, 2) the initial ratio of nutrient:carbon in sinking particles is low, or 3) the resistance for backtransport of carbon to the photic zone by physical mixing processes is higher than for nutrients (see further discussion in text).

At equilibrium, the storage of carbon in the interior of the ocean is determined by the concentration below the photic zone, not by the rate at which it is exchanged between the photic and aphotic zone . We will thus refer to an 'efficient' pump as one that has an equilibrium state characterized by high difference in C-concentration between the aphotic and photic zones (high C, see below). This is not necessarily equivalent to a high rate of export production (high P_0) from the photic zone. The essential features of the mechanisms of the soft- tissue pump may be illustrated by an extremely simplified two-layer model of the ocean as outlined in Fig. 1.

At equilibrium, the transports of any element into and out of the aphotic zone must balance. Assuming the interface with the photic zone to be the only import/export boundary of the aphotic zone, the equilibrium conditions for carbon and limiting nutrient give: $P_0=k_C C$, and $rP_0=k_N N$, respectively, where P_0 is the exported production from the photic to the aphotic zone in carbon units, r is the nutrient:carbon-ratio of the exported material, C is the concentration difference of dissolved carbon between the two layers, and N is the concentration of dissolved nutrients in the aphotic layer. Nutrient concentration in the photic zone is assumed to be 0. The transport from the aphotic to the photic layer is assumed to be proportional to the concentration differences with specific rate constants k_C and k_N for carbon and nutrient respectively. The inverse of k_C and k_N would be analogous to a 'resistance' towards the backtransport of carbon and nutrient to the photic zone. C is an expression of the efficiency of the pump and is obtained by eliminating P_0:

$$C = (k_N/k_C)\bullet N/r, \qquad (1)$$

while export production is given by:

$$P_0 = k_N \bullet N/r. \qquad (2)$$

While export production thus is proportional to the rate constant k_N of nutrient transport into the photic zone, the storage capacity depends on the ratio k_N/k_C between the rate constants for nutrient and carbon transport. Since dissolved carbon and nutrients are transported back to the photic zone by the same physical mixing processes, one would, in the simple two layer model of Fig.1, tend to set $k_C=k_N$, and hence $k_N/k_C=1$.

In a more realistic description with a more detailed vertical resolution, any mechanism that leads to a physical separation of C and N along the depth profile will affect the 'resistance' towards backtransport of carbon and nutrient to the photic zone with a change in the efficiency of the carbon pump as a result. Such a separation will be reflected in the ratio $\zeta = Z_N/Z_C$ between the e-folding depths for nutrient and carbon respectively (the depths where a fraction 1/e of C or N remains in the sinking particles). Due to its more direct biological relevance, ζ is used in the subsequent discussion rather than k_N/k_C. Also, we will refer to the total nutrient content N_T of the deep ocean, rather than the spatially varying concentration N. From this we can define three classes of biological mechanisms that may potentially influence the efficiency of the 'soft tissue pump':

Class A: Mechanisms that influence the total nutrient content N_T

Class B: Mechanisms that influence the ratio ζ of e-folding depths for limiting nutrient and carbon.

Class C: Mechanisms that influence the nutrient:carbon ratio r of export production.

Class A: Microbial processes influencing N_T:

The ocean's content of fixed nitrogen depends on the balance between terrestrial and atmospheric inputs in combination with nitrogen fixation on one side, and loss by denitrification on the other. Changes in the ocean's content of fixed nitrogen over long time scales is therefore a possibility to be considered. Climate- related factors such as temperature and circulation-dependent oxygen content may affect the nitrogen fixation and denitrification processes, resulting in possible feedback effects in the climate system (Codispoti,1989). Also the P-content of the ocean may have varied over geological time scales due to massive deposition of phosphorites (Froelich et al.,1982). Interestingly, there is geological evidence, based on Cd/Ca-ratios and $\delta^{13}C$ in sediment profiles, for periods of increased phosphate concentrations in the deep ocean (e.g. Boyle & Keigwin, 1987). These observations are, however, probably more easily interpreted as changes in the oceans circulation patterns (e.g. Boyle & Keigwin, 1987) than as indications of changes in total nutrient content of the ocean. As discussed later, it may be that not only the total content of limiting nutrient, but also the type of limiting element (N,P or other ?) may be important. The processes mentioned above clearly have the potential to lead to shifts between N and P limitation, but the large scale balancing of N- and P- content is poorly understood. The argument has been used that the phosphorus content is the ultimately controlling factor and that the amount of fixed nitrogen will adjust to the phosphorus through the stoichiometry determined by the biological requirements (Redfield,1958). If this is correct, long- time shifts in P-level would shift the total nutrient level N_T, even in a nutrient- limited ocean, but may not lead to qualitative shifts between N and P limitation of primary production.

Class B: Microbial processes influencing ζ.

The transformation of sinking particles is a complex process likely to be influenced by numerous bacterial- and zooplankton- activities. The bacterial and zooplankton activities are not independent. There is both a potential 'top-down' relationship through the predatory food chain from zooplankton to bacteria, and a potential 'bottom-up' effect through the bacteria-zooplankton competition for the particulate organic matter. Material can be transformed from sinking particles to material passively following the water movement in at least four different ways:

1) Physical disintegration of the material into smaller, non-sinking, particles of identical chemical composition.
2) Enzymatic hydrolysis of the macromolecules in the particle, releasing oligomers, monomers or ions to the surrounding water.
3) Consumption of the material by colonizing microorganisms which, apart from degrading the organic carbon, may excrete or assimilate nutrients to/from the surrounding water.
4) Colonizing microorganisms may be eaten by predators, go into lysis, or leave the sinking particle.

These processes may in principle lead to different couplings between the vertical fluxes of C, N, and P. Processes of type 1 would affect the e-folding depths of both carbon and nutrients, but force the ratio between them towards 1 since the material is transferred into non-sinking forms before separation of carbon and nutrients occur. The other processes (types 2, 3 and 4) may, however, have different effects on ζ. While carbon is a part of all organic material, including macromolecules such as nucleic acids, proteins, lipids and carbohydrates, nitrogen is a major constituent of proteins and nucleic acids, and phosphorus is mainly found in nucleic acids and phospholipids. The hydrolysis of these macromolecules are catalyzed by different enzyme systems (Table 1).

Table 1: Some bacterial enzyme systems of potential importance in releasing carbon and nutrients from sinking particles

Macro-molecule	Elements contained	Hydrolytic enzymes	Action	Comments
Structural carbohydrates	C	Glucosidases	Polymers to oligo- and monomers	Rate dependent on sugar moities, type of linkages (α,β), branching, water solubility of carbo-hydrates
Proteins	C,N,	Proteases	Polymers to oligo- and monomers	
Nucleic acids	C,N,P	DNA'ses, RNA'ses,	Polymers to oligo- and monomers	
		Nucleo-tidases,	Orthophosphate from monomers	Primarily cellbound to bacteria
		Alkaline-phosphatases	Orthophosphate from monomers	Inhibited by free orthophosphate

If type 2 processes dominate, C, N, and P release from the falling particles could not a priori be expected to be tightly coupled. Recalcitrant carbohydrates could be remaining in particles stripped for e.g. phosphate through the action of enzymes such as nucleases and nucleotidases. It is, however, not only the type of polymer attacked by hydrolyzing enzymes that should be considered; the mechanism of hydrolysis may also be important. Polymer-hydrolyzing enzymes may either split off monomers from the end of the polymer (exohydrolases) or split the polymer somewhere internally (endohydrolases). The enzymes may also be bound to the cell surface or excreted into the surrounding medium. As discussed by Azam & Smith (1991), the type of enzymes produced by the particle-colonizing bacteria would be expected to determine whether hydrolysis is coupled to consumption of the products by the particle-attached bacteria, or whether the products diffuse into the surrounding medium before consumption.

A dominance of cell-bound exohydrolases would presumably favor degradation processes physically attached to the sinking particles, while a dominance of free endohydrolases would favor a process where the main role of the attached bacteria would be to dissolve the particle. The main bacterial consumption of the released material would then be by free living bacteria as supported by an investigation of Cho and Azam (1988) comparing the activity of free versus that of particle-bound bacteria. As pointed out by Azam & Smith (1991), a rapid solubilization of sinking particles with few attached bacteria could be argued as an adaptive strategy in the competition/predation-relationship between bacteria and metazoan zooplankton. Processes of type 3 could also affect the ratio between the e-folding depths. Bacteria seem to have a relatively high content of phosphorus relative to carbon (Vadstein et al.,1988), as would be expected for small, relatively fast growing organisms with a high content of nucleic acids and a low content of structural carbohydrates. Consumption or remineralization of nutrients by bacteria is a question of the nutrient:carbon ratios of the substrate and of bacterial biomass, and of the respiration coefficient of the bacteria (Thingstad,1987). When colonizing carbon-rich particles phosphate, and in more extreme cases also nitrate, would be expected to be assimilated from the surrounding water by the particle-bound bacteria. Bacterial colonization could therefore be expected to increase ζ when this is originally low. In cases where C-rich material is exported from the photic zone, bacterial colonization would therefore be expected to decrease the efficiency of the biological pumping, not primarily because of the release of carbon by bacterial respiration, but because of the potential increase in the ratio between the e-folding depths of nutrients and carbon caused by the colonization. If, on the other hand, the material is originally enriched in nitrogen and phosphorus, bacterial remineralization would be expected to preferentially release nutrients faster than carbon. Studying the degradation of algal assemblages of different initial nutrient:carbon ratios, Tezuka (1989) have demonstrated how degradation have an equalization effect by tending to decrease high values of particulate nutrient:carbon-ratios while low values increase.

Processes of type 4 will of course further complicate this picture. Bacteria may leave the sinking particles. One interesting physiological mechanism here is the change in hydrophobicity occurring as some bacteria go into a state of starvation (Kjelleberg & Hermansson, 1984), suggesting that bacteria may regulate their attachment/detachment according to degree of starvation. Viral lysis of particle-bound bacteria may be more than a theoretical possibility since high frequencies of viral infection of organisms in sedimenting material has recently been demonstrated (Proctor & Fuhrman,1991). Both of these processes would transfer both carbon and nutrients in bacterial biomass from sinking to non-sinking forms, and therefore decrease both of the e-folding depths. By selectively acting on the nutrient-rich bacterial component of the sinking particles, these processes could also decrease ζ and thereby increase the pump efficiency. Protozoan grazing on bacterial biomass will release carbon through protozoan respiration, but also release nutrient in an amount dependent upon respiration ratio, nutrient:carbon ratio of prey, and nutrient:carbon ratio of predator biomass. With an expected lower nutrient:carbon ratio in protozoa than in bacterial biomass, predation on bacteria by particle-bound protozoa would may also help to decrease ζ. A clear demonstration that release rates for carbon and nitrogen are different in nature may be found in the data set from the VERTEX study in the northeast Pacific (Martin et al.,1987).

Another potentially important aspect is that some of the processes discussed above may also lead to a difference in the e-folding depths of nitrogen and phosphorus. The evidence is that phosphorus is preferentially released from particles relative to both nitrogen and carbon (Bishop et al.,1977; Martin et al.,1979; Garber,1984). If so, the shape of the nitrate and the phosphate profiles in the deep ocean would not be identical. Redfield et al. (1963) noted that such differences could be observed, and ascribed the difference to a more rapid release (i.e. a smaller e-folding depth) for phosphorus than for nitrogen. The implication of a smaller e-folding depth for P than for N would be that the biological pump could work with greater efficiency in a system with phosphorus-limited primary production than in one with nitrogen limitation. While nitrogen is assumed to be the limiting nutrient in the majority of todays oceans, this aspect may be of importance in freshwater influenced areas such as Norwegian fjords (Sakshaug & Olsen,1986), and potentially also in other areas such as the Mediterranean Sea (Bethoux & Copin-Montegut,1986; Krom et al.,1991). Freshwater discharge to the ocean has undergone large changes in both amount and geographical distribution during geological times (Fairbanks,1989). Since major freshwater releases naturally were linked to periods of deglaciation, fresh-water induced P-limitation seems, however, difficult to invoke as an explanation for increased oceanic CO_2- storage in glacial periods. From a more practical modelling point of view, however, the potential importance of a difference in e-folding depths may be worth having in mind when choosing between phosphorus- and nitrogen-based geochemical models of the ocean.

While it is the biological processes of the aphotic zone that will primarily affect , the biology of the photic zone has an indirect effect by determining the chemical quality of the exported material. A food web in the photic zone exporting material with part of the carbon bound in slowly degradable structural carbohydrates would thus be expected to lead to a more efficient pump than a food web exporting material with the same r-value, but with the carbon contained in easily degradable proteins, lipids or storage carbohydrates. The important effect of a more turbulent upper ocean may not be the potential increase in export production, but rather the shift from a 'microbial' food chain in the photic zone towards a different food chain with larger organisms producing qualitatively different sinking material (see e.g. Legendre and LeFèvre,1991).

From a modelling point of view, the importance of the spatial separation of carbon and nutrients by processes forcing to values below 1 depends on the pattern of turbulence in the upper layer. If turbulence is rapidly decreasing downwards, the effect of a low will be enhanced since the nutrient released at shallow depths will be easily transported back to the photic zone while the carbon released at larger depths will be isolated from the surface.

Class C: Microbial processes influencing the Redfield ratio r of export production.

The material exported from the photic zone shows large seasonal variations both in magnitude and in composition (Bathmann et al.,1990). To properly model the storage capacity for carbon in deep waters, the model for the biological processes in the photic zone must therefore reflect an appropriate understanding of the factors determining the quality of the exported material. It should be remembered that the magnitude of the exported production (P_0) in nutrient limited areas is not a function of the model chosen for the food web in the photic zone, but is determined through a combination of the model of the physical mixing processes and the model for the biological processes determining the e-folding depth for the limiting nutrient in the aphotic zone. The model chosen for the photic zone food web will, however, determine the amount of regenerated production corresponding to any given magnitude of the exported production. Empirical relationships between exported and total production have been demonstrated (e.g. Wassmann,1991), showing a high degree of correlation between the two. With export production fixed as described above, total and regenerated production can therefore be estimated from these relationships. Empirical relationships contain, however, in themselves no understanding of the underlying mechanisms and may be dangerous to use for predictive purposes in a changing system.

At equilibrium, the total concentration (sum of all particulate, dissolved organic and inorganic forms) of nutrient remaining in the photic zone (N_E) must also in some way be subject to the requirement that import must equal export.

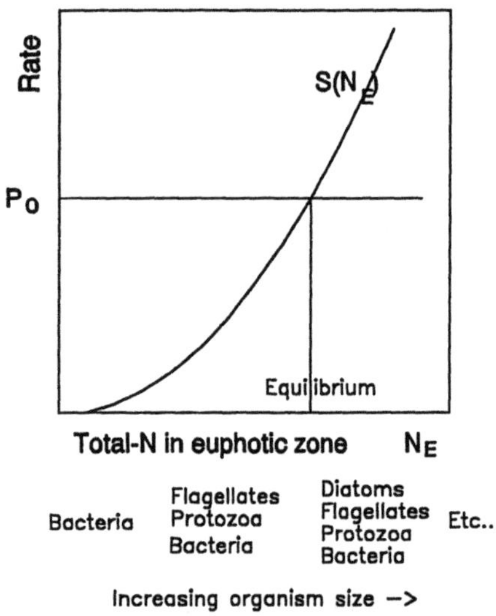

Figure. 2. Coupling between physical mixing, export production, and food web structure in the photic zone. Physical mixing and degradation processes in the aphotic zone determine the input of nutrients to the photic zone and therefore the new production (horizontal line P_0). Over time, this must be balanced by the sedimentary loss (illustrated by curve S). Over time, the photic zone food web therefore has to adjust to a structure which produces the required sedimentation. If large organisms generally produce more sedimenting material than small, this implies a change in food web structure with increasing P_0 towards dominance of large organisms. The figure suggests an equivalence between increased nutrient enrichment of the photic zone (N_E) and a food chain dominated by large phytoplankton species and/or including larger predators.

To illustrate this, assume that there is a one-to-one relationship between the equilibrium structure of the photic zone food web and N_E. The equilibrium value for N_E is then determined by the crossing point between the horizontal line representing P_0 in Fig.2, and the curve S describing how sedimentation loss from the food web changes as food web structure changes with increasing N_E. Assuming S to be monotonically increasing with increasing N_E (not necessarily true in nature), increasing P_0 will lead to a shift of the equilibrium point to the right, corresponding to a new food web structure with a higher sedimentary loss.

A good model of this relationship would require a good description of the form of the curve S, i.e. it would require a firm biological understanding of how food web structure changes with increasing amounts of total available nutrient. The present qualitative understanding of this relationship is a general trend towards increasing dominance of larger organisms as one moves from left right in Fig.2 with the biomass dominated by bacteria at the extreme left (Cho and Azam,1990) and more 'classical' food chains dominated by diatoms, copepods and fish to the right (Thingstad and Sakshaug,1991 and references therein). Systems with a high input of nutrients due e.g. to unstable hydrographic conditions, can only reach equilibrium through a compensating loss. If this can only occur from large organisms, the food chain structure must somehow adjust to include large auto- or heterotrophic organisms. A recent discussion of the connections between size relationships and hydrographic conditions may be found in Legendre & LeFèvre (1991).

The main implications for the storage of carbon in deep waters would again be if there are qualitative changes in the nutrient:carbon ratio or the ratio of e-folding depths for the exported material along the curve S. Using the argument that smaller organisms as a general tendency contain more nucleic acids and less structural carbohydrates, the possibility exists of changes in both the nutrient:carbon ratio (r) and in the ratio ζ of the e-folding depths leading to higher efficiency of carbon storage as the equilibrium moves from left to right with increased mixing.

Dissolved organic carbon.

Production of organic material is not necessarily linked to a stoichiometric consumption of nitrogen and phosphorus. Both phytoplankton (Myklestad,1977) and bacteria (Dawes & Senior,1973) may produce polymeric carbohydrates or other C-rich compounds under N- or P- limited conditions. These may be either stored intracellularly or excreted to add to the pool of DOC produced by other mechanisms. If advective or diffusive processes transport dissolved organic compounds out of the photic zone before degradation, this may add to the storage capacity through a mechanism very different from the 'sinking particle'-process discussed above. The potentially important biogeochemical consequences of the 'new' DOC measured by Sugimura & Suzuki (1988) has recently been discussed by Toggweiler (1989). For DOC-production to be an active process involved in the storing of C in the deep sea, the DOC must be produced and consumed, but the time scale of degradation in the photic zone must be comparable to or longer than the characteristic time scales of the physical mixing processes. The depth profiles presented by Sugimura & Suzuki (1988) suggest this to be the case. Kirchman et al. (1991) have recently shown that a substantial fraction of the 'new' DOC may be removed by bacteria within a time scale of days. Bacterial degradation of the material in the photic zone may be retarded by the chemical structure of the compounds produced, which is probably the case for a large fraction of the DOC accumulating in seawater. Ecological factors such as predation and competition for limiting nutrients between phytoplankton and bacteria would, however, also be expected to interfere with the degradation even of readily assimilable bacterial substrates (Pengerud et al.,1987). Since climatic effects could change the area covered by cold surface water, an understanding of the effect of temperature on DOC degradation would be important. This effect can probably only be understood within an ecosystem context. From theoretical considerations the response of a dynamic food web to a lowered temperature may as well be a higher concentration of dissolved organic bacterial substrates, as a lowered bacterial growth rate. While the DOC may be produced with C:N:P ratios very different from the Redfield ratio (corresponding to different r-values for exported particles), the transport of dissolved compounds would not be expected to spatially separate nutrients and carbon as was discussed for the particulate material (corresponding to $\zeta=1$). The concentrations of dissolved organic N and P are known to be substantial (Duursma,1961) and the inclusion of dissolved organic phosphorus in simulation models has been shown to change the oceanic distribution for orthophosphate to a more realistic pattern (Najjar et al,1992). From a modelling point of view, this would indicate the requirement of a proper description of the mechanisms producing and consuming DOC,DON and DOP. This would include the linkage of production to ecological factors such as food web structure and nutrient limitation of phytoplankton and bacteria, and it would require an understanding of the ecological factors regulating the production and activity of enzymes hydrolyzing polymeric DOC, DON, and DOP polymers (Table 1).

Concluding remarks.

Microbial processes interact with a series of mechanisms determining the efficiency by which carbon is stored in the oceans interior. The total nutrient content, the spatial separation of carbon and nutrients by the microbial degradation of sinking particles, and the degradation of dissolved organic compounds in the photic zone are all processes obviously closely connected to the activity of marine bacteria, and changes in their activity would be expected to change the C-storage in the ocean. A much more difficult and complex question is whether these bacterial processes may change in a significant manner under the influence of a changing climate; i.e. whether feedback effects to the climate are possible via the atmospheric greenhouse effect of CO_2. Cause and effect relationships in this field may be simple, such as direct effects of temperature on growth rate, or there may be complex cause and effect chains. One candidate for such a complex chain is from changing river runoff, changing meltwater layers or changed frequencies in low pressure passages via changing turbulence in the upper mixed layer to food web structure and further to subsequent changes in the quality of sinking material, another candidate is the potential change in nutrient conditions or oxygen content as a response to large scale changes in ocean circulation. Whether these couplings are sufficiently strong to alter atmospheric CO_2, or whether the bacterial processes can be regarded as constants, can only be analyzed theoretically by combining a proper understanding of the biological processes involved with a proper quantification of their interactions with hydrography through mathematical models. Interestingly, some information on past changes may be obtained from marine sediments. Since the 'soft-tissue' pump discussed here will lead to a fractionation of carbon isotopes, while carbonate formation will not, geological $\delta^{13}C$ values contains in principle information on the relative importance of these two pump-mechanisms. Recent investigations on this suggests that changes in the strength of the 'soft-tissue' pump cannot alone have been responsible for the reduction in atmospheric CO_2 in glacial times (Leuenberger et al.,1992).

References

Azam, F, Smith, DC (1991) Bacterial influence on the variability in the ocean's biogeochemical state: A mechanistic view. In: Demers, S. (ed.) Particle analysis in oceanography. Springer Verlag, Berlin

Barnola, JM, Raynaud, D, Korotkevitch, YS, Lorius, C (1987) Vostok ice core: A 160,000 year record of atmospheric CO_2. Nature, **329**:408-414

Bathmann, UV, Peinert, R, Noji, T, Bodungen, B (1990) Pelagic origin and fate of sedimenting particles in the Norwegian Sea. Prog. Oceanog., **24**:117-125

Bell, W, and Sakshaug, E (1980) Bacterial utilization of algal extracellular products 2. A kinetic study of natural populations. Limnol. Oceanogr. **25**:1021-1033

Bethoux, JP, Copin-Montegut, G (1986) Biological fixation of atmospheric nitrogen in the Mediterranean Sea. Limnol. Oceanogr. **31**:1353-1358

Bishop, JKB, Collier, RW, Ketten, DR, and Edmond, JM (1977) The chemistry, biology,and vertical flux of particulate matter from the upper 400m of the equatorial Atlantic ocean. Deep-Sea Res. **24**:511-548

Boyle, EA, and Keigwin, L (1987) North Atlantic thermohaline circulation during the past 20,000 years linked to high-latitude surface temperatures. Nature, **330**:35-40

Cho, BC, Azam, F (1988) Major role of bacteria in biogeochemical fluxes of the ocean's interior. Nature, **332**:441-443

Cho, BC, Azam, F (1990) Biogeochemical significance of bacterial biomass in the oceans euphotic zone. Mar. Ecol. Prog. Ser. **63**:253-259

Codispoti, LA (1989) Phosphorus vs. nitrogen limitation of new and export production. p.377-408. In: Berger, W et al. (eds.) Productivity of the ocean:Present and past. Dahlem Conf., Wiley

Dawes, EA, Senior, PJ (1973) The role and regulations of energy reserve polymers. Adv. Microbial Physiol. **10**:135-266

Ducklow, H, and Fasham, M (1991) Bacteria in the greenhouse: Modeling the role of oceanic plankton in the global carbon cycle. pp 1-32 in Mitchell, R (ed.) New concepts in environmental microbiology. Wiley-Liss.

Duursma, EK (1961) Dissolved organic carbon, nitrogen, and phosphorus. Neth. J. Sea. Res. **1**:1-147

Fairbanks, RG (1989) A 17,000-year glacio-eustatic sea level record: influence of glacial melting rates on the Younger Dryas event and deep ocean circulation. Nature(London). **342**:637-642

Froelich, PN, Bender, ML, Luedtke, NA, Heath, GR, De Vries, T. (1982) The Marine phosphorus cycle. Am. J. Sci. **282**:474-511

Garber, JH (1984) Laboratory study of nitrogen and phosphorus remineralization during the decomposition of coastal plankton and seston. Estuar. Coastal Shelf Sci. **18**:685-702

Kirchman, D, Suzuki, Y, Garside, C, and Ducklow, HW (1991) High turnover rates of dissolved organic carbon during a spring phytoplankton bloom. Nature, **352**:612-614

Kjelleberg, K, Hermansson, M (1984) Starvation-induced effects on bacterial surface characteristics. Appl. Environ. Microbiol. **48**: 497-503

Krom, MD, Kress, N, Brenner, S (1991) Phosphorus limitation of primary productivity in the eastern Mediterranean Sea. Limnol. Oceanogr. **36**:424-432

Legendre, L, LeFèvre, J (1991) From individual plankton cells to pelagic marine ecosystems and to global bigeochemical cycles. In: Demers,S.(ed.) Particle analysis in oceanography. Springer Verlag, Berlin

Leuenberger, M.,Siegenthaler, U, Langway, CC (1992) Carbon isotope composition of atmospheric CO_2 during the last ice age from an Antarctic ice core. Nature. **356**:488-490

Maier-Reimer, E, Bacastow, R (1990) Modelling of Geochemical tracers in the ocean. pp.233-267 in: Schlesinger, ME (ed.) Climate-Ocean Interaction. Kluwer Academic Publishers, the Netherlands

Martin, JH, Knauer, GA, Bruland, K (1979) Fluxes of particulate carbon,nitrogen and phosphorus in the upper water column of the northeast Pacific. Deep-Sea Res. **26**:97-108

Martin, JH, Knauer, GA, Karl, DM, Broenkow, WW (1987) VERTEX: carbon cycling in the northeast Pacific. Deep-Sea Research, **34**:267-285

Myklestad, SM (1977) Production of carbohydrates by marine planktonic diatoms.II. Influence of N/P ratio in growth medium on the assimilation ratio, growth rate, and production of cellular and extracellular carbohydrates by *Chaetocheros affinis* var. Willei(Gran) Hustedt and *Skeletonema costatum* (Grev.) Cleve. J. exp. mar. Biol. Ecol. **29**:161-179

Najjar, RG, Sarmiento, JL, Toggweiler, JR (1992) Downward transport and fate of organic matter in the ocean: Simulations with a general circulation model. Global Biogeochemical Cycles **6**:45-76

Pengerud, B, Skjoldal, EF, Thingstad, TF (1987) The reciprocal interaction between degradation of glucose and ecosystem structure. Studies in mixed chemostat cultures of marine bacteria, algae, and bacterivorous nanoflagellates. Mar. Ecol. Prog. Ser. **35**:111-117

Proctor, LM, and Fuhrman, JA (1991) Roles of viral infection in organic particle flux. Mar.Ecol.Prog.Ser. **69**:133-142

Redfield, AC (1958) The biological control of chemical factors in the in the environment. Amer. Sci. **46**:205-221

Redfield, AC, Ketchum, BH, Richards, FA (1963) The influence of organisms on the composition of sea-water. In: Hill,M.N. (ed.) The sea. Vol 2, pp.26-77

Sakshaug, E, Olsen, Y (1986) Nutrient status of phytoplankton blooms in Norwegian waters and algal strategies for nutrient competition. Can. J. Fish. Aquat. Sci., **43**:389-396

Sarmiento, JL, Toggweiler, JR (1984) A new model for the role of the oceans in determining atmospheric CO_2 . Nature, **308**:621-624

Siegenthaler, U, Wenk, T (1984) Rapid atmospheric CO_2-variations and oceanic circulation. Nature, **308**:624-626

Sugimura, Y, Suzuki, Y (1988) A high-temperature catalytic oxidation method for the determination of non-volatile dissolved organic carbon in seawater by direct injection of a liquid sample. Mar. Chem. **16**:83-97

Taylor, AH, and Joint, I (1990) A steady-state analysis of the 'microbial loop' in stratified systems. Mar. Ecol. Prog. Ser. **59**:1-17

Tezuka, Y (1989) The C:N:P ratio of phytoplankton determine the relative amounts of dissolved inorganic nitrogen and phosphorus released during aerobic decomposition. Hydrobiologia, **173**:55-62

Thingstad, TF (1987) Utilization of N,P, and organic-C by heterotrophic bacteria. I. Outline of a chemostat theory with a consistent concept of maintenance metabolism. Mar. Ecol. Prog. Ser. **35**:99-109

Thingstad, TF, Sakshaug, E (1991) Control of phytoplankton growth in nutrient recycling ecosystems. Theory and terminology. Mar. Ecol. Prog. Ser. **63**:261-272

Toggweiler, JR (1989) Is the downward dissolved organic matter (DOM) flux important in carbon transport. pp.65-83 in Berger, W, Smetacek, VS, and Wefer, G: Productivity of the Ocean: Present and Past. Wiley & Sons

Vadstein, O, Jensen, A ,Olsen, Y, Reinertsen, H (1988) Growth and phosphorus status of limnetic phytoplankton and bacteria. Limnol. Oceanogr. **33**:489-503

Volk, T, Hoffert, MI (1985) Ocean carbon pumps: Analysis of relative strengths and efficiencies in ocean driven CO_2 changes. pp. 99-110 in: Sundquist, ET, Broecker, WS (eds) The carbon cycle and atmospheric CO_2: natural variations archean to present. AGU Monograph 32, Amer. Geophys. Union, Washington DC

Wassmann, P (1991) Relationship between primary and export production in the boreal coastal zone of the Atlantic. Limnol. Oceanogr. **35**:464-471

DISSOLVED ORGANIC MATTER IN BIOGEOCHEMICAL MODELS OF THE OCEAN

David L. Kirchman
University of Delaware
College of Marine Studies
Lewes, Delaware 19958
USA

C. Lancelot[1], M. Fasham[2], L. Legendre[3], G. Radach[4], M. Scott[5]

I. Introduction

Recent developments in high temperature catalytic oxidation techniques (Sugimura and Suzuki, 1988) have raised a number of questions about dissolved organic carbon (DOC). Several key aspects of the DOC pool are still unknown, including its size (Martin & Fitzwater, 1992; Ogawa & Ogura, 1992) and turnover rates (Kirchman et al., 1991; Kepkay & Wells, 1992; Bauer et al., 1992). Even the low estimates of DOC concentrations, however, imply that DOC is large enough (700 x 10^9 tons C) to rival the terrestrial biomass and atmospheric CO_2. Thus the dissolved carbon pool needs to be included in biogeochemical models. Yet uncertainty about concentrations and many key aspects of DOC production and consumption does limit our ability to include DOC in these models. Furthermore, the lack of data allows a model to evoke DOC with few restrictions in order to explain otherwise unresolved issues. Because of these uncertainties, the DOC working group spent more time discussing what is known about DOC than about modelling it. What emerged from that discussion are some recommendations for future

[1]Universite Libre de Bruxelles, GMMA, Boulevard du Triomphe, B-1050 Bruxelle, Belgium
[2]James Rennell Centre for Ocean Circulation, Chilworth Research Centre, Southampton, UK
[3]Universite Laval, Departement de Biologie, Ste-Foy, QC G1K, Canada
[4]Universitat Hamburg, Institut fur Meereskunde, 2000 Hamburg 54, Germany
[5]University of Glasgow, Department of Statistics, Glasgow G12 8QW, UK

NATO ASI Series, Vol I 10
Towards a Model of Ocean
Biogeochemical Processes
Edited by G T Evans and M J R Fasham

experimental research in order to provide data that would aid in modeling more accurately the role of DOC in biogeochemical processes.

DOC is traditionally defined as the organic carbon that passes through a Whatman GF/F filter. We now know that this definition includes several classes of particulate carbon, such as inert colloids, bacteria and viruses, which all pass through GF/F filters to varying degrees. These particles are important in determining DOC fluxes, such as catabolism by bacteria of DOC to CO_2 or coagulation of DOC mediated by colloids, but small particles are a small fraction of the DOC pool, and their inclusion in the DOC pool is less of a problem than other uncertainties.

One conclusion of the DOC working group was that models need to consider at least three pools of DOC: 1) a labile pool with turnover times of days or less; 2) a refractory pool with extremely long turnover times (centuries or greater); and 3) a "semi-refractory" pool that varies on a seasonal time scale. These pools have different roles in biogeochemical processes and thus would seem to require that models differentiate among them. The labile pool, and to a lesser extent the semi- refractory pool, fuel bacterial production. As an important part of the microbial loop, the DOC-bacteria pathway is a major mechanism by which primary production is respired to CO_2 or routed to higher trophic levels and larger organisms. The refractory DOC, on the other hand, is available for export to the deep ocean. In addition, the presence of refractory DOC in surface waters implies that carbon is being stored in the oceans independent of export to the deep layers. Finally, the semi-refractory pool may include intermediates of compounds that eventually become part of the refractory, long-lived DOC.

Molecular size is another important characteristic of DOC. As a first approximation, two sizes of dissolved compounds are important to consider: less than and greater than 500 Daltons. Low molecular weight compounds, i.e. <500 Daltons, can be assimilated directly by microorganisms without any modifications. The cutoff at 500 Daltons is based on extensive work with oligopeptides and proteins (Law, 1980), although it may not be applicable to all compounds. Dissolved compounds roughly greater than 500 Daltons, on the other hand, must be hydrolyzed to smaller compounds before transport into the cell and further catabolism is possible. This extracellular hydrolysis is mediated by enzymes that are bound to the outer membrane of bacteria or are part of other extracellular polymers extending from the outer membrane

(Vives-Rego et al., 1985); in special cases such as with bacteria attached to particles, the hydrolytic enzymes may be excreted by bacteria. The distinction between low and high molecular weight material may be important in modeling DOC because low molecular weight material generally turns over faster than high molecular weight material and because the synthesis of these extracellular hydrolytic enzymes may delay the microbial response to changes in >500 Dalton DOC. Also, the high molecular weight DOC is more likely than low molecular weight DOC to participate in abiotic reactions, such as adsorption and coagulation to even larger compounds and particles.

What follows is a short review of some of the major problems faced by modelers trying to include DOC in biogeochemical models. Most of the "facts" presented here may be disputable but, like a good model, hopefully this discussion will stimulate further research and thought. Figure 1 depicts only a few aspects of dissolved organic matter (DOM) that should be included in biogeochemical models.

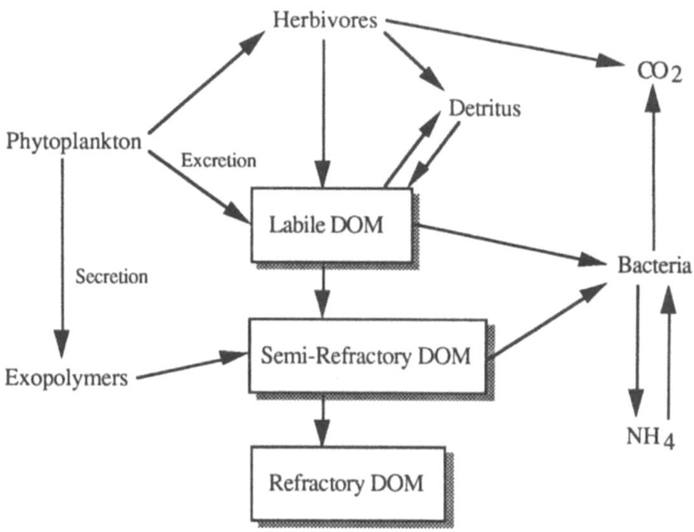

Figure 1 Three major pools of dissolved organic matter (DOM) and the processes contributing to them.

2. Problems in Modeling DOC

Three general problems in modeling DOC were identified:

1. Mechanisms of DOC production

The major problem in examining DOC production is that there are several (at least 5) possible mechanisms, and the existing data do not allow us to ignore any of them. In addition, even less is known about the mechanisms by which refractory DOC is produced. The relationships between labile and refractory DOC seem particularly important, but our ignorance makes it difficult to include them in biogeochemical models.

2. Trophic interactions

Trophic interactions are very important in determining DOM concentrations, fluxes, and distributions because several components of the plankton produce DOM; only bacteria are important consumers of DOM (see below). Since these interactions are technically outside the scope of this working group, they were not discussed in great detail.

3. Coupling of C and N during DOM degradation

Although the dissolved material is measured frequently as carbon, and for several questions about global change we are most interested in carbon, DOM is not just DOC. Furthermore, the use of specifically DOC is tied to the supply of appropriate inorganic nutrients, most importantly ammonium, and regeneration of these nutrients, again most importantly ammonium, is dependent on the C:N ratio of DOM used by bacteria (Goldman et al., 1987). This coupling of C and N (P has not been considered) during DOM use causes some problems in biogeochemical modeling, especially models using solely C or N units, but is now being critically examined (e.g. Anderson, submitted).

Because of these important connections between carbon and nitrogen, it is seems more useful to use "DOM" rather than just "DOC", even if our main interest is in carbon.

3. Production of Labile DOM

The production of labile DOM is a large fraction of the material flux in pelagic environments as evident from the observation that the ratio of bacterial production to primary production is high. (Most bacteria are free-living, so their growth must be supported by DOM.) Assuming a 50% growth efficiency the flux of DOM via bacteria is roughly 50% of primary production (Cole et al., 1988; Ducklow & Carlson, 1992). Although the concentration of labile DOM is low and thus not important in carbon export to the deep ocean, this high DOM flux seems an important component of trophic dynamics of pelagic ecosystems and needs to be considered by biogeochemical models.

Most models of oceanic processes probably can safely ignore the input of labile DOM from terrestrial and probably even estuarine sources. Although there may be reason to consider non-oceanic sources of refractory DOM for balancing oceanic carbon budgets (Sarmiento & Sundquist, 1992), the plankton are the only important source of labile DOM in the ocean. We can list several mechanisms by which labile DOM is produced, ranging from those most closely coupled to phytoplankton (the ultimate source of oceanic organic C) to those most distant. Five mechanisms of labile DOM production were identified:

1. Excretion and secretion of DOM directly from phytoplankton

Excretion of low molecular weight compounds by phytoplankton has been examined extensively, yet we still know little about its mechanism and controlling factors (Williams, 1990). It appears to be a small fraction of primary production (<10%) (Lancelot, 1984). Currently, three mathematical formulations for modeling DOM excretion have been suggested: a) a constant percent of primary production (Fasham et al., 1990); b) a constant percent of biomass or some internal soluble pool (Bjornsen, 1988): c) the difference between maximum photosynthesis allowed by light and net primary (biomass) production under nutrient limitation (Bratbak & Thingstad, 1985) .

Secretion of extracellular polymers, such as polysaccharides that make up the gelatinous matrix of colonial algae, may be important in selected environments (Lancelot & Mathot, 1988). When the colonies are disrupted, these polymers are released into the water and become available for free-living bacteria. A portion of these polymers may flocculate and sink rapidly to below the euphotic zone before being degraded. Unlike the small molecular weight compounds excreted by phytoplankton, these extracellular polymers are not degraded rapidly by bacteria and may contribute eventually to the refractory DOM. Benner et al. (1992) argued that much of oceanic DOM is made up of polysaccharides. Perhaps these polysaccharides were originally secreted by phytoplankton. Thus, it seems important to distinguish secretion of high molecular weight material from excretion of low molecular weight DOM.

2. Viral attack and spontaneous autolysis of microorganisms

Recent studies have found that a high proportion of both phytoplankton and bacteria are infected with viruses. When lysed, these infected organisms release all their cellular constituents. What's interesting about this mechanism is that it would help explain the existence of high molecular weight DOM in seawater. It also suggests that the chemical composition of DOM might closely mirror that of the lysed phytoplankton or bacteria, although both biotic and abiotic reactions may modify this DOM beyond recognition, depending on the relative rates of production and modification.

Lysis of bacteria and release of DOM is a circular loop (DOM---> bacteria --->DOM) and thus does not explain how primary production makes its way to the DOM pool. However, it may help explain the chemical composition and molecular weight distribution of DOM. Lysis of phytoplankton, on the other hand, would obviously be a mechanism that connects primary production and the DOM pool. Lysis induced by viruses and possible production of DOM is only now being examined (Bratbak et al. in press).

Also, spontaneous autolysis of phytoplankton is another mechanism that would release DOM (van Boekel et al., 1992). The difference between viral-induced and spontaneous autolysis may be too subtle for most models, but there may be some important distinctions to retain. For example, population density may be critical in determining the importance of viral-induced lysis because of the need for viruses to contact their "prey". Autolysis is probably only important at the end of a bloom when phytoplankton are stressed by low nutrient concentrations.

3. Sloppy Feeding

DOM may be released when zooplankton graze on phytoplankton and physically damage the cell before complete ingestion (Roy et al., 1989). Little is known about sloppy feeding. It shares some similarities with DOM production via lysis in that sloppy feeding would release the entire cell contents, which would impact the composition of the DOM pool.

Sloppy feeding is usually attributed to only zooplankton such as copepods that "handle", with filter or raptorial appendages, their prey before ingestion. It seems unlikely to be important with phagotrophic grazers such as protozoa, because the entire prey is engulfed by the grazer with no pre-ingestion contact. In the open ocean, therefore, this mechanism of DOM production is probably insignificant because much of primary production is by picophytoplankton which is grazed on by phagotrophic grazers.

4. Excretion of DOM by Herbivores

Many heterotrophic organisms are capable of releasing DOM but the most important would be grazers of phytoplankton. Some studies have reported release of urea and amino acids by copepods and microflagellates. Although these studies show the capacity of these organisms to release DOM, the importance of this release in situ and relative to other mechanisms is unclear. Grazers of bacteria may also release DOM (specifically amino acids; Nagata and Kirchman 1991), although this is a circular loop of protozoa, bacteria, and DOM.

5. Degradation of fecal material and other detritus

The general rule is that about 33% of carbon ingestion by zooplankton goes towards zooplankton biomass production, another 33% to respiration, and the final 33% goes to excretion, including fecal pellets. The form of this released organic matter, i.e. particulate fecal pellets vs. DOM, is obviously important, but the factors governing when grazers release pellets vs. DOM have not been examined. When released as fecal matter, some studies suggest that bacteria are necessary to degrade that particulate material which in the process releases DOM. Alternatively, reingestion by other grazers is possible, and also DOM may abiotically leach from pellets. These processes are still very poorly understood.

The amount and C:N ratio of the excreted organic matter has been modelled; these models do not distinguish between DOM excretion and pellet production. One approach is to examine the C and N content of the prey (e.g. phytoplankton). Another way of examining this problem is similar to that suggested by Jumars et al. (1989). Briefly, excretion of organic carbon and nitrogen is possible when ingestion, or the engulfing of a prey item, is faster than the digestion process, i.e. the mechano-chemical breakdown of the prey.

The important point is that herbivores maximize digestion and thus may allow high ingestion and excretion; when food supply is high ingestion efficiency (organic matter assimilated into the herbivore/prey ingested) is not necessarily maximized and may not be high. There is some experimental support for this hypothesis (Nagata & Kirchman, 1991).

4. Production of Refractory DOM

A large fraction of DOC (>50%) even in surface waters has turnover times greater than a thousand years (Williams and Druffel, 1987; Bauer et al., 1992). It had been suggested that even a small amount of runoff from terrestrial sources could contribute refractory DOM, such as lignin-cellulose, that builds up sufficiently over the years to account for the observed concentrations in seawater. However, isotopic data (^{13}C and ^{14}C) and the lack of lignin signatures in DOM argues against the terrestrial sources. Thus, refractory DOM is produced somehow from plankton- derived compounds. But the compounds that comprise plankton (e.g. polysaccharides, proteins, lipids and nucleic acids) are fairly labile. So the question becomes, how is refractory DOM produced from these labile compounds?

There are two general classes of mechanisms for producing refractory DOM from labile compounds (Hedges, 1988): 1) condensation of small molecular weight labile DOM to larger, refractory DOM; 2) modification of large molecular weight material, which initially is relatively labile (e.g. protein), but becomes more refractory after modification. Part of the problem in determining the relative importance of these two mechanisms is that we know little about the chemical composition of DOM. Roughly 30% of this organic matter has been chemically identified at a molecular level. The conclusion that the refractory DOM is "humic material" says little about its origin as humic and fulvic acids are not well-defined compounds with specific structures. Also, the importance of humic material may have been exaggerated because previous methods for isolating DOM from seawater would tend to pull out hydrophobic humic material and miss the hydrophilic compounds that may dominate the DOM pool (Benner et al., 1992).

5. Degradation Processes

Heterotrophic bacteria are the predominate organisms responsible for the degradation of DOM in pelagic ecosystems. Eukaryotic organisms may assimilate DOM, but this can be safely ignored by biogeochemical models, although it may be important for the organism. Several abiotic transformations of DOM have been suggested, some of which may be important in specific cases: 1) condensation and coagulation of DOM to particulate organic matter (POM); and 2) photochemical lysis. The latter reaction may be important in transforming refractory DOM to DOM that can be more readily attacked by bacteria.

Uptake of individual components of the DOM pool can be modelled by standard Michaelis-Menten kinetics. That is, the uptake of the entire DOM pool (or some component of the pool) is a function of two parameters, the maximum or saturated rate (V_{max}) and the half-saturation constant (K_m), and the DOM concentration (S). The standard equation is

$$v = V_{max}/(K_m + S) \qquad (1).$$

More work on these kinetic parameters may be needed, especially in waters with very low concentrations of specific compounds (amino acids, for example) as measured by modern high performance liquid chromatography methods. However, current models appear to be rather insensitive to changes in these parameters, although the steady- state concentration of DOM does vary with different kinetic parameters (Billen et al., 1980; see below).

It may be adequate to describe uptake as a first-order process, i.e. $v = \lambda * S$ where λ is a constant. The advantage of this simplification is that it may be easier to incorporate into a model several components of the DOM with differing λ's, i.e. turnover rates. This is the basic assumption behind the "multiple G" model which has been applied to the degradation of organic matter in sediments (Berner, 1980). The multiple G model has not been used to model DOM degradation in the water column. Incidentally, the simplest multiple G model, which has been shown to describe organic matter degradation quite well in sediments, has three organic compounds with different λ's, similar to our suggestion here for the DOM pool.

Perhaps the biggest uncertainty in examining bacterial uptake of DOM is the growth efficiency. This parameter is critical in understanding total DOM uptake and its relationship to bacterial biomass production. It sets whether carbon is respired or is transferred to higher trophic levels. Although most models assume a constant and fairly high growth efficiency (30-50%), some studies indicate that it could vary and be as low as 10% (Kirchman et al., 1991). Usually the growth efficiency is modelled by comparing the C:N ratio of the DOM used by bacteria (DOM_s) vs. the C:N ratio of the bacteria ($C:N_b$) (see Goldman et al., 1987). The problem with this approach is that DOM_s is not known. The effect of other factors, such as temperature (see below), on growth efficiency is also largely unknown.

6. What information is needed to test models with a DOM compartment?

The current methodological problems in measuring DOC and dissolved organic nitrogen (DON) obviously need to be resolved before we can test any DOM model. Several other authors have argued that dissolved organic phosphorus (DOP) also needs to be considered if we are to completely understand the dissolved organic pool. We also need more information about the chemical composition (including C:N and molecular weight) of the DOM pool: presently roughly only 30% has been identified (Benner et al., 1992). This information may give us some clues about its origin. For example, the molecular size distribution and chemical composition would help us distinguish excretion vs. lysis by viruses; only small compounds can be excreted whereas at least the fresh DOM released by cell lysis would reflect the chemical composition of the phytoplankton cell. Additional measurements, however, would be needed to estimate the C:N ratio of DOM_s. The C:N ratio of exported DOM is also an important parameter, although it seems even more difficult (if not impossible) to estimate this parameter.

Although invaluable, data about composition and concentrations tell us little about rates of production and degradation. These rates will be estimated by different methods, depending on which of the three major pools of DOM interest the investigator. As a first approximation the production of labile DOM must equal consumption over the day to week time scale. We can estimate consumption fairly easily with radiotracers. Also, production can be estimated by isotope dilution techniques (Fuhrman, 1987). Degradation rates of total labile DOM can be estimated from: DOM degradation= bacterial production/growth efficiency. At the other extreme, production and consumption rates of refractory DOM appear to be impossible to measure directly considering that the time scale is on the order of hundreds of years. Perhaps estimates can be derived by examining relationships among apparent oxygen utilization, inorganic nutrient concentrations and DOM concentrations in deep waters, although the degradation of sinking particles complicates these relationships.

Data about temporal changes in DOC and DON concentrations would be very informative for examining sources and turnover rates of the semi-refractory DOM with timescales of weeks to months. The limited seasonal data from oceanic studies suggest that this pool may be large fraction (*ca.* 50%) of the total DOM in surface waters (see discussion in Kirchman et al. (1991)).

A more precise estimate of the size of the semi-refractory pool could be made with more extensive seasonal studies. Also, we need to compare DOC concentrations with other microbial variables such as bacterial and phytoplankton biomass and production over a seasonal cycle. Samples need to be taken at least monthly, and sampling over a diel cycle (for a few days) also would be very informative. In addition, it would be useful to compare total DOC concentrations with the concentrations of specific components such as dissolved combined amino acids (DCAA) and dissolved combined monosaccharides (DCMS), i.e. proteins and polysaccharides. Concentrations of DCAA and DCMS are typically 5- to 10-fold higher than that of dissolved free amino acids (DFAA) or monosaccharides. Studies in coastal areas prove the utility of these data (e.g. Ittekkot et al., 1981), but similar data are needed for the open ocean. Data on the seasonal variation in total DOM, DCAA and DCMS concentrations could be used to estimate turnover times of these pools. These estimates of turnover times for DCAA and DCMS could be compared to estimates based on radiotracers or fluorescent analogs (Billen, 1991; Keil & Kirchman, 1992). Unlike low molecular weight compounds, such as DFAA, the proper tracer for high molecular weight compounds is not obvious. Thus, there is a need to have methods for estimating DCAA and DCMS turnover rates that are not sensitive to the choice of tracer.

It seems that even models with a DOM component have largely ignored the semi-refractory DOM pool. Trophic dynamic models that include bacteria using a DOM pool do not include the refractory nor the semi-refractory DOM (e.g. Fasham et al., 1990); DOM in this type of model is labile, by definition. At the other extreme, geochemical models have evoked refractory DOM to help explain distributions in nutrients (e.g. Bacastow & Maier-Reimer, 1991; Najjar et al., 1992). This DOM has turnover times of years, longer than the timescale of our semi- refractory DOM pool, although less than the thousand year turnover time of the most refractory DOC (Druffel & Williams, 1987). Clearly there is a need for models to include the semi-refractory pool, which may be a large fraction of the total.

7. DOM and Global Change

One important goal of biogeochemical models is to predict changes in ecosystems in response to global changes in climate caused by anthropogenic forcing, i.e. increases in CO_2 and other radiative gases. Of the several predicted changes in global climate, temperature seems to be the most important to be examined by, and to include in oceanic DOM models. The critical questions concern how temperature affects relative rates of production and consumption of DOM.

The response of bacteria, which largely determines DOM consumption, to temperature appears to differ from that of primary production, which is one component of DOM production. On one hand, rates of bacterial activity and primary production appear to share the same statistical relationship with temperature (Li & Dickie, 1987). On the other hand, research in arctic and subarctic waters suggests that bacterial activity is suppressed relative to primary production by low water temperatures (Pomeroy et al., 1991; Kirchman et al., 1992). These data imply that the relative importance of the microbial loop is low in cold waters, which in turn implies that the entire trophic structure of the ecosystem varies with temperature. Some of these temperature effects may be peculiar to waters around 0 ^{o}C where even a one degree increase may have a large impact. But temperature effects appear to be important also for ca. 10 ^{o}C waters (Kirchman et al., 1992).

Temperature may affect quasi-steady state DOM concentrations and thus would change storage of carbon even in surface layers of the ocean. The simple model of Billen et al. (1980) indicates that the steady-state DOM concentration (S) is described by

$$S = \frac{K_m}{(V_{max}Y)g^{-1}} \qquad (2)$$

where K_m and V_{max} are the kinetic parameters of DOM uptake (see equation 1), Y the growth efficiency, and g is a cell specific grazing term. Note that the steady-state DOM concentration is not dependent on DOM production, according to this formulation. Half-saturation constants (K_m) of enzymatic reactions decrease and V_{max} of uptake (Robarts et al. 1991) increase with increasing temperature; both effects would lead to lower DOM concentrations.

The impacts of temperature on grazing and growth efficiency are less clear. Laboratory studies indicate that bacteria require higher DOM concentrations for growth as temperature decreases (Wiebe et al., 1992). Currently it is not clear if this higher DOM requirement is because bacterial uptake systems are less efficient at low temperatures or because bacterial growth efficiencies are low in cold waters. The former would imply that DOM concentrations should be high in cold waters, and there is some evidence to support this (Pomeroy et al., 1991).

8. Conclusions

The challenge in including DOM in biogeochemical models is exemplified by examining potential effects of temperature on DOM fluxes and concentrations. Changes in temperature may affect DOM production, for example, even if primary production is not affected, because several organisms and processes potentially produce DOM and are susceptible to temperature shifts. In short, virtually any change in ecosystem structure and function potentially impacts the DOM pool. Also, since we know so little about the production of refractory DOM, it is hard to predict how temperature would affect that DOM. One guess is that an increase in temperature would lead to a decrease in refractory DOM concentrations because of increased microbial activity, but there are no data to prove or disprove this hypothesis.

Our ignorance of key aspects of the DOM pool should not stop models from including it, although caution is advised. We need to include DOM because it is a large pool and is probably more dynamic than traditionally depicted (Kirchman et al., 1991; Kepkay & Wells, 1992). But because of this ignorance, perhaps DOM models have a special obligation to help identify those processes and parameters about which we need information most urgently. Also, some type of model may be essential for extracting information about concentrations and turnover rates of important components of DOM (e.g. semi-refractory and refractory pools) from spatial or seasonal studies of total DOM concentrations. It is already clear that models are necessary to help interpret correlations among concentrations of DOC, DON, and inorganic compounds such as oxygen and nitrate (e.g. Toggweiler, 1989).

References

Anderson, TR (1992) Modelling the influence of food C:N ratio and respiration on growth and nitrogen excretion in marine zooplankton and bacteria. J Plankton Res 14:1645-1671

Bacastow R, Maier-Reimer, E (1991) Dissolved organic carbon in modeling oceanic new production. Global Biogeochem Cycles 5: 71-85

Bauer, JE, Williams, PM, Druffel, ERM (1992) ^{14}C activity of dissolved organic carbon fractions in the north-central Pacific and Sargasso Sea. Nature 357: 667-670

Benner R, Pakulski JD, McCarthy M, Hedges JI, Hatcher PG (1992) Bulk chemical characteristics of dissolved organic matter in the ocean. Science 255: 1561-1564

Berner, RA (1980) Early diagenesis: a theoretical approach. Princeton University Press Princeton

Billen G (1991) Protein degradation in aquatic environments. In: Chrost, RJ (ed) Bacterial Enzymes in Aquatic Environments. Springer Verlag, Berlin, p 122

Billen G, Joiris C, Wijnant J, Gillain G (1980) Concentration and microbiological utilization of small organic molecules in the Scheldt Estuary, the Belgian Coastal Zone of the North Sea and the English Channel. Est Coast Mar Sci 11: 279-294

Bjornsen, P.K. (1988). Phytoplankton exudation of organic matter: *Why* do healthy cells do it? Limnol Oceanogr 33: 151-154

van Boekel, W.H.M., Hanse, F.C., Riegman, R., Bak, R.P.M. (1992). Lysis-induced decline of a *Phaeocystis* bloom and coupling with the microbial food-web. Mar Ecol Prog Ser 81: 269-276

Bratbak G, Thingstad TF (1985) Phytoplankton-bacteria interactions: an apparent paradox? Analysis of a model system with both competition and commensalism. Mar Ecol Prog Ser 25: 23-30

Bratbak, G., Heldal, M., Thingstad, T.F., Riemann, B., Haslund, O.H. (1992). Incorporation of viruses into the budget of microbial C-transfer. A first approach. Mar Ecol Prog Ser In press

Cole JJ, Findlay S, Pace ML (1988) Bacterial production in fresh and saltwater ecosystems: a cross-system overview. Mar Ecol. Prog Ser 43: 1-10

Ducklow HW, Carlson CA (1992) Oceanic bacterial production. Advances in Microbial Ecology. In press

Fasham MJR, Ducklow HW, McKelvie SM (1990) A nitrogen-based model of plankton dynamics in the oceanic mixed layer. J Mar Res 48: 591-639

Fuhrman JA (1987) Close coupling between release and uptake of dissolved free amino acids in seawater studied by an isotope dilution approach. Mar Ecol Prog Ser 37: 45-52

Goldman JC, Caron DA, Dennett MR (1987) Regulation of gross growth efficiency and ammonium regeneration in bacteria by substrate C:N ratio. Limnol Oceanogr 32: 1239-1252

Hedges, JI (1988) Polymerization of humic substances in natural environments. In: Frimmel, FH, Christman, RF (eds) Humic substances and their role in the environment. John Wiley & Sons Limited.

Ittekkot V, Brockmann U, Michaelis W, Degens, ET (1981) Dissolved free and combined carbohydrates during a phytoplankton bloom in the northern North Sea. Mar Ecol Prog Ser 4: 299-305

Jumars PA, Penry DL, Baross JA, Perry MJ, Frost BW (1989) Closing the microbial loop: dissolved carbon pathway to heterotrophic bacteria from incomplete ingestion, digestion and absorption in animals. Deep-Sea Res 36: 483-495

Keil RG, Kirchman DL (1992) Bacterial hydrolysis of protein and methylated protein and its implications for studies of protein degradation in aquatic systems. Appl Environ Microbiol 58: 1374-1375

Kepkay, P.E., Well, M.L. (1992). Dissolved organic carbon in North Atlantic surface waters. Mar Ecol Prog Ser 80: 275-283

Kirchman DL, Suzuki Y, Garside G, Ducklow HW (1991) High turnover rates of dissolved organic carbon during a spring phytoplankton bloom. Nature 352: 612-614

Kirchman DL, Keil RG, Simon M, Welschmeyer NA Biomass and production of heterotrophic bacterioplankton in the oceanic subarctic Pacific. Deep-Sea Res. In press

Lancelot, C. (1984). Extracellular release of small and large molecules by phytoplankton in the Southern Bight of the North Sea. Est Coast Shelf Sci 18: 65-77

Lancelot C, Mathot S (1988) Dynamics of a *Phaeocystis*-dominated spring bloom in Belgian coastal waters. I. Phytoplanktonic activities and related parameters. Mar Ecol Prog Ser 37: 239-248

Law BA (1980) Transport and utilization of proteins by bacteria. In: Payne, JW (ed) Microorganisms and nitrogen sources. Wiley, New York

Li WKW, Dickie PM (1987) Temperature characteristics of photosynthetic and heterotrophic activities: seasonal variations in temperate microbial plankton. Appl Environ Microbiol 53: 2282-2295

Martin, J.H., Fitzwater, S.E. (1992). Dissolved organic carbon in North Atlantic surface waters. Nature 356: 699-700

Nagata T, Kirchman DL (1991) Release of dissolved free and combined amino acids by bacterivorous marine flagellates. Limnol Oceanogr 36: 433-443

Najjar, R.G., Sarmiento, J.L., Toggweiler, J.R. (1992). Downward transport and fate of organic matter in the ocean: simulations with a general circulation model. Global Biogeochem Cycles 6: 45-76

Ogawa H, Ogura N (1992) Comparison of two methods for measuring dissolved organic carbon in sea water. Nature 356: 696-698

Pomeroy LR, Wiebe WJ, Deibel D, Thompson RJ, Rowe GT, Pakulski JD (1991) Bacterial responses to temperature and substrate concentration during the Newfoundland spring bloom. Mar Ecol Prog Ser 75: 143-159

Robarts RD, Sephton LM, Wicks RJ (1991) Labile dissolved organic carbon and water temperature as regulators of heterotrophic bacterial activity and production in the lakes of Sub-Antarctic Marion Island. Polar Biol 11: 403-413

Roy, S., Harris, R.P., Poulet, S.A. (1989). Inefficient feeding by *Calanus helgolandicus* and *Temora longicornis* on *Coscinodiscus wailesii*: quantitative estimation using chlorophyll-type pigments and effects on dissolved free amino acids. Mar Ecol Prog Ser 52: 145-153

Sarmiento, J.L., Sundquist, E.T. (1992). Revised budget for the oceanic uptake of anthropogenic carbon dioxide. Nature 356: 589-593

Sugimura Y, Suzuki Y (1988) A high-temperature catalytic oxidation method for the determination of non-volatile dissolved organic carbon in seawater by direct injection of a liquid sample. Mar Chem 24: 105-131

Vives-Rego, J., Billen, G., Fontigny, A., Somville, M. (1985). Free and attached proteolytic activity in water environment. Mar Ecol Prog Ser 21: 245-249

Wiebe WJ, Sheldon WMJ, Pomeroy LR (1992) Bacterial growth in the cold: evidence for an enhanced substrate requirement. Appl Environ Microbiol 58: 359-364

Williams PJ (1990) The importance of losses during microbial growth: commentary on the physiology, measurement and ecology of the release of dissolved organic material. Mar Microb Food Webs 4: 175-206

Williams PM, Druffel ERM (1987) Radiocarbon in dissolved organic matter in the central North Pacific Ocean. Nature 330: 246-248

MODELLING PARTICLE FLUXES

P. Tett
Marine Sciences Department.
University College of North Wales,
Menai Bridge,
Gwynedd, LL59 5EY

G. Jackson[1], F. Joos[2], P. Nival[3], J. Rodriguez[4], U. Wolf[5]

1. Introduction

Marine particles play an important rôle in the chemistry of the oceans, not only because they provide sources of, and sinks for, solutes, but also, and especially, because they move relative to water. Examples of the resulting fluxes can be seen in the vertical transport of organic carbon and nitrogen, calcium carbonate, or trace metals. The components of most oceanic particles were originally formed as a result of light-driven biological processes in the euphotic zone. The vertical flux of particles decreases with depth and is controlled by a range of biological, physical, and chemical processes.

The passive settling of particles is a dominant mechanism of vertical transport. For low Reynolds numbers (Re<1), settling velocity w (relative to that of the surrounding water) is described by Stokes' Law in terms of the balance between frictional and gravitational forces. In the sphere, the equation is:

$$w = \frac{2}{9} \pi \, \Delta\rho \, g \, r^2 \mu^{-1}$$

where $\Delta\rho$ is the density difference between the particle and the fluid, r is the particle radius, and μ is the fluid's viscosity (which depends on temperature). This

[1]Texas A&M University, Dept. of Oceanography, College Station, Texas 77843, US
[2]University of Bern, Physics Institute, 3012 Bern, Switzerland
[3]Station zoologique, BP 28, 06230 Villefranche, France
[4]University of Malaga, Dept. of Ecology, 29071 Malaga, Spain
[5]Institut fur Meerskunde, JGOFS Buro, 2300 Kiel 1, Germany

NATO ASI Series, Vol I 10
Towards a Model of Ocean
Biogeochemical Processes
Edited by G T Evans and M J R Fasham

equation can be modified by means of a form factor (itself related to the fractal dimension of the particle surface) when considering non-spherical particles and porous aggregates. Furthermore, the value of $\Delta\rho$ can change with particle type and with physiological state in the case of a living particle. A more general form of the equation is

$$w \propto \text{(density excess).(volume).g /(effective radius).}\mu$$

where the effective radius includes the effect of shape on friction.

Sinking particles differ in their density excess, volume and effective radius, and some may be able to influence the viscosity of water around them. Given information on these variables, it should be possible to transform the spectrum of particle types into a spectrum of particle fluxes.

2. Observed categories of particles

Phytoplankton. In some circumstances, such as after the spring bloom in the North Atlantic (Billet *et al.,* 1983) algal cells are a dominant component of the sinking flux. It is generally held that nutrient-depleted and dead diatom cells sink at up to 10 times the rate of living, nutrient-replete cells (Reynolds, 1984). Settling rates can be increased by collection of individual cells into aggregates. Thus, sedimentation of *Phaeocystis* colonies has been invoked as an important carbon loss from high-latitude and North Sea spring blooms. Probably only the larger planktonic algae contribute substantially to this flux.

Faecal pellets. Zooplankton defecation is a source of relatively large, dense, particles. Salp faeces, for example, sink at up to 1000 m d^{-1}.

Cast-offs and corpses. Some sedimenting material comes from detached parts of living organisms, including algal coccoliths, the houses of appendicularians, the mucus feeding webs and the shells of pteropods, the molted exoskeletons of crustaceans, the skeletal parts of radiolarians, and the tests of foraminifera. Although the lives of many planktonic organisms end when the animals are eaten, some die in one piece. For example, carcasses of gelatinous animals may make a significant contribution to the flux of large particles.

Mineral particles. These arrive in the sea from atmospheric (aeolian) and riverine sources. Although they can be important in relation to micronutrients such as iron, we will not consider them further, except in so far as they become part of the final category.

Aggregates. Aggregates are collections of a variety of materials, including cast-off feeding structures, faecal pellets, and micro-organisms. Small algae and microheterotrophs may make an important contribution. 'Marine snow' is another name for large aggregates.

3. Processes

3.1. Formation.

Organic material in sedimenting particles originates in primary production, although it may be transformed or laterally transported before sinking from the photic zone. Sedimenting particles can be categorized into those (i) of direct origin from primary production and (ii) that are formed after one or more transformations, such as feeding followed by defecation, or aggregation. Chemical composition and elemental ratio depend on the origin as well as on the subsequent transformation of particles.

3.2. Transformations

Transformations include physical aggregation, breakdown,and reaggregation, and the more thorough chemical transformation and repackaging following the capture of a sinking particle by a meso- or bathypelagic animal.

Aggregation. Aggregation can repackage small particles into larger, more rapidly sinking clumps. Important controlling factors include the rate of contact between particles and the probability of sticking on contact. Contact is a nonlinear process with a probability depending on the square of the particle concentration and a mechanism-dependent power of the particle size (Jackson, 1990). Organisms can influence aggregation through surface features such as stickiness and spininess; the former property is strongly affected by surface microbial activity. Veil-feeding dinoflagellates may sometimes play an important role in the aggregation of diatoms (Kennaway *et al.* ,ms). 'Cast-offs', especially feeding nets, can form the cores of aggregates. Adsorption of colloids onto aggregates or air bubbles offers a pathway for the removal of dissolved organic matter.

Disaggregation. Disaggregation is defined as the transformation of large to small particles by physical processes such as shear-induced turbulence, or chemical processes such as remineralisation.

Defecation. Major producers of faecal pellets include crustaceans and salps.

Midwater capture. Sinking fluxes may be harvested by midwater feeders, such as pteropods, which spin feeding webs at depths in the water column between 300 and 1000 m. If these organisms are widespread and abundant, they must play an important role in the transformation of sinking particles.

3.3. Consumption.

Sinking organic particles are, according to the strict logic of the scheme proposed here, *consumed* (as opposed to transformed) only by dissolution and mineralization. Nevertheless, consumption by zooplankton may be best considered as a consumption processes, and subsequent defecation as a new production, since the alternative is to treat deep-water zooplankters merely as another type of particle.

Some inorganic components of detritus, including phosphates and silicates, dissolve continuously in seawater. Carbonate, however, does not dissolve until the particle reaches the appropriate carbonate compensation depth. Microbial remineralization, perhaps aided by mechanical and enzymatic reworking in the guts of zooplankters, is one of the main routes of breakdown of organic material. The process depends on detrital composition (and, in particular N:C ratio) as well as surface:volume ratio, water temperature, and oxygen concentration at the particle surface. According to some views, zooplanktonic metabolism is very important in mineralization. The production of dissolved organic matter during these processes may also contribute to the sequestering, as well as the mineralization, of organic material in deep water.

3.4. Vertical transport

Water motions. As already discussed, particles sink relative to water. An important question is, however, whether local vertical motions of the water due to advection and turbulence are significant in relation to particle sinking speed. It seems likely that this is not the case in the deep ocean. In the diel and seasonal thermoclines, however, local upwelling and downwelling can make a significant contribution to particle transport. These transports are best seen as advective on short time-scales and turbulent on longer timescales, where 'short' and 'long' depend also on the spatial scale of the physical processes. In shelf seas, turbulence and associated bottom stress may resuspend particles for lateral transport to regions of lower turbulence.

Migration. Zooplankton vertical migration may provide substantial transport of organic production in the guts and bodies of the zooplankters, before the latter defecate this material or themselves are eaten. This process could be important if the amplitude of the vertical migration is larger than the thickness of the seasonal thermocline and if zooplankters retain a significant fraction of their food in their guts throughout the migration. Although more information is needed on retention times in vertically migrating oceanic zooplankton, it is known that copepod gut retention times are typically only 20 minutes, whereas the migration occurs over several hours. Finally, vertical migration periodically increases near-surface grazing and thus the production of faecal pellets and other particulate material.

Sinking. Notwithstanding the other transports, we believe that sinking of large particles is the dominant pathway by which organic material reaches the deep ocean.

4. Variation in space and time

Timescales. The seasonal timescale is important in production. As already mentioned, timescales for consumption and transformation must be considered in relation to speeds of particle sinking, animal migration, and water motion. Spectra of particle size are unlikely to be the same as spectra of particle flux, because large particles normally sink faster than small particles. Degradation mechanisms may depend on size and sinking rate.

Space scales. In general, vertical gradients of particle distributions and processes are more significant than horizontal gradients. Some water columns exhibit obvious subsurface maxima of particles. Production in the photic zone is horizontally heterogeneous on basin- and meso-scales: we know little about the horizontal distribution of particles in deeper waters. Settlement of local photic zone production onto the continental slope may be augmented by lateral transport of resuspended material generated on the continental shelf.

Oceans and shelf seas. There are important differences between processes affecting particles in shelf seas and those in the ocean. Resuspension and lateral transport, contributing to enhanced sedimentation flux at the shelf break, are thought to be characteristic of tidally-stirred shelf seas, a hypothesis to be investigated by the LOICZ programme ('Land Ocean Interactions in the Coastal Zone' of IGBP). Differences in organisms and life cycles distinguish production processes in the two regions, and some categories of consumers are absent in the shelf seas.

5. Modelling

5.1 Introduction and overall description of particle dynamics

Particulate matter is so diverse in size, shape and composition that it seems impermissible to combine all types in the one category "detritus".
A minimum description could use 3 classes defined by sinking speed:

- non sinking particles
- slow sinking
- fast sinking

For most purposes however, we suggest using size to scale other properties. The continuum of size should be divided into 3 to 20 intervals: the smaller number corresponding to the minimum description by sinking speed, and the larger number being the maximum feasible with present numerical methods. Thus a model should describe the rate of change of the concentration D_{iz} at depth z of the ith particle size class (i from 1 to n), as in the following equation

$$\frac{dD_{iz}}{dt} = \text{formation - consumption} \pm \text{transformation} - \frac{\partial}{\partial z}[(w_{iz}+w_w)D_{iz}] + \frac{\partial}{\partial z}[k_z \frac{\partial D_{iz}}{\partial z}]$$

where w_{iz} is the sinking speed of the ith particle size class at depth z, w_w is vertical water velocity (which might, additionally, be considered a function of z), and it is assumed that vertical turbulent diffusivity, parameterized by k_z, applies equally to all sizes of particles. In this equation z is taken as increasing downwards, and hence the vertical velocities w are positive for particle sinking or downwards water motion.

If D_{iz} stands for the concentration of carbon per m^3 of seawater in the form of particles of category i at depth z, there may need to be additional arrays of state variables for nitrogen, silicon, etc. Given sufficient information to describe the terms, the set of equations can be evaluated for the sinking fluxes at chosen depths. They can also be used for scaling analysis to identify the important components of the fluxes or the main processes driving the fluxes.

Our ability to model the different types of process is uneven. It may be useful to develop detailed models of each process, perhaps included in a single comprehensive sedimentation model, in order to aid the bulk parameterization of simplified implementations in GCMs. An important aspect of both the detailed and the GCM-coupled models will be the incorporation of adequate temporal and spatial – both vertical and horizontal – resolution of flux variability. The transformation term may be especially important, and must include, in appropriate cases, the collection of small by large particles as well as the simple aggregation of small particles into those in the next larger size category.

Small particles can be considered as more or less homogeneous bodies for which rates depend mainly on volume ($\sim r^3$) and surface ($\sim r^2$). Large aggregates are usually porous and should be seen as micro-ecosystems, including bacteria, flagellates and ciliates, in which nutrients are regenerated. Such aggregates can accumulate material. A simulation using a Lagrangian ensemble method (e.g. Woods & Onken, 1982) might allow better understanding of the behaviour of such complex particles.

5.2. Process equations

Equations for particle dynamics have been developed by Jackson (1990) and Riebessell & Wolf-Gladrow (1992) and the process equations suggested below draw on this work. B refers to living, potentially photosynthetically active, phytoplankton biomass and Z to living biomass of meso- and macro- zooplankton, thus clarifying the description of D as relating to "detritus" , which may however include microheterotrophs and inactive phytoplankters. To avoid complex subscripts, z is omitted. i is used to refer to the size class of phytoplankton as well as of detrital particles. j refers to type of zooplankton, in size categories corresponding to those of the other particles. The a are (dimensionless) efficiencies, introduced because some processes give rise to detrital particles with an efficiency less than 100%; rates are $TIME^{-1}$ unless otherwise stated. The rate variables in particular are likely to depend on more variables than just size. The time rate of change of D_i is given by

$$\frac{\partial D_i}{\partial t} = \Delta_1 + \Delta_2 + \Delta_3$$

where Δ_1,Δ_2 , and Δ_3 are the time rate of change brought about by the processes of formation, consumption, and transformation respectively and are defined below.

Formation:

Δ_1	=	$a_{1i}m_iB_i$ +	$\Sigma e_{ij}Z_j$ +	$a_{2j}m_jZ_j$
		phytoplankton. mortality	zooplankton defaecation	zooplankton mortality

where the m are mortality rates, and the summation in the defaecation term is over all zooplankton types making faeces in size category i. The defaecation rates e will depend on the feeding history of the zooplankon, and on temperature, as well as on size.

Consumption:

Δ_2	=	- $\Sigma a_{3j}k_{ji}D_iZ_j$	- s_iD_i
		zooplankton grazing	dissolution

where k_{ji} is the volume clearance rate (LENGTH^3 (zooplankton biomass)$^{-1}$ TIME^{-1}) which will depend on the feeding history of zooplankton and the present concentration of all types of zooplankton prey. The grazing term is summed over all zooplankton types grazing on detrital particles in size category i. s is a dissolution rate which will depend on temperature, pressure and particle chemistry as well as size.

Transformation:

$\Delta_3 = + a_{4(i-1)}b_{(i-1)}(D_{(i-1)})^2/2$
aggregation gain from smaller particles
$-a_{4i}b_i(D_i)^2/2$
aggregation loss to larger particles
$- b'_iD_i$
disaggregation loss to smaller particles
$- b'_{(i+1)}D_{(i+1)}$
dissagregation gain from larger particles

In this simplified model, aggregation and disaggregation involve interactions only with the next smaller or larger size categories. The aggregation efficiency a_4 is related to particle stickiness. The encounter rate b (LENGTH^3 (particle biomass)$^{-1}$ TIME^{-1}) is likely to depend on water velocity shear, related to particle size, for particles sinking at the same rate. In a model that includes the "snowball" effect of large particles sweeping up smaller ones as they overtake them, it will depend on the relative sinking rates of the particles. Finally, the disaggregation rate will depend on local velocity shear (which determines the forces tending to break a particle up), and on mineralization, since this results in particles shrinking.

6. Validation and data needs

The working group had insufficient time to review this topic in detail. There was some discussion of sediment traps and particle sizing techniques for validating models of sinking particles. Biochemical tracers like fatty acids or photosynthetic pigments, and indexes derived from them can be used to estimate fluxes or age of particles captured by traps. Natural radionuclides have been used to estimate the depletion rate of particulate matter. It was suggested that there was a need for novel equipment involving bulk optical and/or imaging techniques to give information on the size and shape characteristics of particulates in a given volume of water. The instruments should be simple and cheap enough to be deployed on a grid array, and should be active for sufficient time to capture strong events like the sedimentation of spring bloom biomass, or episodic rains of large faecal pellets, and record their duration as well as their intensity. Work with ship-lowered instruments should concentrate on improving knowledge of vertical profiles of particulates, especially in key depth zones such as the base of the turbocline or the region of a deep chlorophyll maximum, and should be intensive during key periods such as the decay of the spring bloom.

References

Billett DSM, Lampitt RS, Rice AL. & Mantoura, RFC (1983) Seasonal sedimentation of phytoplankton to the deep sea benthos. Nature, London, 302: 520-522.

Jackson GA (1990) A model for the formation of marine algal flocs by physical coagulation processes. Deep Sea Research, 37: 1197-1211.

Kennaway GMA, Tett P, Lucas IAN (Ms) Dinoflagellates as grazers of diatoms in a Scottish sea-loch. Submitted to Marine Ecology - Progress Series.

Reynolds CS (1984) The ecology of freshwater phytoplankton. Cambridge University Press.

Riebessell Ú, Wolf-Gladrow DA (1992) The relationship between physical aggregation of phytoplankton and particle flux: a numerical model. Deep-Sea Research, 39A:1085-1102.

Woods JD, Onken R (1982) Diurnal variation and primary production in the ocean- preliminary results of a Lagrangian ensemble model. J Plank Res, 4:735-756

THE SIGNIFICANCE OF INTERANNUAL VARIABILITY

John H. Steele and Eric W. Henderson[1]
Woods Hole Oceanographic Institution
Woods Hole, MA 02543
USA

Introduction

The attention to climate change has aroused interest in the effects of longer-term physical variability on ecological systems. Can ecological changes be simply related to physical trends; or are the changes so modified by the biological dynamics that simple physical consequences will not be observable?

In the ocean we do not have simultaneous, long time series of physical and biological measurements that allow straightforward statistical tests. The Continuous Plankton Recorder (CPR) data provide the best space and time coverage of plankton (but no physics) available for any ocean region (Colebrook, 1982). These data show spatially coherent decadal trends in abundance of many species of plankton in the North Sea and North Atlantic. It has been suggested that the patterns of wind direction over the North Atlantic are similar to the plankton variability in the North Sea and could be related to upwelling changes (Aebischer *et al.*, 1990; Bakun, 1990). It is more difficult to provide causal explanations in terms of ecological processes.

This problem is exacerbated when the data are much more sparse and sporadic in their space and time distribution. The available process studies are often for other species in different locations. Further, the physical and biological data are usually collected separately, on different time scales, and the

[1]Marine Laboratory (SOAFD), PO Box 101, Victoria Road, Aberdeen AB9 8DB, Scotland

NATO ASI Series, Vol. I 10
Towards a Model of Ocean
Biogeochemical Processes
Edited by G. T. Evans and M. J. R. Fasham

physical observations are normally much more dense, and are collected over longer periods, than the biological data.

For these reasons, inferences about the causes of longer-term biological variability depend upon the circumstantial evidence derived from modelling studies that can patch together physical data and process studies, test the models against a few biological observations, and infer ecological consequences on decadal time scales.

In this paper, physical variability is introduced as annual cycles with superimposed variance. The general effects are illustrated using a simple phytoplankton-herbivore (P-H) model. Then a N (nutrient)-P-H model is applied to the Sargasso Sea using long-term data on mixed layer depth (MLD).

Physical Variability

The red noise character of variability in the ocean has been noted frequently. Davis (1976) pointed out the difference between atmospheric white noise and the longer-term scales of variance in sea surface temperature. Wunsch (1981) gives a full description of frequency distributions of variance (power spectra) for several parameters. In particular, using the 30-year Bermuda data set, he shows that deep water temperature has a power spectrum that is red over decadal time scales.

Hasselmann (1976) simulated this transfer of variance to longer time scales by a first-order autogressive function with a relaxation time of five months for the surface ocean response to atmospheric forcing. Lemke (1977) extended this approach to deeper layers by proposing an additive set of stochastic relations with successively longer time scales for deeper ocean layers.

More recently, this stochastic approach has been used by Wigley and Raper (1990) to estimate the long-term variance of the climate system with and without superimposed trends. Their stochastic model of temperature in the mixed layer has relaxation time scales of 2–8 years. However, these kinds of models do not take account of any interannual variations in mixed layer depths.

These simple simulations are in marked contrast to the massive numerical models that attempt to capture the deterministic processes in space and time. The stochastic models lack the physical realism but they permit much more exploration of parameter space. This greater flexibility makes the stochastic approach attractive for study of the possible ecological consequences of longer-term physical variability. Earlier work (Steele and Henderson, 1984; Steele, 1985) used truncated Fourier series with random phases as stochastic forcing; these forcing functions were used in single population models to represent physical inputs of variance with red noise characteristics. Here, simple P-H models are used to illustrate the effects at different time scales. For this general approach, annual cycles have superimposed variance derived from autoregressive functions.

The major problem, obviously, is to determine which changes in different physical processes will affect the biological rates and abundances. A unique set of hydrographic data off Bermuda (WHOI and BBSR, 1988) provides a 34-year sequence for the vertical density structure.[2] The seasonal pattern (Fig. 1) shows that the main variability is in the depth of the winter mixed layers. By taking the deepest value in each winter, a sequence is obtained (Fig. 2) that can be used in conjunction with biological process models.

As a specific application to the Sargasso Sea, a seasonal cycle of incident

---

[2]The depth of the mixed layer was estimated from each profile as the depth below 25 m where the temperature decreased by more than 0.2°C. For the deepest layers this was checked by inspection and some adjustments were made.

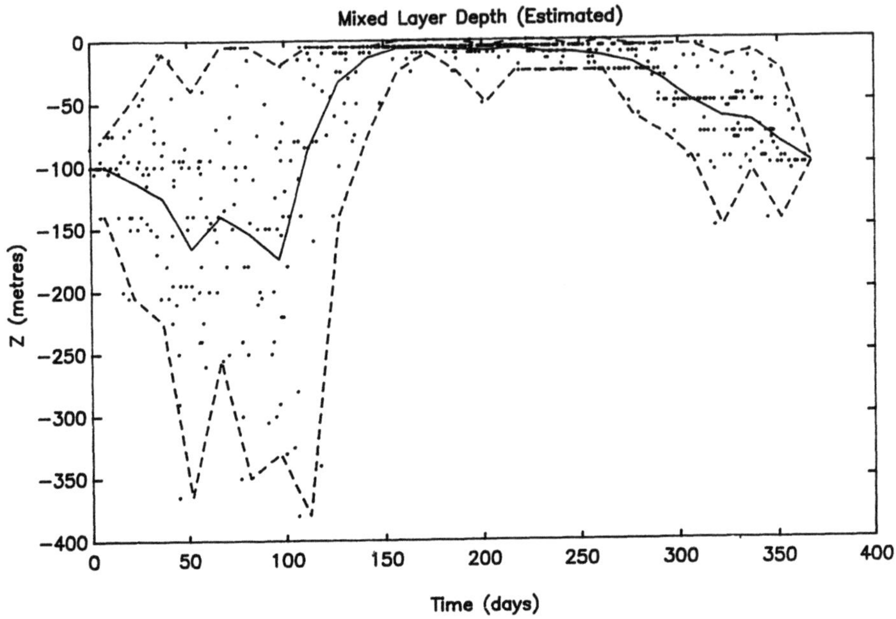

Fig. 1. Estimated mixed layer depths from Station "S" off Bermuda, 1954–1987. (Solid line is 15-day average; dashed line is range.)

radiation is used to derive phytoplankton photosynthesis. A two-layered system will be used with the mixed layer depth cycling seasonally between Z_{min} and Z_{max}. The value of Z_{max} is obtained from the Bermuda data.

Biological Processes

The simplest plant-herbivore models have the form (Steele and Henderson, 1981)

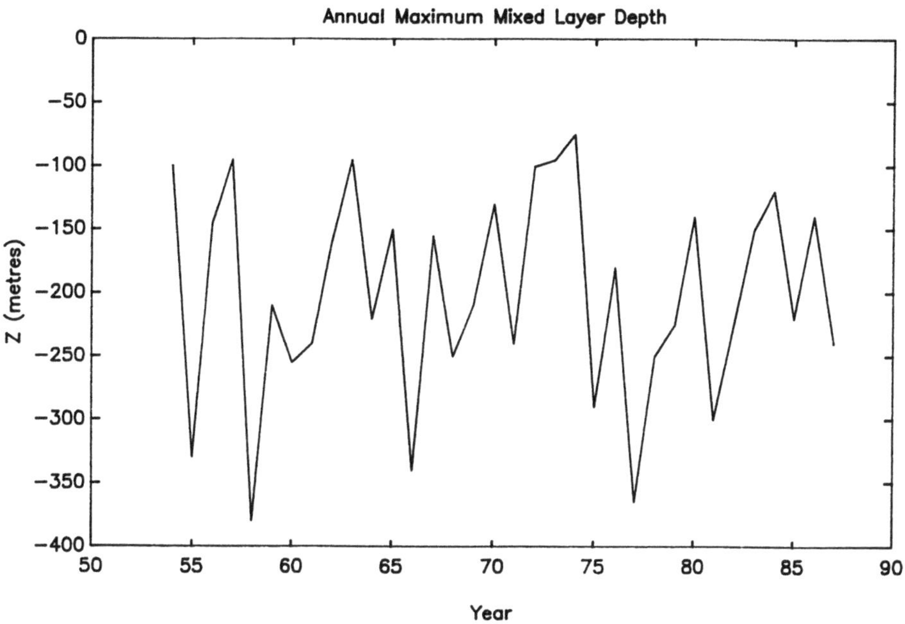

Fig. 2. Estimated maximum mixed layer depth at Station "S," for each year, 1954–1987.

$$dP/dt = f(P) - g(p) \cdot H \tag{1}$$

$$dH/dt = e \cdot g(P) \cdot H - h(H) \tag{2}$$

where the production function f(P) usually takes the logistic form

$$f(P) = a \cdot P(1 - P/c) \tag{3}$$

and

$$g(P) = b \cdot P^n / (d + P^n) \tag{4}$$

where $n = 1$ gives the hyperbolic form and $n = 2$ gives the S-shaped or threshold form. The form $n = 2$ is used here. The closure term $h(H)$ usually takes the form $h(H) \propto H$, but for reasons given in Steele and Henderson (1992) we use $h(H) = m.H^2$.

The variability is introduced in this simple illustration as with

$$a = \bar{a} + a_1 \sin 2p\pi t + v(t)$$

with

$$v(t) = (1 - \alpha)\cdot v(t - 1) + \beta \cdot u(t)$$

where u(t) has a Gaussian distribution with zero mean and unit variance.

With $\bar{a} = 0.3$. $a_1 = 0.15$, and $p = 1/365$, the growth rate of P cycles between 0.2 and 0.6 per day. For simplicity, $b = d = 1$. The value of $c = 10$ is used on the assumption that carrying capacity is large relative to half-saturation grazing rate. A growth efficiency, $e = 0.3$, is assumed. For $a = 1$, the system of equations (1), (2) exhibits a bifurcation near $m = 0.3$, and this value was used so that the deterministic run exhibits large amplitude oscillations in P and H. The stochastic runs in Figure 3 were produced by varying α and selecting β so that the total variance remained the same. Thus $\alpha = 1$ gives white noise; and $\alpha = 0.1, 0.01$, and 0.001 correspond to ten, one hundred, and one thousand days' relaxation times.

It is apparent that this simple "planktonic" system, even with multiple steady states, is not significantly affected by relatively high frequency variance of the order of one week or less relaxation time (Fig. 3a,b). However, relaxation periods of order one year do significantly affect the seasonal cycles (Figs. 3c,d). Thus we are interested in physical processes with longer-term variability at interannual time scales. The Bermuda data (Figs. 1 and 2) provide an excellent example, but their use requires a more detailed model emphasizing interannual changes with specific relevance to the Sargasso Sea. Such a model

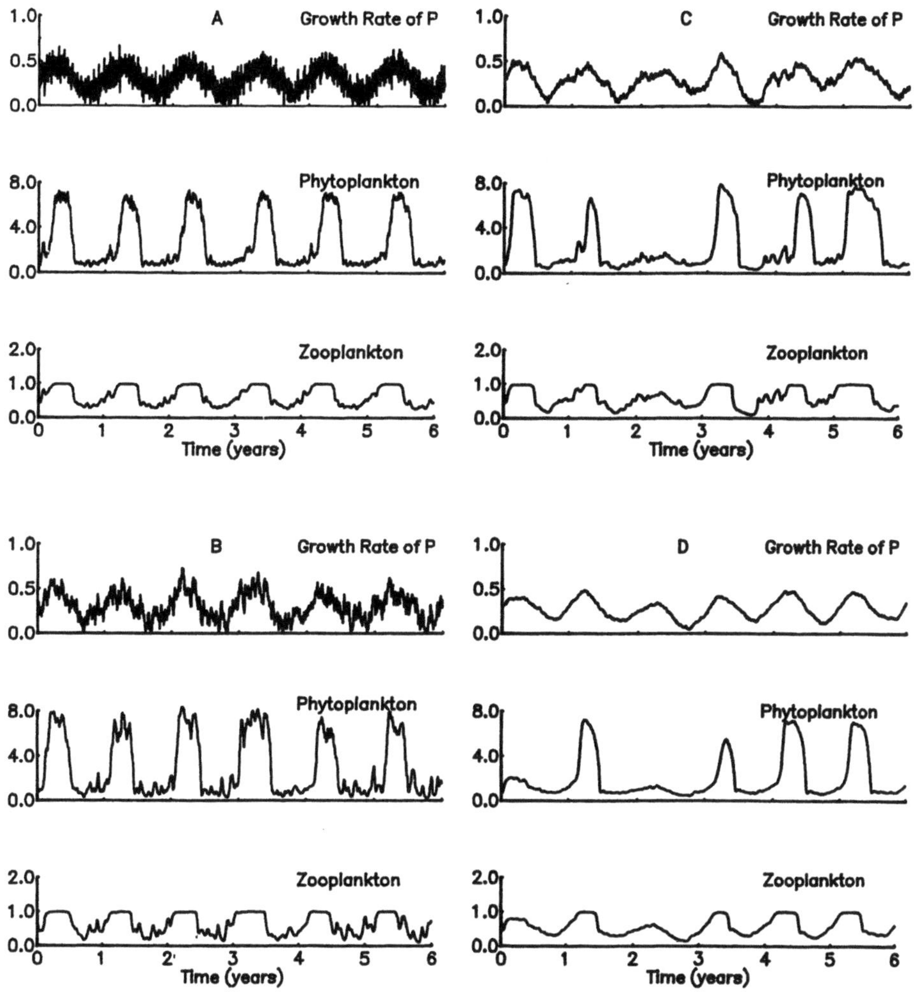

Fig. 3. Response of a simple P-H model to cyclical forcing (see text for details); a) response with white noise; b) response with high frequency noise (α = 0.1); c) response with α = 0.01; and d) response with α = 0.001.

is developed in the next section. A similar model was developed by Evans and Pepin (1989), who conclude that it is difficult to predict plankton and fish concentrations from mixed layer depth history.

Bermuda Simulation

Plankton production off Bermuda was studied intensively about three decades ago (Menzel and Ryther, 1960), and these data form the basis for later analyses (Steele and Menzel, 1962; Musgrave *et al.*, 1988), and especially for models (e.g., Fasham *et al.*, 1990). The model used here is, essentially, the N-P-H model in Steele and Henderson (1981), with seasonal irradiance (RI) and a variable mixed layer depth (MLD). Details of the formulation are given in the Appendix. The main feature is the adaptation and simplification of the data in Figures 1 and 2 to introduce seasonal and interannual variation in MLD. Formally the latter is achieved by varying only the winter maximum (Fig. 4) based on the data in Figure 2.

A fixed vertical distribution of nutrient (nitrate) below the mixed layer is assumed

$$N = 0.13(Z - 20)\ mmol.m^{-3}, for\ Z{>}20\ m \tag{5}$$

When the thermocline deepens, nutrient is entrained. It is assumed that $P = 0$ below the mixed layer. (This, obviously, is false in the summer; but it is appropriate for the winter and spring sequences, which are the focus of this model.) Herbivores, H, are assumed to migrate through the mixed layer.

There are two quite different kinds of variability associated with this model: (1) interannual variability arising from the changes in mixed layer depth; and (2) variability that can be introduced by changing parameter values. These two aspects are considered separately (Figs. 5 and 6). The "standard" parameter values are given in Table 1 in the Appendix.

The estimated maximum MLDs (Fig. 2) provide the interannual forcing. (A random sequence could have been used.) It is clear that the model output (Fig. 5) is closely related to the physical changes. This is the obvious

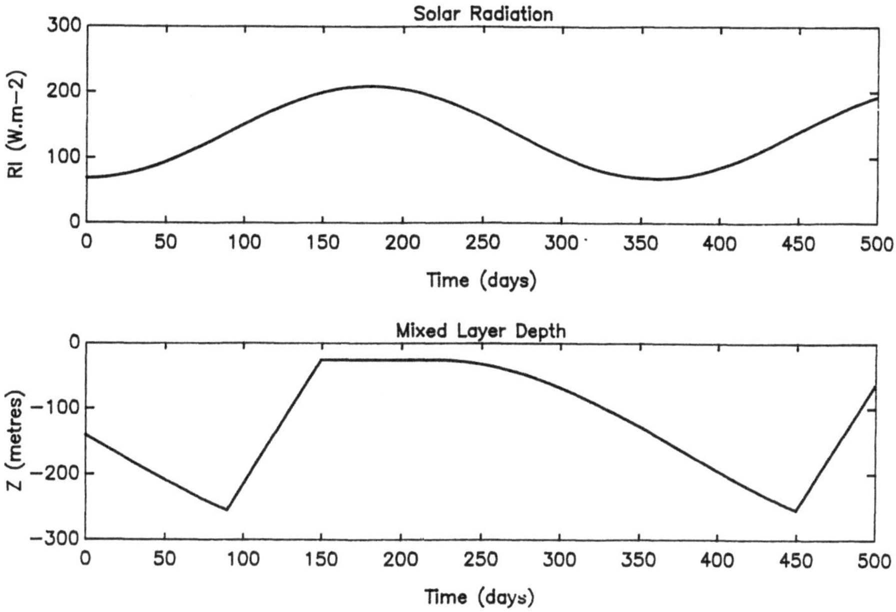

Fig. 4. Simplified representation of radiation and MLD used in the model. The maximum annual depth is taken from Figure 2.

consequence of larger nutrient addition with increased winter MLD, and is seen in the winter nutrient maxima (Fig. 5a).

In the model, the nutrient concentration below the mixed layer increases linearly with depth (equation 5) so the amount of nutrient mixed into the upper layer during the winter is approximately proportional to $(Z_{max})^2$, emphasising the significance of varying MLD for new production. On the other hand, much of the phytoplankton production depends on nutrients recycled within the mixed layer. For this reason, model phytoplankton and zooplankton abundances will have the same interannual ranking but need not correspond closely to winter MLD values.

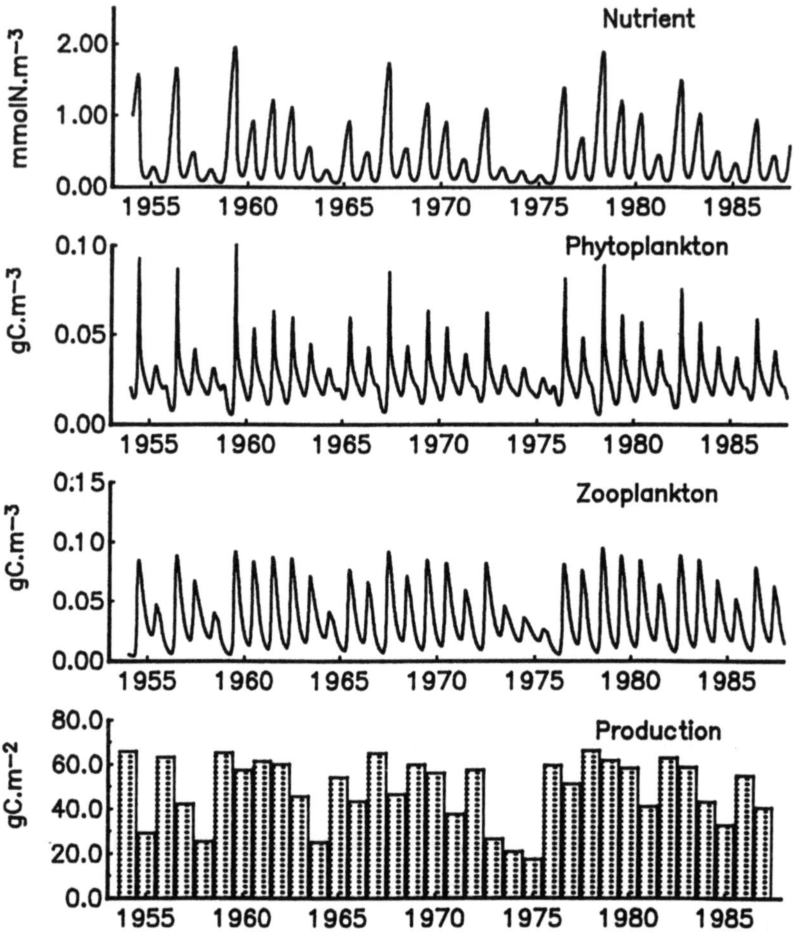

Fig. 5. Model output with the standard parameter values: a) nutrient (as nitrogen equivalent); b) phytoplankton (x 20 to chlorophyll equivalent); c) herbivores (x 2.5 to give dry wt); d) integrated annual net primary production.

These features are reflected in the yearly integral values for net primary production (Fig. 5d). There is significant interannual variation, from about 20 to 60 $gC.m^{-2}.yr^{-1}$, but the ratio (max/min) is much smaller than the range in winter nutrient levels. If we calculate the yearly integral of flux out of the

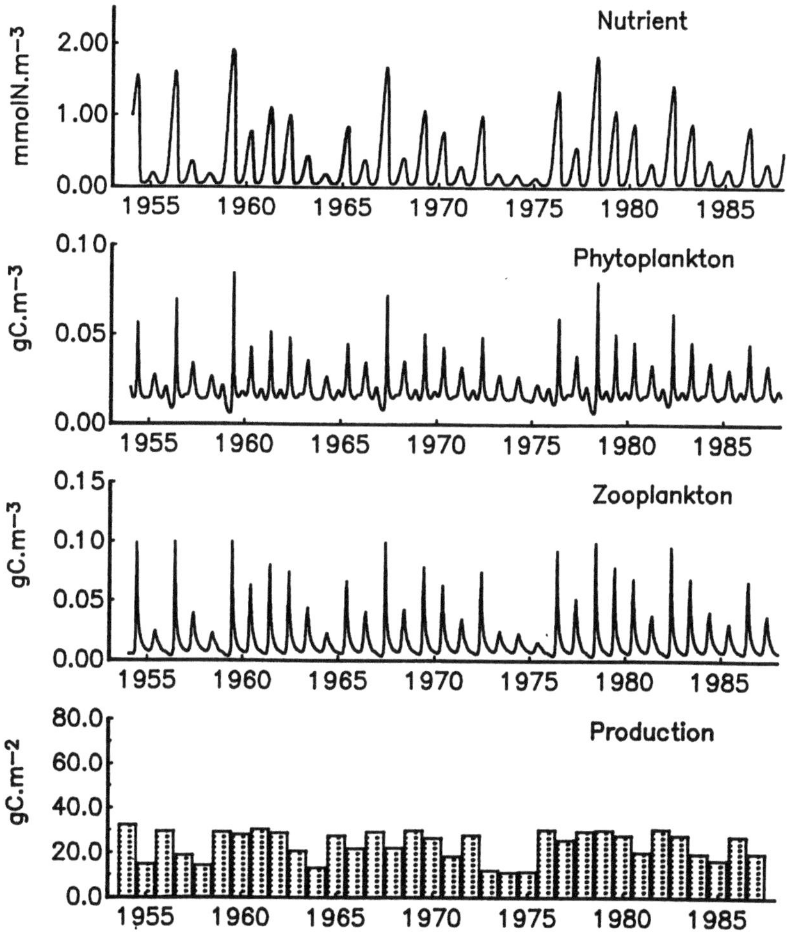

Fig. 6. Model output with changed values for herbivore growth efficiency and predator consumption rate (see text).

mixed layer (Fig. 7), then the range is, proportionately, similar from 7-21 $gC.m^{-2}.yr^{-1}$. Note that the ranges of net production and of flux overlap, so that flux in some years can be greater than production in others. This may be relevant to disagreements on the magnitude of flux out of the mixed layer (Jenkins and Goldman, 1985).

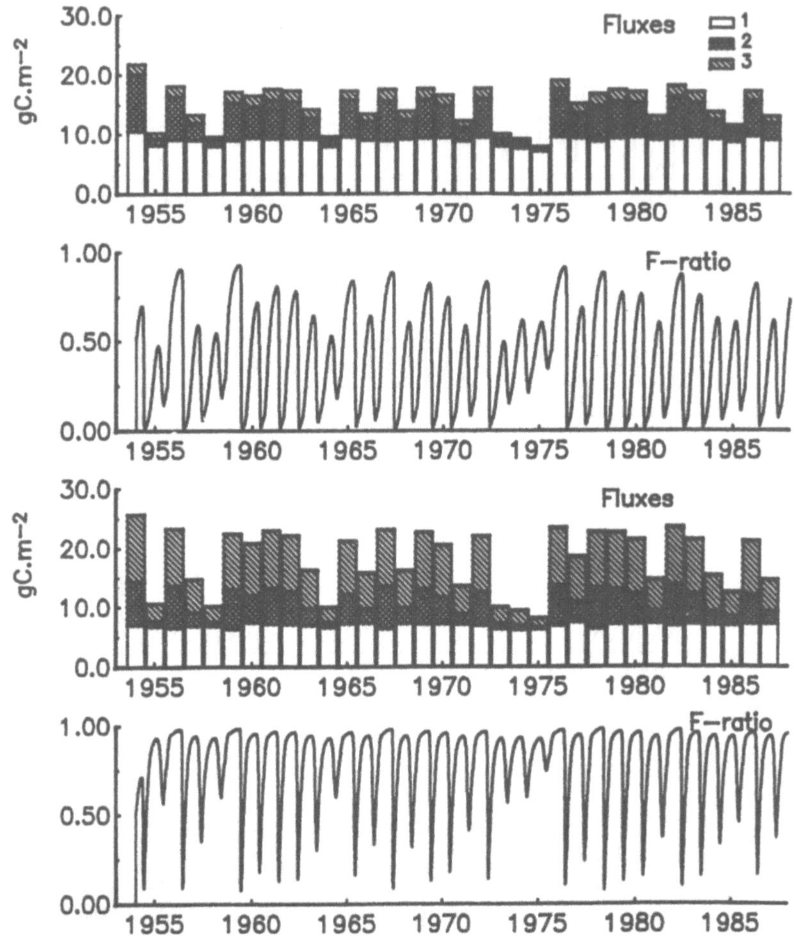

Fig. 7. Yearly fluxes and F-ratios; for (top) standard run (Fig. 5) and (bottom) changed efficiency (Fig. 6). Flux (1) sinking and mixing out of P; (2) loss during spring thermocline shallowing; (3) removal by carnivores (see Appendix).

This application of the model illustrates the consequences of interannual differences in the physics. But the output of the model depends also on the reliability of the parameter values (Table 1). These are put in three groups.

The first group refers to phytoplankton growth. There is considerable agreement on these factors, and the parameters used here are close to those of Fasham *et al.* (1990) and Platt *et al.* (1991). The second group refers to the conditions off Bermuda and can be documented with the exception of the mixing rate, which, as usual, is used for smoothing. It can be halved without significant change.

This consensus does not hold for the third group, the zooplankton parameter values. There is very little experimental or observational basis for choosing any of these values for the Sargasso Sea. In particular, the choice of herbivore growth efficiency is critical, since the difference between grazing and growth can go to provide regenerated nutrients. In the "standard" run used here, growth efficiency (C2) is 20%, and so 80% of ingested nutrients are recycled. This is based on the assumption of predominantly microzooplankton with some internal "loops" resulting in a relatively low overall growth efficiency and a large recycling component. Fasham *et al.* (1990) use a very high growth efficiency of 75%, which, as they say, is more typical of a temperate copepod. Their value was used here, Fig. 6, for a re-run (C2 = 0.75). The predator regeneration was also changed to a comparable efficiency (C3 = 0.25). These changes require adjustment in the predator coefficient (D3), which, in any model, is an essentially free choice, used for tuning (D3 = 1.0).

These changes (Fig. 6) alter the seasonal patterns of P and H but not the general range. There is no significant change in the winter maxima in N but a slight decrease in the summer minima. The major changes are in the annual production values, which are less than half those of the standard run. This is a consequence of the greatly decreased recycling of nutrients and is reflected in the F-ratios (Fig. 7) expressed as the fraction of nutrient uptake that derives from recycled nutrient. (New and recycled nutrient are tracked separately in the model.) In the standard run there is significant interannual variation in the seasonal cycle, whereas the low regeneration regime usually has F values close to unity.

The annual flux of particulate carbon and nitrogen out of the mixed layer is necessarily dependent on upward mixing of inorganic nutrient (i.e., NO_3). Thus there is relatively little change in total flux between the two regimes. There are, however, major changes in the composition of this flux (Fig. 7). In particular, low regeneration from carnivores means that flux from this component can be dominant. Also, as Fasham *et al.* (1990) point out, a large and variable part of the flux in the model (A_2) arises from entrainment of P into the lower layer during the rapid spring shallowing.

The changes in flux result from the larger zooplankton mortality; i.e., that component of zooplankton production that is not recycled. Thus this part of the variability in the system—the downward flux of particulate carbon—is critically dependent on the choice of the three zooplankton parameters, the values for which are poorly known.

In the long term, the flux out of the system must balance the input (remember that a fixed C/N ratio is used). It would appear that fluxes to deeper water are more closely linked to nutrient mixing than to primary production. For flux calculations, a knowledge of vertical mixing and entrainment, and of nutrient concentration, are the real requirements. The description of primary production is needed to define chlorophyll or ocean "colour," since this is the variable that can be compared with satellite observations at the ocean basin scale (Sarmiento *et al.*, in press).

Part of this problem concerns the processes responsible for the flux. In broad terms, flux in the model comes in two categories: the direct products of photosynthesis sinking or mixing out; and the products from secondary production in the form of particulate excreta, dead animals, or through vertical transfers by predators. These different mechanisms can produce quite separate patterns in the depth and timing of regeneration of nutrients, but an investigation of these factors would require detailed vertical structure in the physics and the biology.

Discussion

The simple models used here have three purposes:

(1) To illustrate the relevance of long-term variability;
(2) To demonstrate the dependence on physical factors; and
(3) To explore the concepts that need further development.

Long-term variability

There is a wide range of time scales associated with ocean physics, from the short scales of near-surface events to the millennial responses of Broecker's (1991) abyssal "conveyor belt." The cumulative effect of these processes is that the dominant time scales in the physics increase with depth. The overall response is increased variance with increasing period. This is in contrast to atmospheric variability, which is nearly white up to century time scales (Steele, 1985).

The impact of different types of variance on a simple prey-predator system (Fig. 3) illustrates the abilities of such systems to absorb the high frequency components but respond to the longer-term, red noise variance. This form of analysis ignores the potential effect of occasional, sporadic large events, such as storms, except as these are contributors to the low-frequency variance.

This simple model provides a reason to focus attention on longer-term changes in the ocean, such as those associated with deep winter mixing, and the consequent ecological responses. Deeper-water, longer-term changes in the physics, combined with the general nature of the ecological response, provide a basis for considering specific features of the marine environment.

Dependence on physical factors

Variation in mixed layer depth (MLD) has been recognised as a dominant feature in oceanic production since Sverdrup (1953). MLD varies widely with latitude and ocean dynamics (e.g., Steele, 1966). Obviously it varies seasonally, but it can be expected to have interannual variations as a consequence of large-scale changes in ocean circulation such as El Niño or the north Atlantic "great salinity anomaly" (Dickson *et al.*, 1990). In the Sargasso Sea isopycnal mixing can bring in water from different latitudes and depths. Thus the interannual variations in vertical structure will have a lateral component and should not be regarded as a purely vertical process, as has been done here.

With these caveats, however, the "Bermuda" model illustrates the close dependence of spring nutrient levels, primary production, and downward flux on physical mixing in the winter. Further, the range in maximum MLD at this latitude (32°10′N) is about 100–400 m, and this corresponds to the *average* values given for 20°N to 55°N (Yentsch, 1990). Thus the interannual variability off Bermuda can be of the same order of magnitude as the spatial variability over a major part of the North Atlantic. On this basis, relatively simple models of "new" production (Yentsch, 1990) or steady-state solutions without horizontal dynamics (Wroblewski *et al.*, 1988) may be adequate to portray average conditions and fluxes. They do not demonstrate the very large interannual variability that is evident at Bermuda.

The data on MLD (Fig. 1) off Bermuda indicate that there are many other scales of physical variability that have been ignored in this model. These can contribute to the biological variance in different ways (see Fig. 8 in Appendix). The first, simple model suggested that the overall long-term effects may not be great, but high-frequency noise can alter the "internal" processes such as nutrient uptake at the bottom of the mixed layer (Goldman, 1988). Also, the Bermuda data cannot represent the horizontal scales of variability in MLD.

Further developments

The advantage of these simple models is the relative transparency of their structure. Particularly, the inadequacies are obvious; thus the two critical factors are the variation in mixed layer depth and the form and magnitude of the grazing and mortality terms. The former has been grossly simplified to a two-layer system, the latter require much more experimental data. There are now a number of physical models that describe the dynamics of mixing through the thermocline (e.g., Price *et al.*, 1986), and these have been applied to average conditions at Bermuda (Musgrave *et al.*, 1988). Can such models be extended to handle interannual variability?

Similarly, there are models that utilise age-structured zooplankton populations (e.g., Steele and Frost, 1977; Davis, 1984). These were intended to simulate simple nutrient-phytoplankton-copepod food chains such as the North Sea or Georges Bank systems. The problem with the micro-planktonic system in the open ocean is the lack of data on critical rates; for example, for growth, grazing, and excretion. If we are to understand—and to measure—the mean and variance in vertical fluxes, then we need much more information on these processes.

Appendix

The equations for the Bermuda simulation are:

$$dN/dt = -F1(N,P) + A6 \cdot P + (1 - C2) \cdot F2(P) \cdot H + F3(N) + (1 - C3) \cdot F5(H)$$

$$dP/dt = F1(N,P) - A6 \cdot P - F2(P) \cdot H - F4(P) - (Vs + Ak) \cdot P/Z$$

$$dH/dt = C2 \cdot F2(P) \cdot H - F5(H)$$

where *N*, *P*, *H* represent nutrient, phytoplankton, and herbivore concentrations. *Z* is the mixed layer depth, and

$$F1 = N \cdot P \cdot Ph / ((R2 + N) \cdot (A4 + A5 \cdot P))$$

where Ph is average photosynthesis in the mixed layer

$$F2 = B1 \cdot P^2 / (B2 + P^2)$$

$$F3 = (Ak + DELZ) \cdot (NZ - N) / Z$$

with *DELZ* = *dZ/dt* for Z increasing; otherwise zero.

$$NZ = N0\ (Z - 20) / 250$$

$$F4 = DELZ \cdot P/Z$$

$$F5 = D3 \cdot H^2$$

The depth averaged rate, *Ph*, is derived from a hyperbolic relation of photosynthesis with photosynthetically active radiation (taken as 0.4 x total radiation) depending on the initial slope, α, and maximum rate, *C1*.

$$Ph = C1 \cdot \log(I + RI/R1)/Z$$

$$R1 = C1/\alpha$$

The rate of flux out of the mixed layer

$$TLFX = A1 + A2 + A3$$

where

A1	=	$-P.dZ/dt$ when *Z* decreasing	(shallowing)
	=	otherwise	
A2	=	$(Ak + Vs).P$	(sinking and mixing)
A3	=	$C3.F5(H)$	(predation)

The two physical driving functions are incident radiation, *RI*, and mixed layer depth, *Z* (see Fig. 4). Nutrient concentration below the mixed layer is assumed to increase uniformly to a concentration N0 = 3 mgat.m^{-3}, at 250 m. This determines incorporation of nutrient into the mixed layer as *Z* increases. There is also a small mixing rate introduced by the exchange coefficient, *AK* = 0.5. The units are gC, m, days. It is assumed that

$$1\ gC \equiv 20\ mg\ Chl \equiv 10\ mg\ at\ N$$

The parameter values are given in Table 1. The best data to test the effects of interannual variability are the winter nitrate levels for the six years 1957–63 (Fig. 8). These show that there are significant differences between years, but also that there are many shorter-term fluctuations not captured by the model, which shows only one winter peak in *N* each year (Fig. 3). As a partial comparison of the variability in the data and the model, the maximum values are plotted against each other (Fig. 9). As discussed in the main text, differences in winter nutrient concentrations are the single most important factor in varying the interannual fluxes.

Parameter	Symbol	Value
Light attenuation due to water	A4	0.04 m^{-1}
Light attenuation by phytoplankton	A5	0.3 $m^2(gC)^{-1}$
Phytoplankton respiration rate	A6	0.05 $(day)^{-1}$
Initial slope of P - I curve	α	0.01 $(W.m^{-2})^{-1}(day)^{-1}$
Maximum phytoplankton growth rate	C1	1.0 $(day)^{-1}$
Sinking rate of phytoplankton	Vs	0.5 $m.(day)^{-1}$
Half-saturation nitrogen uptake	R2	0.02 $gC.m^{-3}$
Maximum incident radiation	R_{max}	210 $W.m^{-2}$
Minimum incident radiation	R_{min}	70 $W.m^{-2}$
Mixing rate for phytoplankton	Ak	0.5 $m.(day)^{-1}$
Nitrate at 250 m	N0	0.3 $gC.m^{-3}$
Maximum grazing rate	B1	1.0 $(day)^{-1}$
Zooplankton half-saturation grazing	B2	0.02 $(gC.m^{-3})^{-2}$
Zooplankton growth efficiency	C2	0.2
Zooplankton mortality	D3	0.1 $m^3.(gC)^{-1}.(day)^{-1}$
Removal by predators	C3	0.2

Table 1. Parameters used in the Bermuda model.

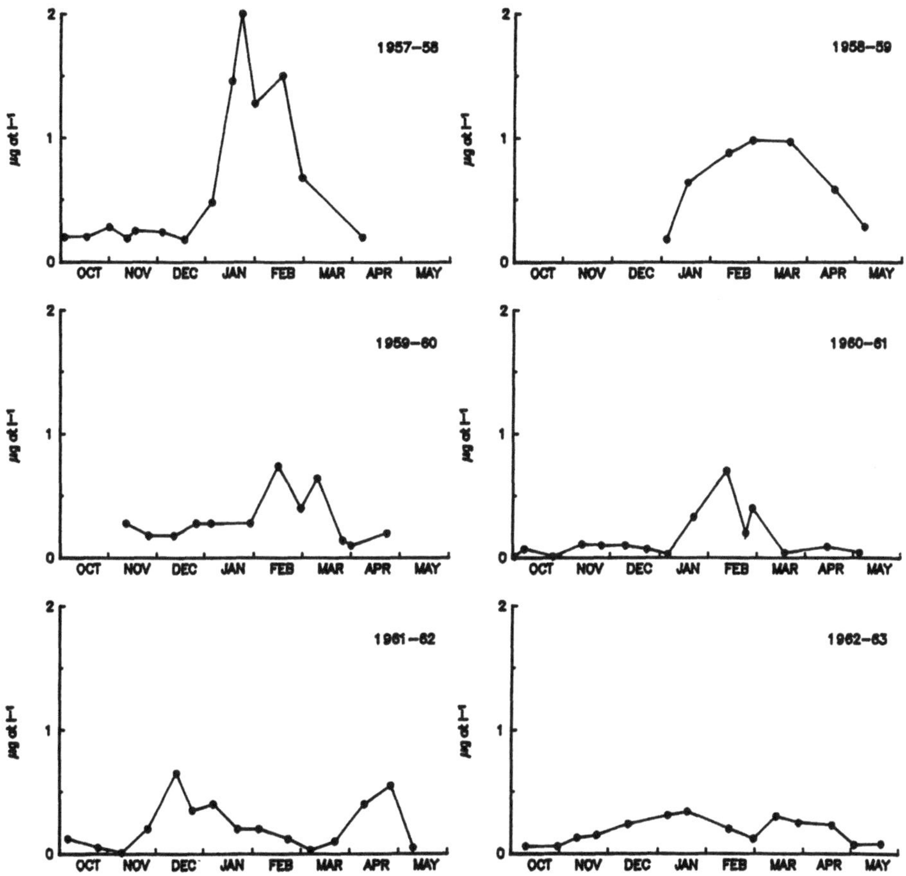

Fig. 8. The average concentration of nitrate in the mixed layer during six winters (from Steele and Menzel, 1962; Ryther and Menzel, 1964).

Acknowledgments

We are grateful to Geoff Evans, Mike Fasham, and Tony Michaels for helpful comments. Support was provided by ONR Grant No. N00014-92-J-1527. This is WHOI Contribution No. 8250 and BBSR Contribution No. 1341.

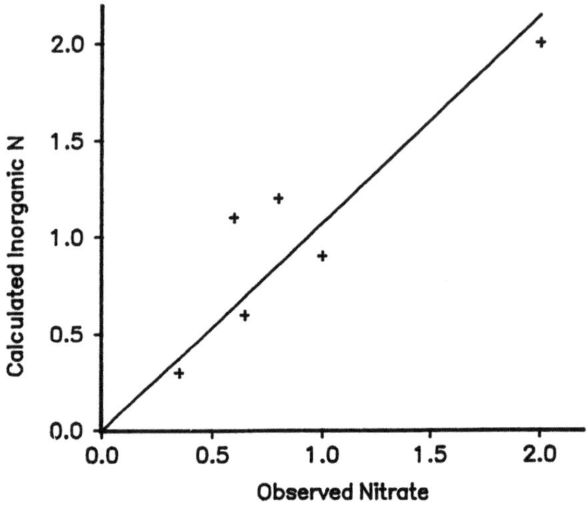

Fig. 9. Comparison of observed nitrate maximum (Fig. 8) and N_{max} in the model (Fig. 5) for the years 1957–1963 (m mol $N.m^{-3}$).

References

Aebischer JJ, Coulson JC, Colebrook JM (1990) Parallel long term trends across four marine tropic levels and weather. Nature 347:753-755.

Bakun A (1990) Global climate change and intensification of coastal ocean upwelling. Science 237:198-201.

Broecker WS (1991) The great ocean conveyer. Oceanography 4(2):79-89.

Colebrook JM (1982) Continuous plankton records: seasonal variations in the distribution and abundance of plankton in the North Atlantic Ocean and the North Sea. J. Plankton Res. 4: 435-462.

Davis CS (1984) Interaction of a copepod population with the mean circulation on Georges Bank. J. Mar. Res. 42:573-590.

Davis RE (1976) Predictability of sea surface temperature and sea level pressure anomalies over the North Pacific Ocean. J. Phys. Ocean. 6(3):249-266.

Evans GT, and Pepin P (1989) Potential for predicting plankton populations (and fish recruitment) from environmental data. Can. J. Fish. Aquat. Sci. 46:898-903.

Fasham MJR, Ducklow HW, McKelvie SM (1990) A nitrogen-based model of plankton dynamics in the oceanic mixed layer. J. Marine Res. 48:591-639.

Goldman JC (1988) Spatial and temporal discontinuities of biological process in pelagic surface waters. In: BJ Rothschild (ed) Toward a Theory on Biological-Physical Interactions in the World Ocean, Kluwer Academic Publishers, Dordrecht, pp. 273-296.

Hasselmann K (1976) Stochastic climate models, part 1. Theory. Tellus 28:473-485.

Lemke P (1977) Stochastic climate models, part 3. Application to zonally averaged energy models. Tellus 29:385-392.

Menzel DW, Ryther JH (1960) The annual cycle of primary production in the Sargasso Sea off Bermuda. Deep-Sea Res. 6:351-367.

Musgrave DL, Chou J, Jenkins WJ (1988) Application of a model of upper-ocean physics for studying seasonal cycles of oxygen. J. Geophys. Res. 93(C12):15,679-15,700.

Platt T, Bird DF, Sathyendranath S (1991) Critical depth and marine primary production. Proc. R. Soc. Lond. B 246:205-217.

Price JF, Weller RA, Pinkel R (1986) Diurnal cycling: observations and models of the upper ocean response to diurnal heating, cooling, and wind mixing. J. Geophys. Res. 91(C7):8411-8427.

Ryther JH, Menzel DW (1964) Physical, Chemical and Biological Observations in the Sargasso Sea Off Bermuda, 1960-1963. WHOI Reference No. 64-8, Appendix II, Progress Report submitted to the US Atomic Energy Commission under contract AT(30-1)-3140. Woods Hole Oceanographic Institution, Woods Hole, MA.

Sarmiento JL, Slater RD, Fasham MJR, Ducklow HW, Toggweiler JR, Evans GT (In press) A seasonal three-dimensional ecosystem model of nitrogen cycling in the North Atlantic euphotic zone. Global Biogeochem. Cycles.

Steele JH (1966) Sublittoral benthic production on a sandy beach. In: C.E. Goldman (ed) Primary Productivity in Aquatic Environments. University of California Press, Berkeley and Los Angeles.

Steele JH (1985) A comparison of terrestrial and marine ecological systems. Nature 313:355-358.

Steele JH, Frost BW (1977) The structure of plankton communities. Phil. Trans. Roy. Soc. Lond. 280:485-534.

Steele JH, Menzel DW (1962) Conditions for maximum primary production in the mixed layer. Deep-Sea Res. 9:39-49.

Steele JH, Henderson EW (1981) A simple plankton model. Amer. Nat. 117(5):676-691.

Steele JH, Henderson EW (1984) Modelling long term fluctuations in fish stocks. Science 22(4652):985-987.

Steele JH, Henderson EW (1992) The role of predation in plankton models. J. Plankton Res. 14(1):157-172.

Sverdrup HU (1953) On conditions for the vernal blooming of phytoplankton. J. Cons. perm. int. Explor. Mer. 18:287-295.

Wigley TML, Raper SCB (1990) Natural variability of the climate system and detection of the greenhouse effect. Nature 344:324-327.

Woods Hole Oceanographic Institution (WHOI), Bermuda Biological Station for Research (BBSR) (1988) Station "S" off Bermuda: Physical Measurements, 1954-1984. WHOI and BBSR: Woods Hole, MA, and Ferry Reach, Bermuda.

Wroblewski JS, Sarmiento JL, Flierl GR (1988) An ocean basin scale model of plankton dynamics in the South Atlantic. 1. Solutions for the climatological oceanographic conditions in May. Global Biogeochem. Cycles 2(3):199-218.

Wunsch C (1981) Low-frequency variability of the sea. In: BA Warren and C Wunsch (eds) Evolution of Physical Oceanography, MIT Press, Cambridge, MA, pp. 342-375.

Yentsch CS (1990) Estimates of "new production" in the Mid-North Atlantic. J. Plankton Res. 12(4):17-734.

SOME PARAMETRIC AND STRUCTURAL SIMULATIONS WITH A THREE-DIMENSIONAL ECOSYSTEM MODEL OF NITROGEN CYCLING IN THE NORTH ATLANTIC EUPHOTIC ZONE

Richard D. Slater
Program in Atmospheric and Oceanic Sciences
Princeton University
Princeton, NJ, 08544
USA

J. L. Sarmiento[1], M. J. R. Fasham[2]

1. Introduction

The Joint Global Ocean Flux Study (JGOFS) is a major international scientific program with the aim of understanding the role of biological processes in the ocean carbon cycle (the "biological pump"). One of the major aims of this project is to be able to model the effect of the biological pump on carbon dioxide drawdown from the atmosphere and to determine its geographical and seasonal variation and what part this pump may play in climatic change (Fasham *et al.*, 1991). This is an ambitious undertaking and it will be some years before we can judge its success. The first requirement is to develop a model of biological production that possesses some geographical generality; the assumption being that the underlying structure of the marine ecosystem is similar throughout the world ocean but that different physical forcing scenarios give rise to different seasonal cycles and the relative concentrations of the individual components of the food web. One of the problems with developing such a generic model is that even simple biological models contain a large number of parameters which are often difficult to measure and, in any event, represent some average value over the diverse species that make up the phytoplankton, zooplankton, and bacterial populations. It is still an

[1] Program in Atmospheric and Oceanic Sciences, Princeton University, Princeton, NJ, 08544, USA

[2] James Rennell Centre for Ocean Circulation, Chilworth Research Centre, Southampton, SO1 7NS, UK

NATO ASI Series, Vol I 10
Towards a Model of Ocean
Biogeochemical Processes
Edited by G. T Evans and M. J R Fasham

open question whether such a model can be produced but in the meantime we should investigate candidate models for their agreement with observations and also determine the senstitvity of the models to the parameter set.

An initial attempt to develop a generic model was made by Fasham *et al.* (1990) and recently this model has been incorporated into a 3-dimensional ocean general circulation model (OGCM) of the North Atlantic basin to investigate how successfully it could predict basin-scale production processes. This model (hereinafter referred to as the standard model) and its results are described in some detail in Sarmiento *et al.* (1993) and Fasham *et al.* (1993). This paper will examine several experiments in which we tried to improve the simulation as compared with our one global, temporal data set: chlorophyll estimated from observations of the Coastal Zone Color Scanner (CZCS).

2. The Standard Model

The ecosystem model is based on the model in Fasham *et al.* (1990) and contains 7 compartments—two dissolved inorganic nutrients, three living forms, and dissolved and detrital particulate organic nitrogen (DON and DPON, respectively). The nutrients are nitrate and ammonium, and the living forms are phytoplankton, zooplankton and bacteria. All ecosystem variables, except DPON, are advected, diffused and convected by the physical forcing determined by the OGCM throughout the model domain. DPON is not allowed to advect, diffuse or convect since we do not explicitly predict its value below the photic layer (see below) and therefore do not have any values to use in determining such fluxes.

Figure 1 shows a flow diagram for the biological model for the upper 123 m (our photic layer). For full details of the equations, see Sarmiento *et al.* (1993). Here we will give a brief overview of the model. Photosynthesis is given by $\overline{J}QP$, where $\overline{J}$ is the nutrient-saturated growth rate which is a function of temperature and light. Temperature dependence comes in the form of the maximum photosynthetic growth rate, $V_p = ab^{cT}$, where a, b and c are constants given in Table 1, and T is water temperature in °C (Eppley, 1972) and the light dependence uses the model of Evans and Parslow (1985). $Q = Q_1 + Q_2$ is a Michaelis-Menten function describing the limiting effect of nitrate and ammonium on the growth rate. The nitrate term Q_1, includes an ammonium inhibition factor after Wroblewski (1977). Part of the photosynthetic uptake, γ_1, is exuded as DON. Zooplankton grazes phytoplankton, bacteria and DPON at rates G_1, G_2,

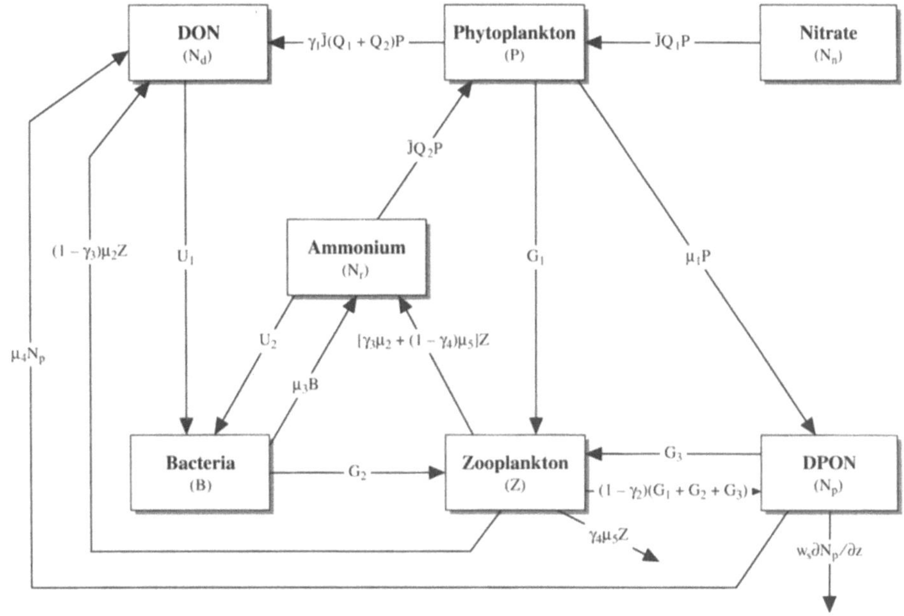

Figure 1: The upper ocean ecosystem model. See text for description of terms. DPON is non-living particulate organic nitrogen and DON is dissolved organic nitrogen.

and G_3, respectively. The assimilation efficiency of grazing is given by γ_2. Bacteria utilize DON and ammonium (the rates U_1 and U_2) via Michaelis-Menten type terms. Phytoplankton mortality is $\mu_1 P$, bacterial excretion is $\mu_3 B$, DPON breakdown is $\mu_4 N_p$, zooplankton excretion is $\mu_2 Z$, and zooplankton mortality is $\mu_5 Z$. Zooplankton excretion is divided between ammonium and DON by the fraction γ_3, and a fraction of the zooplankton mortality, γ_4, is parameterized as being exported immediately from the photic layer by higher trophic levels. Finally, DPON sinks at a constant velocity w_s.

Below 123 m, the flux of particles out of the photic layer (from DPON sinking, and parameterized export by higher trophic levels, the $\gamma_4 \mu_5 Z$ term) is instantly remineralized throughout the water column to ammonium based on an empirical relation given by Martin *et al.* (1987). In contrast, any phytoplankton, zooplankton, bacteria or DON that is advected or diffused below the photic zone is converted to ammonium with a 10 d time scale, and then this ammonium is decayed to nitrate also with a 10 d time scale. No flux of any ecosystem component is allowed

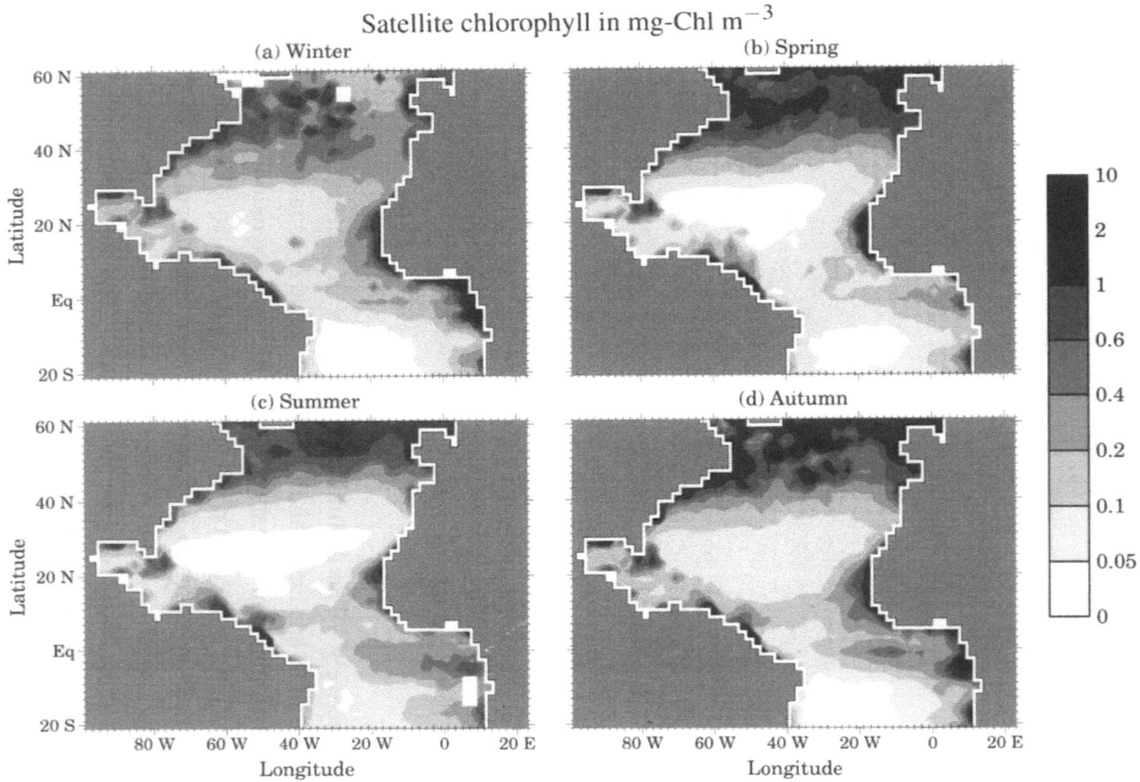

Figure 2: Climatological, seasonal mean chlorophyll concentrations from satellite CZCS observations in mg-Chl m^{-3} (Feldman *et al.*, 1989).

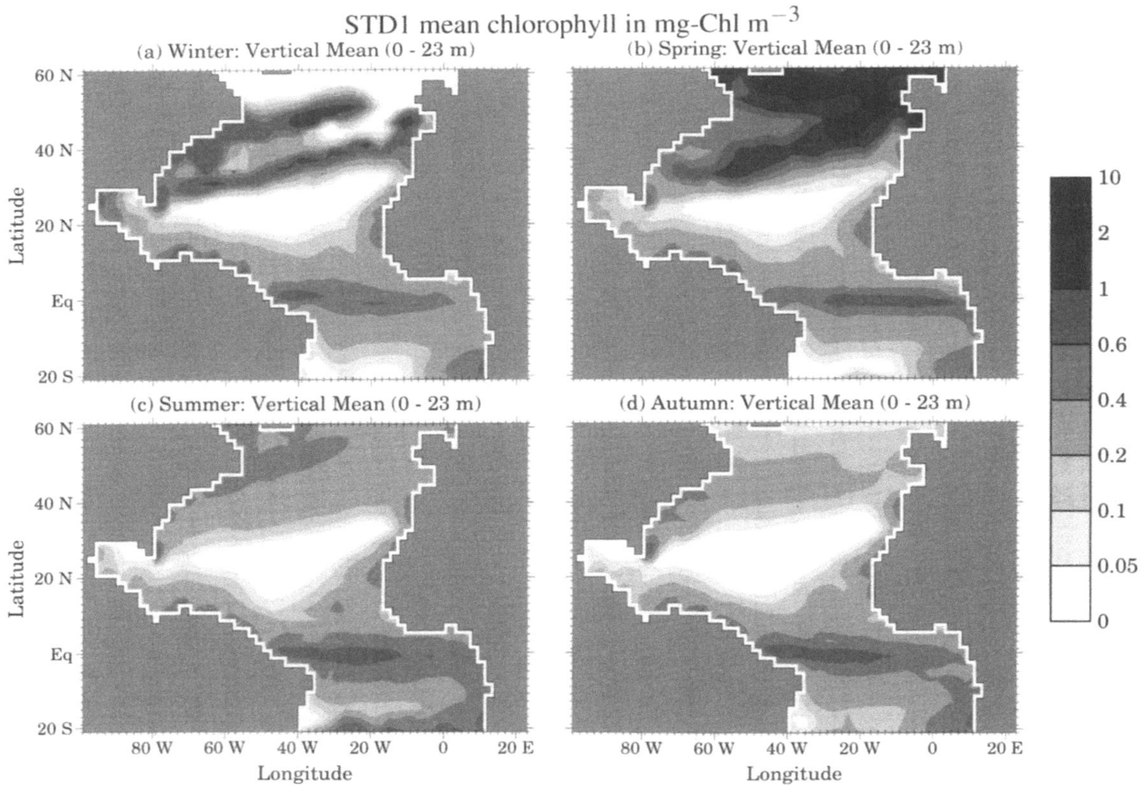

Figure 3: Seasonal mean chlorophyll concentrations from simulation STD1 from model phytoplankton standing crop for the upper 23 m (2 layers) in mg-Chl m^{-3}.

through the sides, bottom (except DPON and ammonium), or surface of the model domain. Any DPON flux, or flux of zooplankton grazed by higher trophic levels, through the bottom (occurring where the depth of the water column is less than or equal to 123 m) is balanced by an equal upward flux of ammonium, thus conserving total nitrogen. Initial conditions for nitrate are from maps produced as described by Kawase and Sarmiento (1986) based primarily on GEOSECS and TTO observations. The other components are initialized to a small value that decays from the surface with a 100 m e-folding scale.

The OGCM is a two-degree, primitive equation model for the North Atlantic from 30°S to 68°N. It is seasonally forced by Hellerman and Rosenstein (1983) winds, and Levitus (1982) temperature and salinity at the surface. There is no flux through the sides, including the open northern and southern ends. The open ends of the basin each have a 10-degree-wide region where temperature and salinity are restored to Levitus (1982) observations throughout the water column (for more details see Sarmiento, 1986). The circulation has been spun up for 1900 years from rest and is used as the initial state for most of the simulations.

The biological simulations were run for two years, with analysis being done during the third year of integration, when the photic zone biology had achieved a near steady-state annual cycle but the nitrate concentrations had not diverged greatly from the initial conditions (Sarmiento *et al.* 1993). Since our model predicts phytoplankton nitrogen, we must convert that to chlorophyll to compare with the satellite data. For this purpose, we used a constant C:N ratio of 6.625, a constant Chl:C ratio of 50, and took the mean of the top two boxes (upper 23 m) as the value for phytoplankton nitrogen. These assumptions mean that no account was taken of any possible geographical or depth-related variations in phytoplankton chemical composition.

Figures 2 and 3 show the surface chlorophyll concentrations obtained from satellite measurements (Feldman *et al.*, 1989) and from our standard simulation (STD1), respectively. In the high latitudes, the model spring bloom began too early (Figures 2a,b and 3a,b), and generally predicted too high phytoplankton concentrations. The model phytoplankton then declined too quickly and had much too low concentrations during the rest of the year (Figures 2c,d and 3c,d). Also, off the North American coast at about 50°N the model showed a trough in the chlorophyll at about 50°N where the bloom started earlier in the winter (Figure 3a) because of the relative stability of the water column (phytoplankton can stay closer to the surface instead of being convected deeper, away from the light) and died off sooner in the spring (Figure 3b) because the nutrients were already depleted and zooplankton were now grazing down the phytoplankton. (Figure 2a appears to show a similar feature to the northwest, but upon examining the monthly plots it does not appear to be related.) This effect was caused by the model Gulf Stream separating from North America too far to the north in the circulation model thereby forcing colder surface waters to stay too far north and shutting down some of the deep convection around 50°N, while causing too deep convection south of that. Figure 4 is a comparison of modeled and observed mixed-layer depths for March showing that the modeled mixed layer depths were too shallow in this area. For more details as to the causes of the discrepencies see Sarmiento (1986).

Figure 5 compares the vertical structure of chlorophyll versus time for the model and data of Menzel and Ryther (1960) at Station “S” at Bermuda (31°N, 65°W from the model). The data shows a deep chlorophyll maximum (DCM) between 80 and 100 m (Menzel and Ryther, 1960; Altabet, 1989), while the model has the DCM much shallower (between 30 and 75 m). Furthermore, there was too little chlorophyll in the model at 100 m. Fasham *et al.* (1993) suggested that the modeled DCM was too shallow because the light-limited growth rate at 50 m was too low to exhaust the large modeled winter concentrations of nitrate (at least twice the observed concentrations) caused by advection from the north-west. Too little phytoplankton (and therefore chlorophyll) occurred at 100 m in the model simply because the net light-limited growth rate was too low to overcome the specified level of phytoplankton mortality and produce a positive net growth rate (Fasham *et al.*, 1993). A similar problem occurred in the center of the subtropical gyre where the model predicted phytoplankton concentrations that were an order of magnitude lower than the CZCS observations resulting in zooplankton and bacterial populations becoming virtually extinct in this area. The reason for this was that the nitrate supply to the photic zone was so low that the resulting nitrate-limited phytoplankton growth rate was again insufficient to produce any net growth.

At the equator, the model predicted much too high concentrations of chlorophyll in the western part of the basin, and somewhat too low concentrations in the eastern basin (Figures 2 and 3). Also, the values off the coast of Africa were much too low because the OGCM does not properly simulate the upwelling there due to its low lateral resolution.

A number of simulations were carried out with modifications to both the ecosystem and circulation models, not as a formal sensitivity analysis, but with

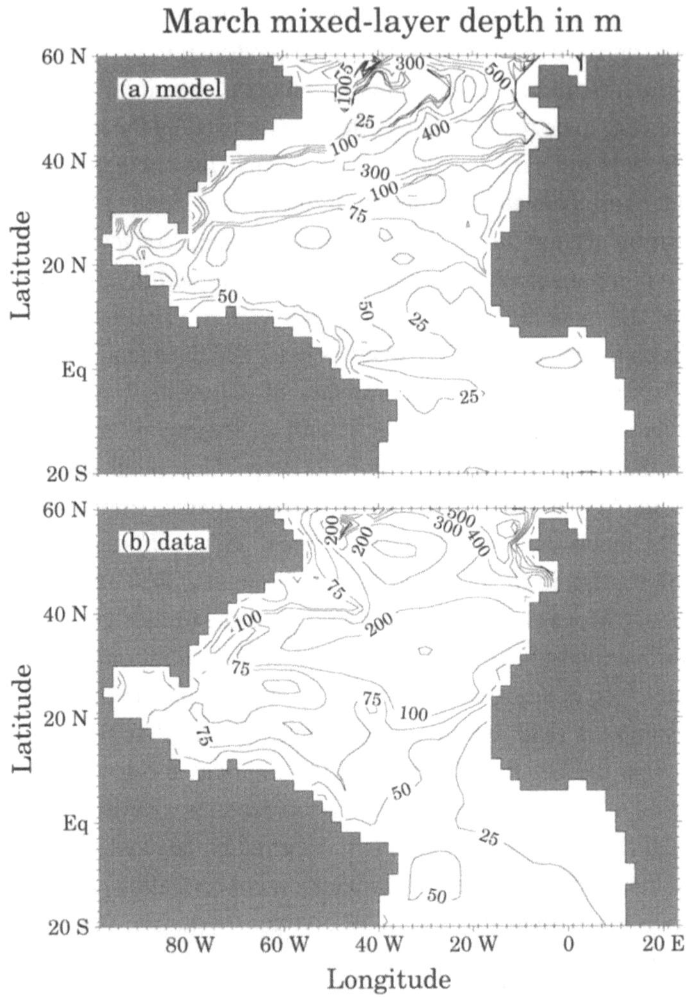

Figure 4: March mixed-layer depths based on a change in σ_t of 0.125 from surface values for (a) the model and (b) data from Levitus (1982). Figure after Sarmiento (1986).

the aim of improving our simulation of the high latitude spring bloom and continued productivity later in the season, as well as to reduce the Equatorial surface phytoplankton and displace it farther to the east, as observed. While neither the ecosystem model nor the OGCM are perfect, our simulations have been aimed more at modifying the ecosystem model, both because it is computationally much

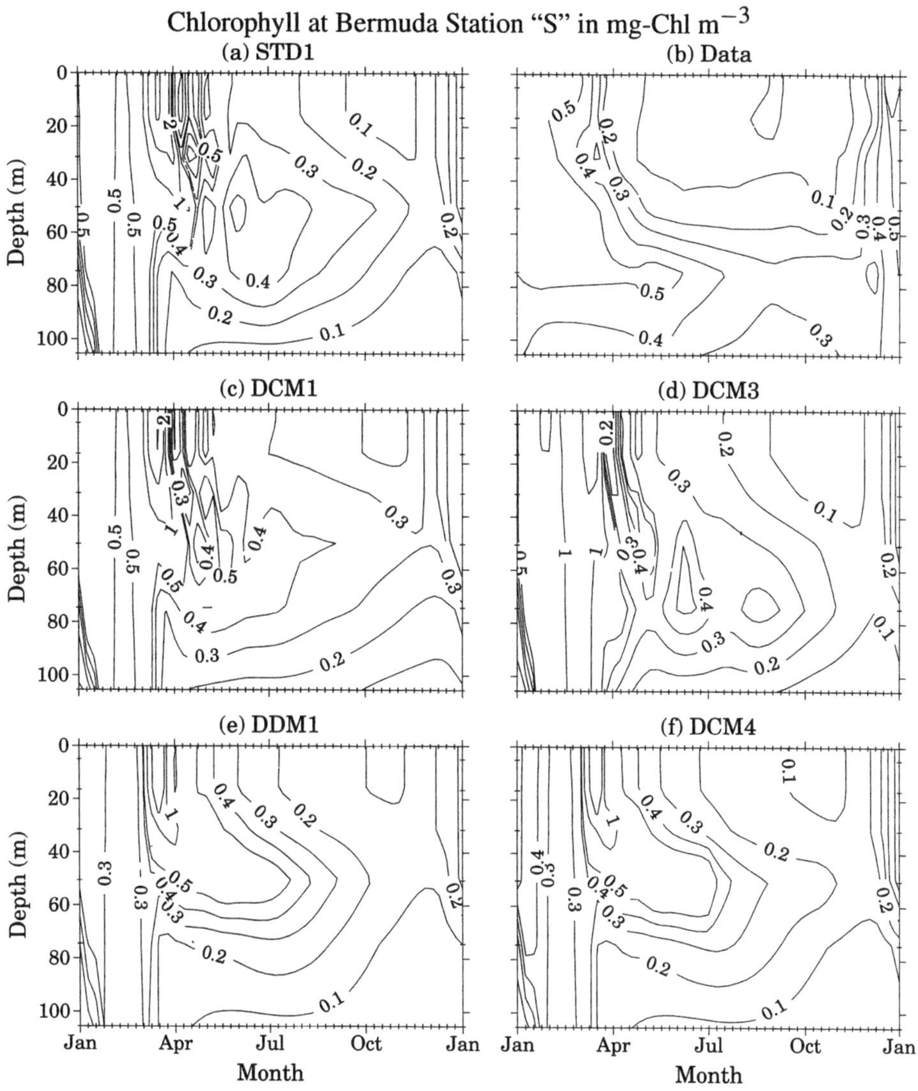

Figure 5: Chlorophyll concentrations for Bermuda Station "S" (31°N, 65°W for the model). Model phytoplankton nitrogen in mmol-N m^{-3} is converted to chlorophyll in mg-Chl m^{-3} by using a constant ratio for g-Chl:mol-N of 1.59. Plot (a) is for simulation STD1, (b) is from data from Menzel and Ryther (1960), (c) is from DCM1, (d) from DCM3, (e) from DDM1, and (f) from DCM4. Contour values are 0.1, 0.2, 0.3, 0.4, 0.5, 1.0, and 2.0. Month labels indicate the middle of the months.

cheaper to do (major changes to the OGCM require an order 1000 year spin-up to reach equilibrium), and because we wish to understand better how changes to the ecosystem model will interact with the OGCM. Although it is difficult to claim that any of these modifications led to a dramatic improvement, many of them did lead to significant changes in the nature of the simulation.

In Section 3 we will describe the simulations and their results, and section 4 gives some conclusions that can be drawn from these simulations.

3. The Simulations

Table 1 gives a brief description of the simulations for future reference. Simulation STD1 is our standard simulation to which we will make comparisons. The parameter set was the same as that used in the mixed layer model of Fasham *et al.* (1990) for Bermuda station "S". Parameter values that were changed in subsequent simulations (with their STD1 values) are: phytoplankton mortality rate, $\mu_1 = 0.04$ d^{-1}; zooplankton excretion rate, $\mu_2 = 0.1$ d^{-1}; zooplankton mortality rate, $\mu_5 = 0.05$ d^{-1}; initial slope of P-I curve, $\alpha = 0.025$ d^{-1} $(W\ m^{-1})^{-1}$; DPON sinking rate, $w_s = -10$ m d^{-1}; maximum zooplankton grazing rate = 1.0 d^{-1}; $\gamma_3 = 0.75$, and $\gamma_4 = 0.33$.

Table 2 gives the annual mean values, averaged over the upper 123 m for the whole model domain, of the seven biological variables for all the simulations. Table 3 gives the annual mean sources and sinks for the biological equations averaged over the same time and volume. Finally, Table 4 gives some derived and computed values. In this table f is the f-ratio (ratio of new production to total primary production); q is, for phytoplankton, the ratio of mortality to the total loss by mortality and grazing; chlorophyll is phytoplankton nitrogen converted using our constant conversion factors; nitrate supply is the net supply of nitrate across 123 m; sedimentation is the loss from the photic layer of DPON by sinking plus the loss of zooplankton by grazing by higher trophic levels; non-nitrate loss is the export by advection, diffusion and convection of all non-nitrate ecosystem components; and change in nitrogen is the time rate of change for the sum of all components. The change in nitrogen should be zero when in steady state.

Figure 6 shows, for STD1, the ratio of the zonal mean of model-predicted to satellite chlorophyll. At latitudes above 30°N, we see that the model over-predicted the magnitude of the spring bloom, and severely under-predicted the chlorophyll concentrations for the rest of the year. The high values shown may be

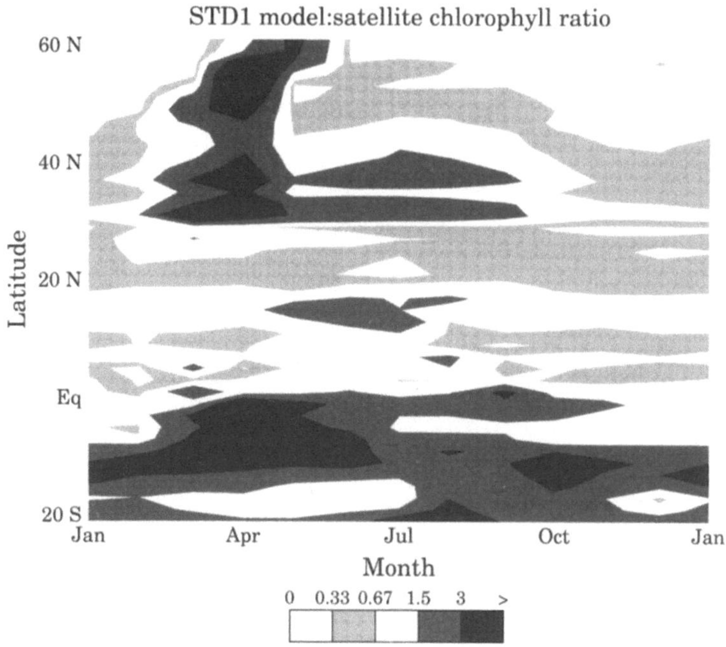

Figure 6: Zonal mean of the ratio of model (upper 23 m) to satellite chlorophyll for simulation STD1. The white area indicates that the model is within ±50% of the observations. Month labels indicate the middle of the month.

due to either having the timing of the bloom wrong or its magnitude. For Bermuda Station "S" (31°N, 65°W in the model), the bloom was about two weeks early and had a maximum chlrophyll concentration that was too high (Fasham *et al.*, 1993).

Simulations STD2 and STD3 examined different sinking velocities for DPON. We also changed the phytoplankton mortality rate in these simulations. In the mixed layer model of Fasham *et al.* (1990), phytoplankton mortality was used as a parameter to tune to the total production value in the literature. Here we used the same values as in that model, namely, for STD2, $w_s = -1$ m d^{-1} and $\mu_1 = 0.08$ d^{-1}, and for STD3, $w_s = -100$ m d^{-1} and $\mu_1 = 0.035$ d^{-1}.

In STD2 we see significant changes. First, in Table 3, we see that new production decreased by about 20% while regenerated production doubled. Phytoplankton mortality increased substantially, but the relative proportions due to mortality and grazing, q, stayed about the same. This simulation was very efficient at recycling nitrogen—the phytoplankton mortality rate has doubled, and the detritus

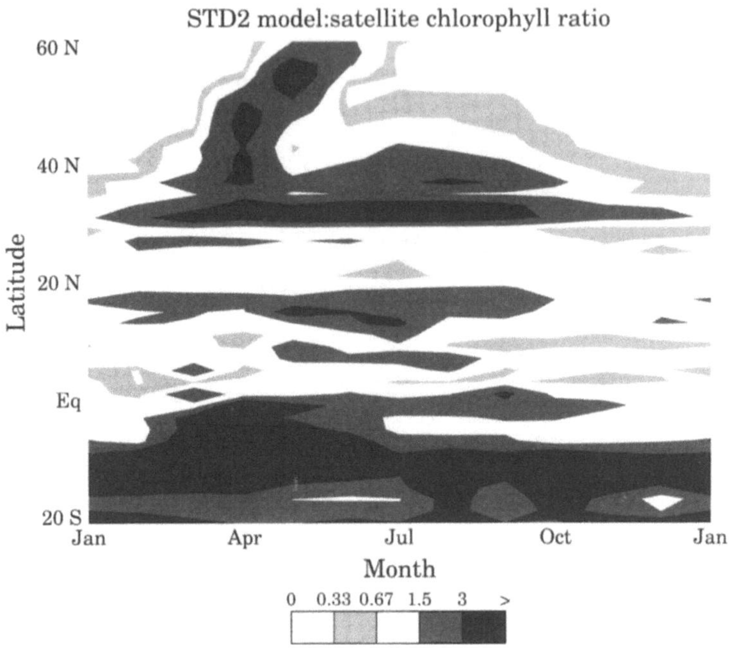

Figure 7: Zonal mean of the ratio of model (upper 23 m) to satellite chlorophyll for simulation STD2.

residence time has increased by a factor of ten due to the slower sinking rate. This provided a large food base for zooplankton (detrital grazing, G_3, increased by an order of magnitude), which was responsible for putting most of the nitrogen into the recycling pool (the left side loop in Figure 1). Note that even though nitrate concentration increased by 25% (Table 2) new production was down (Table 3) because ammonium concentrations, and therefore inhibition of nitrate uptake, had increased greatly. Zooplankton and bacteria populations rose very sharply. The increased zooplankton grazed down the phytoplankton so that global values of phytoplankton were lower. At high latitudes the bloom was later (Figure 7), but the die-off afterwards was larger, and the summer and autumn chlorophyll levels were smaller than in STD1. Finally, the STD2 simulation failed to produce any DCM at Bermuda (Fasham *et al.*, 1993).

In STD3 the results were similar to STD1, but, because of the increased sinking speed for particles (now -100 m d^{-1}), there was little recycling of nutrients. The smaller amount of recycling produced a reduction of zooplankton and bacteria populations, and a halving of the ammonium concentration. It is interesting that with the halving of the ammonium concentration the regenerated production only decreased by about 30%. Perhaps this indicates that the standard simulation produced excess ammonium that was not being fully utilized. Also, the proportions of phytoplankton lost to mortality and grazing were similar in STD3 and STD1. The standing crop of detritus in the photic layer decreased by an order of magnitude. It is, perhaps, somewhat surprising that this simulation produced such similar results to STD1, in terms of phytoplankton and primary production, given the significant changes in the other biological quantities. Further discussion of this issue is given in Fasham *et al.* (1993).

In TDP1, we examined the effect of temperature dependence on photosynthesis by removing this dependence from the maximum growth rate. We used a maximum growth rate of 2.9 d^{-1} throughout the whole region, which is equivalent to a temperature of 24.65°C with the STD1 parameters.

This did not produce much of a change at high latitudes, largely due to the correlation between irradiance and temperature there; the colder months are darker and so little photosynthesis occurs. In the lower latitudes we might not expect a significant change because the temperatures vary less seasonally. Indeed, the results (not shown) are very hard to distinguish from STD1 and the global results given in the tables are much the same as for STD1.

DDM1 replaced the constant phytoplankton and zooplankton specific mortality rates with density-dependent rates (Steele and Henderson, 1992). The specific mortality rates took the form of a Michaelis-Menten function $\mu C/(k + C)$, where C is the concentration of either phytoplankton or zooplankton and k is a half saturation constant. Fasham (1992) found that the density-dependent forms gave a greatly improved simulation of observations at Ocean Weather Ship (OWS) India (59°N) where winter mixed-layer depths are very deep. If a constant mortality rate was used the zooplankton concentrations reached such low winter values that the population values did not recover until August and so had no grazing effect on the spring bloom. The winter zooplankton levels in the standard 3-D model were more reasonable (apparently because of advection of zoopankton from the south) but, even so, the DDM1 simulation seemed to improve the high latitudes results. The parameters used were, for phytoplankton, $\mu = 0.05$ d^{-1} and $k = 0.2$ mmol m^{-3}, and for zooplankton, $\mu = 0.3$ d^{-1} and $k = 0.2$ mmol m^{-3}. This new zooplankton mortality replaced the former terms for mortality and excretion, and was partitioned between ammonium, DON and grazing by higher trophic levels (immediate export from the upper 123 m) in the proportions of 0.7, 0.2 and 0.1, respectively. DDM2 was the same as DDM1 but with a sinking velocity of –5 m d^{-1}.

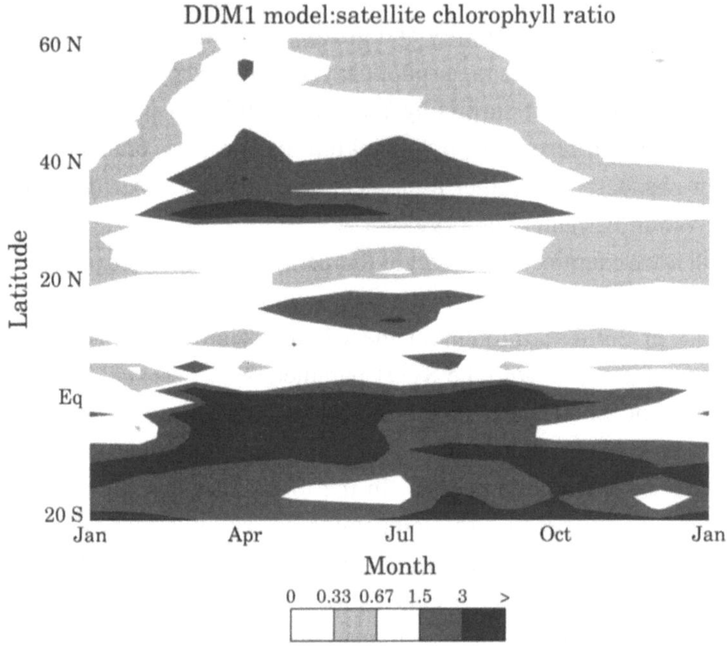

Figure 8: Zonal mean of the ratio of model (upper 23 m) to satellite chlorophyll for simulation DDM1.

Simulation DDM1 showed the most dramatic improvement over the standard model. The ratio of model to satellite chlorophyll is given in Figure 8 for DDM1. This simulation was in better agreement with the magnitude of the bloom at high latitudes than STD1. The onset of the bloom was later than in STD1 and agreed better with the CZCS observations (based on monthly plots), and the bloom began almost simultaneously north of 40°, whereas in STD1 there was a lag at higher latitudes. The bloom was still too intense, and the die-off afterwards too extreme (see also Figures 9 and 2), but it was an improvement in both ways over STD1. The change was primarily due to the lower mortality of zooplankton during the winter. This allowed for more zooplankton to survive the winter and therefore respond more quickly to the phytoplankton bloom in the spring. This reduced the peak of the bloom and allowed the fixed nutrient supply that was convected into the photic layer during the winter to support a longer phytoplankton growing season.

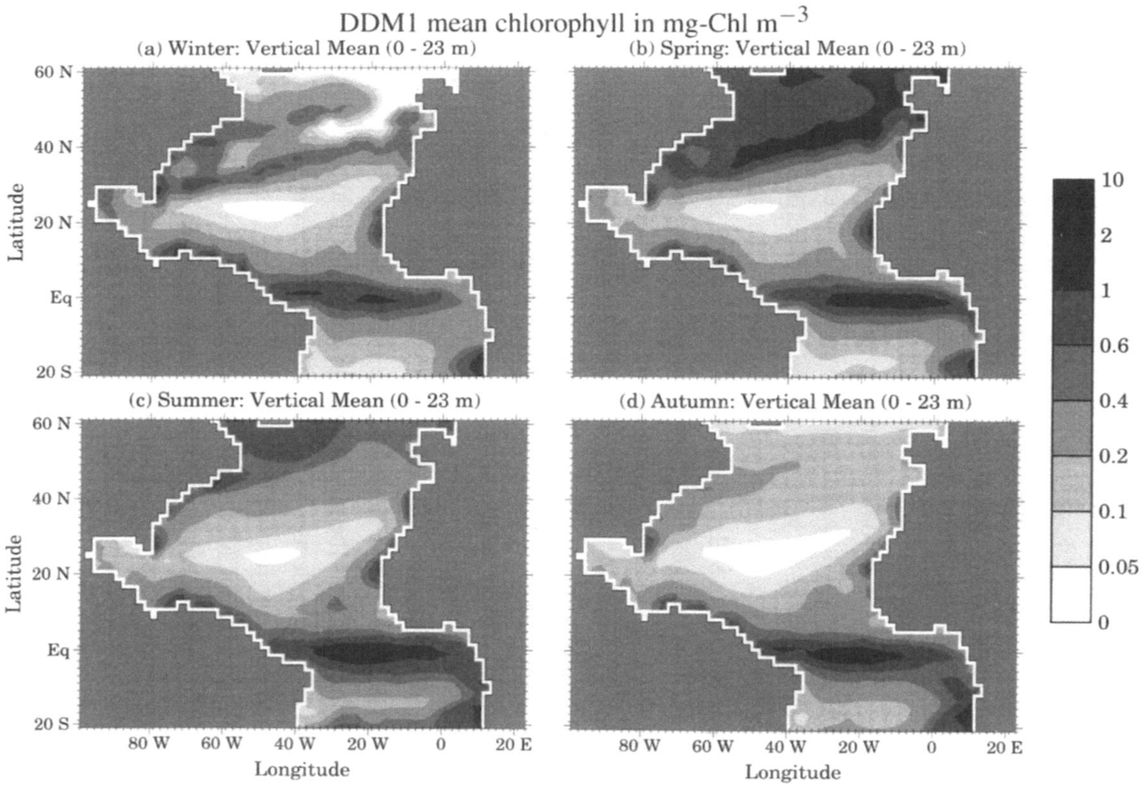

Figure 9: Seasonal mean chlorophyll concentrations from simulation DDM1 from model phytoplankton standing crop for the upper 23 m (2 layers) in mg-Chl m^{-3}.

The major change in standing crop was a 20% increase in zooplankton, and a 13% increase in ammonium. The total loss of zooplankton by mortality, excretion and grazing was almost double that of STD1, due to the increased maximum loss rate in the density dependent term. Note also that, as pointed out by Fasham (1992), the use of the density-dependent mortalities decreased the ratio of phytoplankton loss to natural mortality vis-a-vis zooplankton grazing (q was 0.17 compared to 0.27 for the standard model (Table 4)). The reduced phytoplankton mortality rates at low concentrations also led to significant improvements in the predicted phytoplankton levels in the subtropical gyre (compare Figure 9 with Figure 3).

Simulation DDM2 showed little difference from DDM1. Because of the slower sinking speed there was more recycling of nutrients and therefore a lower f-ratio, but the chlorophyll patterns were very similar. If we look at the tables, and compare the changes from STD1 to STD2 (where we reduced the sinking velocity of DPON), and from DDM1 to DDM2, we see that in every case, except the standing crop of phytoplankton, the same trend obtains, but with smaller magnitudes for the smaller change in DPON sinking velocity (DDM1 to DDM2). These trends are the result of changing the sinking speed. In STD2, the standing crop of phytoplankton was most likely lower than in STD1 because of the doubling of its mortality rate, whereas there was no change in the parameters for the phytoplankton mortality rate in the DDM simulations.

The simulations DCM1, DCM2, and DCM3 were undertaken with the aim of developing a deeper DCM at Bermuda (see Figure 5). In such oligotrophic areas, as the bloom progresses the phytoplankton consume the available nutrients at a given depth and move deeper to where nutrients still exist until finally the phytoplankton become light-limited. As already discussed, Fasham *et al.* (1993) proposed that the modeled DCM was too shallow because the net phytoplankton growth was insufficient to consume all of the winter-supplied nitrate at the shallower depths. To address this, we attempted to increase this net growth rate by reducing phytoplankton mortality (DCM1), reducing the grazing rate by zooplankton (DCM2), and increasing the initial slope of the P-I curve (DCM3). DCM1 halved the phytoplankton mortality rate to 0.02 d^{-1} while DCM2 reduced the maximum zooplankton grazing rate by 10% to 0.9 d^{-1}, and DCM3 doubled α to 0.05 d^{-1} $(W\ m^{-2})^{-1}$.

Simulation DCM1 failed to produce a deeper phytoplankton maximum, but it did produce a higher concentration of phytoplankton at the deepest part of the photic layer (Figure 5c) thereby supporting the contention of Fasham *et al.* (1993)

that it was the net phytoplankton growth rate being too low that was the cause of the absence of phytoplankton at this depth.

The assumption of a constant Chl:N ratio is not the best that could be made since the Chl:N ratio is a function of light and nutrient concentration (Laws and Bannister, 1980; Sakshaug *et al.*, 1989). Under nutrient saturated conditions a lower light level gives a higher Chl:N ratio (Laws and Bannister, 1980) and the inclusion of this effect might improve the comparison of model with data, since model phytoplankton at greater depths would have higher Chl:N ratios than those shallower thereby deepening the DCM in Figure 5a.

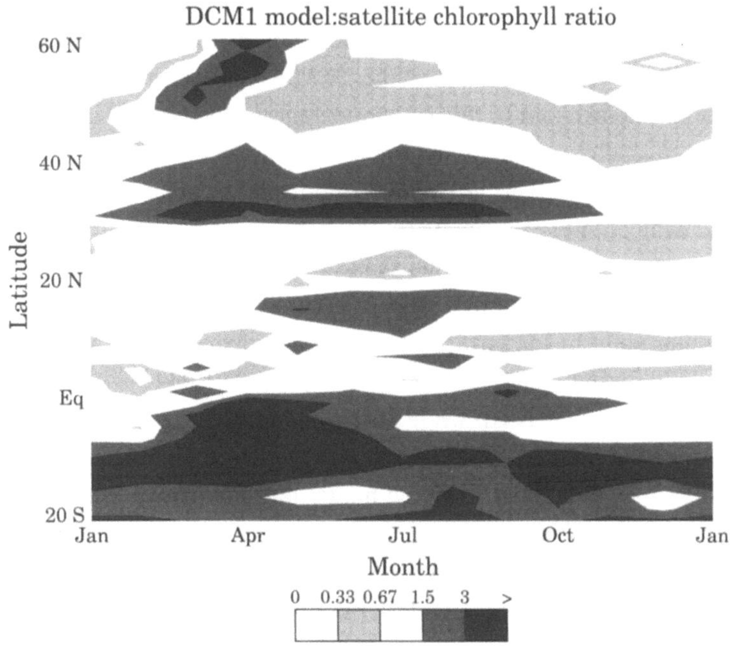

Figure 10: Zonal mean of the ratio of model (upper 23 m) to satellite chlorophyll for simulation DCM1.

Figure 10 shows the ratio of model to satellite chlorophyll ratios for DCM1. The model chlorophyll in DCM1 during the bloom was not as high as STD1, but the phytoplankton still died off too much during the summer and fall. In Table 2 we see that the global phytoplankton population has increased significantly compared to STD1, as has zooplankton. From Table 3 we see that the total production has gone up relative to STD1, with the increase coming from regenerated production due to the increase in phytoplankton grazing and zooplankton excretion. The drop in q from 0.27 for STD1 to 0.15 for DCM1 in Table 4 shows a shift from mortality to zooplankton grazing in the loss of phytoplankton. The phytoplankton-nitrogen loss more readily enters the ammonium recycling loop from zooplankton grazing than from detritus (which may be produced from phytoplankton mortality) thus explaining the increased regenerated production when phytoplankton mortality is reduced.

DCM2 produced a small shift in phytoplankton loss towards mortality (see q in Table 4) and therefore reduced the recycling of nitrogen and lowered the regenerated production (see Table 3). In most respects the results were very similar to those of the standard case, including the vertical structure at Bermuda.

Figure 5d shows, for DCM3, that the DCM has deepened to almost 80 m. The larger nutrient uptake rate in this simulation allows the phytoplankton to exhaust the nutrient at shallower depths and forces them deeper. Although we believe that the problems with the DCM at Bermuda are from too great a supply of nutrients during the winter-time (Fasham *et al.*, 1993), this does point out that the model can produce the desired DCM when the shallower nutrients are exhausted. A plot of the model to satellite chlorophyll ratios is very similar to that for STD1. Figure 11 shows similar patterns for surface chlorophyll compared to STD1 (Figure 3), but the magnitude during winter and spring is increased. Table 2 shows that plankton and bacteria are all increased, while nitrate and ammonium have decreased, over STD1. All of the rates in Table 3 have also increased significantly, showing the increased biological activity from the higher nutrient uptake rate, but the f-ratio and q are about the same.

Simulation DCM4 examined what effect a non-multiplicative form of photosynthesis would have on the DCM at Bermuda. The photosynthesis term was changed to $V_p \cdot \min(\overline{J}/V_p, Q)P$, from $\overline{J}QP$. Also, the density-dependent mortality of DDM1 was used. The terms $\overline{J}$ and Q each used the same parameter values as STD1, and the mortality parameters were the same as DDM1. It was thought that at greater depths the model would be light-limited, and this form would allow for more photosynthesis to occur because $V_p \cdot \min(\overline{J}/V_p, Q) \geq \overline{J}Q$ for $\overline{J}/V_p$ and Q both less than one. When the model is light-limited, nutrient is taken from the nitrate and ammonium pools by the ratios Q_1/Q and Q_2/Q, respectively.

This simulation did not produce a better DCM at Bermuda (see Figure 5f) because at the depths of the model's DCM, the nutrient uptake term, Q is nearly

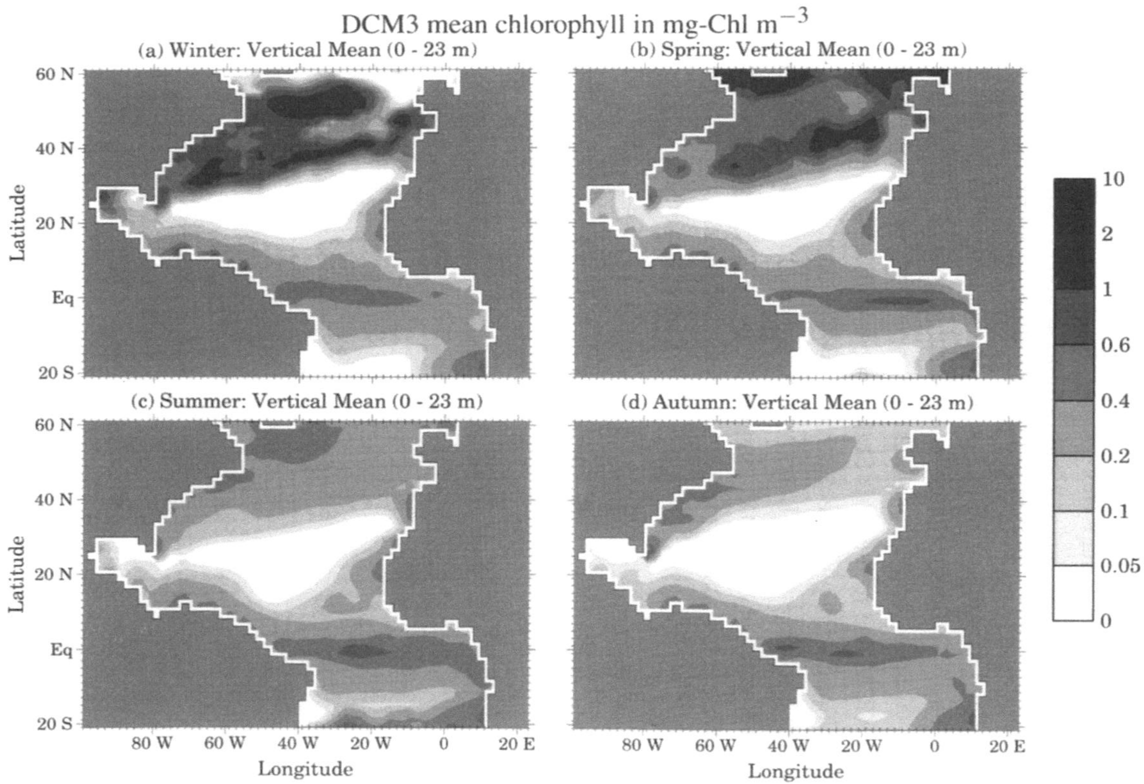

Figure 11: Seasonal mean chlorophyll concentrations from simulation DCM3 from model phytoplankton standing crop for the upper 23 m (2 layers) in mg-Chl m^{-3}.

one, and the two forms for photosynthesis are nearly equal. The only changes are near the surface, where there is somewhat higher primary production, and the seasonal surface chlorophyll maps (not shown) are very similar to DDM1 but with slightly higher values in the summer and autumn at higher latitudes. Figure 5e,f shows very similar results for DDM1 and DCM4 at Bermuda. The living standing crops in Table 2 and the rates in Table 3 are all several percent larger than DDM1, with the nutrients proportionately smaller, indicating more biological activity.

Simulation MLD1 attempted to address the erroneous mixed-layer depths produced in the OGCM due to the Gulf Stream separating from the U.S. coast too far north (see Figure 4). Notice the much too shallow depths at about 50°N, and too large depths just to the south. To try and rectify this, in MLD1 the convection for the biological variables was determined from the mixed-layer depths based on density (σ_t) differences from Levitus (1982) instead of the model predictions.

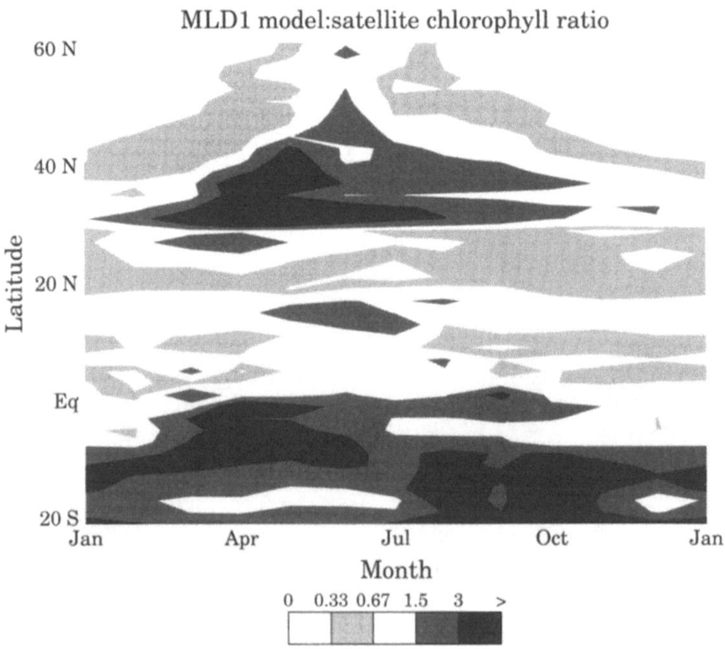

Figure 12: Zonal mean of the ratio of model (upper 23 m) to satellite chlorophyll for simulation MLD1.

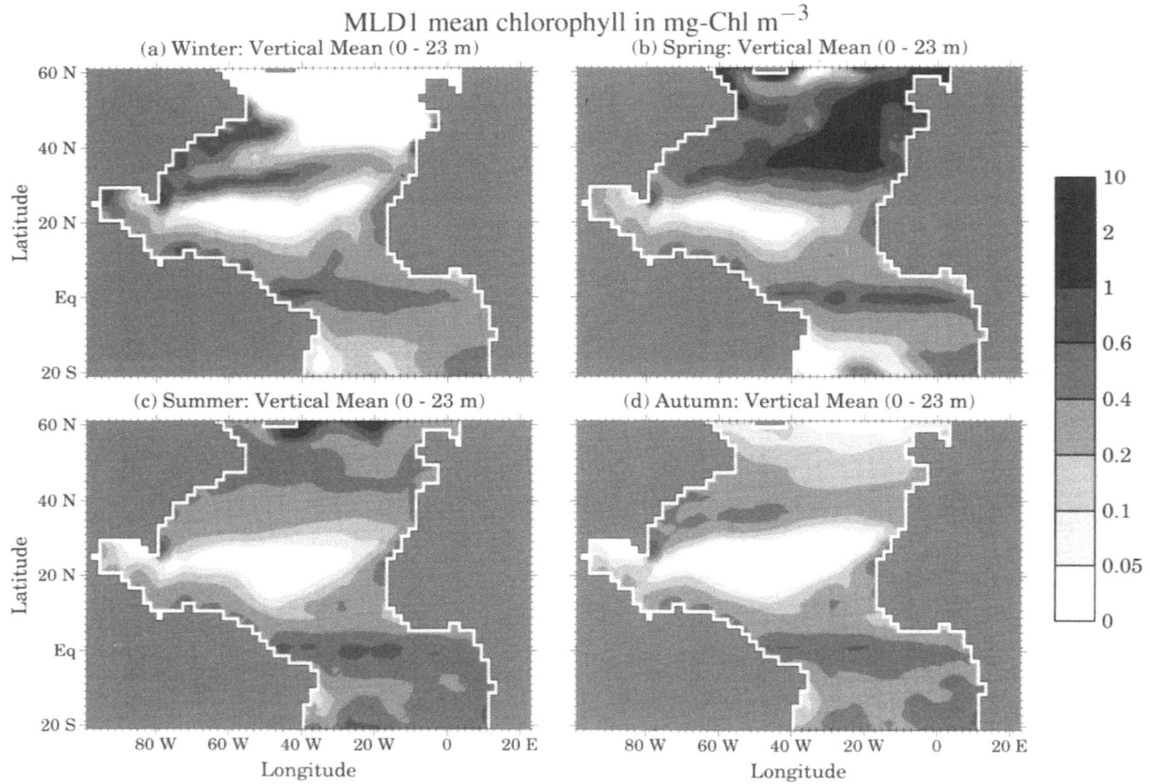

Figure 13: Seasonal mean chlorophyll concentrations from simulation MLD1 from model phytoplankton standing crop for the upper 23 m (2 layers) in mg-Chl m^{-3}.

Figures 12 and 13 give the chlorophyll results. The bloom was pushed back about two months at the high latitudes (the high ratios in Figure 12 show the peak of the model bloom), and was lower in magnitude. The trough of low production has gone (see Figure 3), but the wintertime chlorophyll was worse than in STD1. At latitudes around Bermuda the simulation was very similar to the standard model. The eastern part of the gyre, however, did get significantly more phytoplankton than in STD1 (Figures 3b and 13b). Overall, the standing crops were all up, especially the living forms and nitrate. Primary production was up by about 10%, mainly from an increase in new production and this simulation gave the highest value for the supply of nitrate.

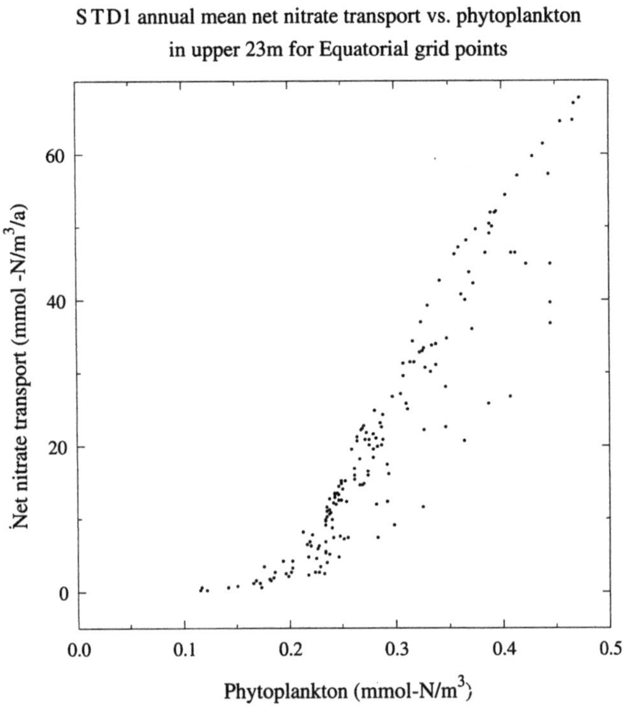

Figure 14: Annual mean plot of net nitrate transport into the upper 23 m versus mean phytoplankton in the upper 23 m for all grid points within 5° of the equator.

The next two experiments examine the model results at the equator. Comparing figures 2 and 3 at the equator, we see that simulation STD1 differed greatly from the satellite observations. The magnitude of chlorophyll concentrations was fairly good in the eastern part of the basin (except along the coast where the model does not resolve the upwelling), but was far too high in the west. If we plot the net nitrate transport (the sum of all advection, diffusion and convection terms) versus the phytoplankton concentration, for the upper 23 m (Figure 14), we find a reasonably good correlation, indicating that it is the nutrient supply that is the main determinant of surface phytoplankton distributions. This implies that the nutrient supply was too high in the equatorial region, which might be because either the equatorial upwelling was coming from too great a depth where nitrate concentrations were high, or organic matter was being regenerated at too shallow a depth and then being advected back to the surface. However, Sarmiento (1986) has shown a similar problem at the equator for the OGCM's heat flux. This leads us to believe that the problems at the equator are probably caused by the circulation.

Simulation EQ1 attempted to fix upwelling water from too great a depth at the equator (Sarmiento *et al.*, 1993). It was thought that perhaps the OGCM had diverged too far from observations, and that a short spin up from rest, initialized with the observed temperature and salinity fields, would allow for a better circulation. This had very little effect on the results. There was a smaller zooplankton population and lower primary production and nitrate supply, but no improvement to the equatorial results. There was some small change at high latitudes, but the circulation at the equator came into equilibrium very quickly in the near surface waters, thus not allowing much change in the biology.

In simulation EQ2 we wanted to see the effects of lower wind-forcing of the circulation to try to reduce the upwelling at the equator. Harrison (1989) has shown that winds at the equator are lower than those specified by Hellerman and Rosenstein (1983); this simulation was run in order to examine how such a change in wind stress would affect the simulation. We reduced the wind stress uniformly by 30%. This did produce significant changes, but they were more a uniform reduction over the whole basin, as can be seen in Figure 15, than an improvement for specific trouble spots. Table 4 shows a 24% reduction in the supply of nitrate to the photic layer. Zooplankton decreased by almost half, which brought down the nutrient recycling and therefore the regenerated production. The high latitudes were not as much affected because there most of the nitrate transport was due to convection rather than advection.

Finally, simulation IC1 used a different nitrate data set for the initial conditions based on a larger dataset than that of Sarmiento and Kawase (1986) (Levitus *et*

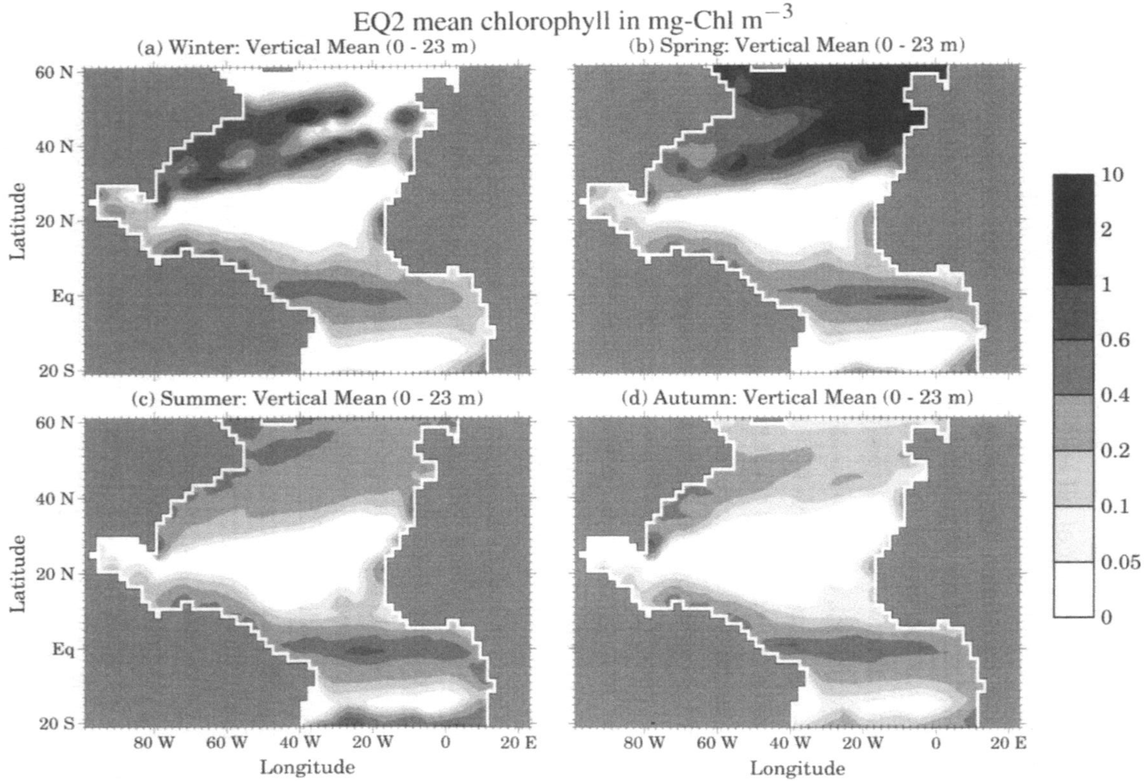

Figure 15: Seasonal mean chlorophyll concentrations from simulation EQ2 from model phytoplankton standing crop for the upper 23 m (2 layers) in mg-Chl m^{-3}.

al., 1992). As we would hope, the results were very similar to STD1. The major change was in the drift of the solution during the third year: it was 30% smaller than STD1, perhaps showing that the new data set was in better agreement with the circulation.

4. Conclusions

Several points have emerged from the results of these simulations. Perhaps the most significant was that the results did not change dramatically for the different parameterizations used. Qualitatively, the results depicted in Figures 3, 9, 13 and 15 all look very similar. The most significant differences occurred when changing parts of the advection and convection fields by using Levitus (1982) mixed layer data for convection of biology in MLD1, and reducing the wind stress in EQ2. Also a major revision of the biological model in DDM1 and DDM2, by using density-dependent mortality for phytoplankton and zooplankton, did give significant improvements at high latitudes.

Increasing the sinking velocity of particles to –100 m d^{-1} had little effect (STD3), whereas decreasing the sinking velocity to –1 m d^{-1} (along with increasing the phytoplankton mortality rate) allowed almost all of the nitrogen to be recycled in the photic layer, thereby lowering the f-ratio (see Table 3 for simulation STD2). In STD2 the vast majority of sedimentation was due to export by higher trophic levels (via zooplankton mortality) rather than the direct sinking of detrital particles produced *in situ* (DPON).

Simulations DCM1, DCM2, and DCM4 did not produce a DCM that more closely resembled the observations at Bermuda, however DCM3 did significantly deepen the DCM there. The higher initial slope for the P-I curve in DCM3 allowed for more nutrient uptake which let the phytoplankton exhaust the nutrients at shallower depths and forced them to move into deeper, more nutrient-rich water. Fasham *et al.* (1993) do not think that the lower nutrient uptake is the cause of the shallow DCM in the model, rather it is from too great a supply of nitrate in the wintertime, but this simulation does show that the model will produce a deeper DCM when the shallower nutrients are depleted. Simulation DCM4 assumed that the maximum growth rate in a nutrient-limited or light-limited situation was the same. However, that may not be a good assumption. Results from Sakshaug *et al.* (1989) seem to indicate that the maximum light-limited growth rate should be greater than the nutrient-limited growth rate (Armstrong, personal communication, 1993).

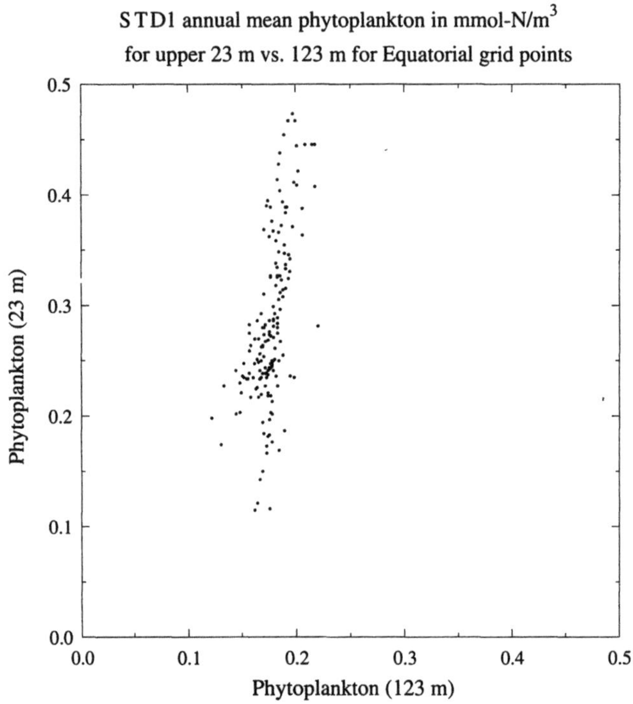

Figure 16: Annual mean phytoplankton concentrations for all grid points within 5° of the equator. The abscissa is the vertical mean for the upper 123 m, and the ordinate is the mean for the upper 23 m (what we assume that the satellite "sees").

The best simulation for high latitudes was DDM1 (see Figures 8 and 9). Phytoplankton standing crop predictions were better than STD1 when compared to CZCS chlorophyll measurements during the spring bloom (but still too high), and they were better for about three months after the bloom. It seems that by allowing a larger biomass of zooplankton to survive the winter we can delay the onset of the bloom, reduce its magnitude and spread the production over a longer period of time by means of increasing the recycling of nutrients within the photic zone.

The chlorophyll to nitrogen ratio that is used could be improved on. The Chl:N ratio is a function of light and nutrient concentration (Laws and Bannister, 1980; Sakshaug *et al.*, 1989). In principle, therefore, we could calculate a variable Chl:N ratio which should improve the comparison of model chlorophyll to data, whether it be CZCS determined or directly measured. However, at least for comparing to CZCS measurements, our judgement is that this would probably not be sufficient to account for all of the differences between models and data.

One change aimed at improving both the timing of the spring bloom in high latitudes and allowing production to occur throughout the summer would be to add multiple size classes to the zooplankton and phytoplankton. Then there could be a large, slow growing phytoplankton that is the major component of the bloom, and a smaller, faster growing plankton that would show up at the end of the bloom and might remain during the summer and fall (Ducklow and Fasham, 1992).

At the equator none of the simulations are satisfactory. In fact, those that showed an improvement in the high latitudes worsened the equatorial results. There are some fundamental problems in the OGCM that need improvement before we can really determine whether the biological model is in error (see Sarmiento, 1986, and Sarmiento *et al.*, 1993). One approach that may produce better results involves solar heating of the water column (Pacanowski, personal communication, 1992), allowing heat to be added deeper than just the surface. This may tend to destabilize the water column and allow for more mixing. The model's subtropical gyre, another problem area, is much too low in chlorophyll (see Figures 2 and 3). It is an area of convergence, so there is very little vertical transport of nutrient. Adding solar heating here may also allow for more nutrient to be mixed into the surface layers.

Attempting to analyze this model shows the great need for more global and temporal biological data sets. The best candidate for this is the CZCS-satellite measurements, but there are potential problems with this in that the satellite only sees the upper levels of the ocean. Figure 16 plots mean phytoplankton in the upper 23 m (a reasonable approximation of what the CZCS would see) versus mean phytoplankton in the upper 123 m, at all grid points within 5 degrees of the equator. The mean phytoplankton in the whole photic layer is quite uniform but the surface values vary by a factor of four. This means that satellite measurements may not be able to tell enough of what is going on in the whole water column.

Table 1: Description of simulations.

Simulation	Feature or parameterization addressed	Description
STD1	Standard simulation	Sinking velocity, $w_s = -10$ m d^{-1}, P mortality rate, $\mu_1 = 0.04$ d^{-1}, Z excretion rate, $\mu_2 = 0.1$ d^{-1}, and Z mortality rate, $\mu_5 = 0.05$ d^{-1}.
STD2	Sinking velocity of DPON	Same parameters as STD1, but $w_s = -1$ m d^{-1} and $\mu_1 = 0.08$ d^{-1}.
STD3	Sinking velocity of DPON	Same parameters as STD1, but $w_s = -100$ m d^{-1} and $\mu_1 = 0.035$ d^{-1}.
TDP1	Temperature dependence of photosyntesis	Same as STD1 but without the temperature dependence in the photosynthesis. The maximum photosynthetic growth rate, V_p, is modified from ab^{cT} where $a = 0.6$ d^{-1}, $b = 1.066$, $c = 1.0$°C^{-1}, and T is temperature, to a constant value of 2.9 d^{-1}. This is equivalent to a constant temperature of 24.65°C.
DDM1	Form of mortality for phytoplankton and zooplankton	Same as STD1 but with a density-dependent (Steele and Henderson, 1992) type mortality used for phytoplankton and zooplankton. This replaces μC by $\mu C^2/(k + C)$, where $\mu_p = 0.05$ d^{-1} and $k_p = 0.2$ mmol m^{-3} for phytoplanton and $\mu_z = 0.3$ d^{-1} and $k_z = 0.2$ mmol m^{-3} for zooplankton. The term for zooplankton replaces the two terms for mortality and excretion, and is divided into N_r, N_d, and export from the photic layer by factors of 0.7, 0.2 and 0.1, respectively.

Table 1: (continued)

DDM2	Form of mortality for plankton and sinking velocity	Same as DDM1 but with $w_s = -5$ m d^{-1}.
DCM1	Deep chlorophyll maximum at Bermuda	Same as STD1 but with $\mu_1 = 0.02$ d^{-1}.
DCM2	Deep chlorophyll maximum at Bermuda	Same as STD1 but with the grazing constant, g, reduced from 1.0 to 0.9 d^{-1}.
DCM3	Deep chlorophyll maximum at Bermuda	Same as STD1 but with α doubled from 0.025 to 0.05 d^{-1} (W m^{-2})$^{-1}$.
DCM4	Deep chlorophyll maximum at Bermuda	Same as DDM1 but with a non-multiplicative form of the photosynthetic uptake term. We used $V_p \cdot \min(\overline{J}/V_p, Q)P$ here instead of $\overline{J}QP$. If photosynthesis is found to be light-limited, then the uptake will be divided for nitrate and ammonium based on the ratios of Q_1 and Q_2.
MLD1	Erroneous mixed layer depths at high latitudes	Same as STD1, but using the climatological monthly mixed layer depths calculated by Levitus (1982) to determine the depth of convective mixing for the biological components.
EQ1	Divergence of model circulation at equator from equilibrium with temperature and salinity data	Same as STD1 but with the model circulation started from rest.
EQ2	Too high equatorial upwelling	Same as EQ1 but with the wind-stress reduced by 30%.
IC1	Initial conditions for nitrate	Same as STD1 but using the Levitus *et al.* (1992) nitrate data set as the initial conditions for nitrate.

Table 2: Annual mean standing crops in the photic layer (upper 123 m) for the complete model domain, in mmol-N m^{-3}. See Figure 1 and text for an explanation of the terms.

Simulation	P	Z	B	N_n	N_r	N_p	N_d
STD1	0.131	0.101	0.114	4.185	0.463	0.065	0.012
STD2	0.104	0.242	0.176	5.125	1.125	0.145	0.019
STD3	0.127	0.072	0.049	3.915	0.210	0.007	0.010
TDP1	0.132	0.107	0.116	4.128	0.468	0.066	0.012
DDM1	0.135	0.121	0.120	4.255	0.524	0.063	0.012
DDM2	0.146	0.154	0.158	4.446	0.749	0.102	0.014
DCM1	0.166	0.125	0.114	4.212	0.480	0.059	0.013
DCM2	0.145	0.095	0.121	4.073	0.453	0.068	0.011
DCM3	0.143	0.124	0.118	3.588	0.410	0.068	0.013
DCM4	0.139	0.127	0.124	4.012	0.458	0.065	0.013
MLD1	0.148	0.112	0.120	4.419	0.478	0.069	0.012
EQ1	0.130	0.094	0.111	4.039	0.457	0.063	0.011
EQ2	0.121	0.057	0.092	3.445	0.405	0.049	0.010
IC1	0.129	0.100	0.112	4.280	0.463	0.064	0.012

Table 3: Annual mean biology rates for the photic layer (upper 123 m) for the complete model domain in mmol-N m^{-3} a^{-1}. See Figure 1 and text for an explanation of the terms.

Simulation	$\overline{J}Q_1P$	$\overline{J}Q_2P$	G_1	G_2	G_3	U_1	U_2	μ_1P^*	μ_2Z^*	μ_3B	μ_4N_p	μ_5Z^*	$w_s\partial N_p/\partial z$
STD1	3.249	4.327	5.194	1.750	0.488	2.460	1.465	1.920	3.689	2.090	1.185	1.845	2.102
STD2	2.527	8.823	7.707	5.338	4.677	5.411	3.223	3.047	8.839	3.220	2.654	4.420	0.145
STD3	3.433	3.051	4.450	0.845	0.006	1.107	0.656	1.621	2.630	0.898	0.135	1.135	2.806
TDP1	3.342	4.531	5.450	1.871	0.544	2.552	1.519	1.934	3.908	2.111	1.211	1.954	2.143
DDM1	3.216	5.016	6.448	2.441	0.637	2.953	1.767	1.294	7.095	2.200	1.152	-	1.885
DDM2	3.075	6.656	7.726	3.906	1.672	4.308	2.580	1.442	9.920	2.883	1.871	-	1.221
DCM1	3.254	5.189	6.672	2.018	0.497	2.621	1.567	1.209	4.566	2.087	1.083	2.283	1.924
DCM2	3.327	4.214	4.957	1.608	0.419	2.450	1.459	2.117	3.470	2.209	1.233	1.735	2.207
DCM3	3.750	5.208	6.236	2.184	0.718	2.786	1.663	2.082	4.529	2.150	1.239	2.265	2.407
DCM4	3.384	5.377	6.894	2.641	0.715	3.119	1.864	1.342	7.630	2.256	1.186	-	2.001
MLD1	3.722	4.648	5.618	1.976	0.663	2.685	1.588	2.165	4.091	2.188	1.262	2.045	2.302
EQ1	3.121	4.071	4.843	1.608	0.437	2.335	1.390	1.901	3.433	2.033	1.144	1.717	2.039
EQ2	2.441	2.723	3.075	0.917	0.216	1.668	0.990	1.773	2.082	1.686	0.902	1.041	1.705
IC1	3.178	4.276	5.111	1.737	0.481	2.428	1.445	1.887	3.653	2.054	1.171	1.826	2.066

*For simulations DDM1 and DDM2 the phytoplankton mortality, zooplankton mortality and zooplankton excretion have density dependent forms and are described in Table 1.

Table 4: Derived quantities: annual mean results for the photic layer (upper 123 m) for the complete model domain. See Figure 1 and text for an explanation of the terms.

Simulation	Primary production	f	q	Chlorophyll	Nitrate supply	Sedimentation	Non-nitrate loss	change in nitrogen
	$g\text{-}C\ m^{-2}\ a^{-1}$			$g\text{-}Chl\ m^{-3}$	$mmol\text{-}N\ m^{-3}\ a^{-1}$			
STD1	74	0.43	0.27	0.21	3.463	2.713	0.484	0.265
STD2	110	0.22	0.28	0.17	2.999	1.603	0.778	0.616
STD3	63	0.53	0.27	0.20	3.572	3.242	0.176	0.155
TDP1	77	0.42	0.26	0.21	3.534	2.789	0.503	0.242
DDM1	80	0.39	0.17	0.21	3.430	2.595	0.568	0.269
DDM2	95	0.32	0.16	0.23	3.345	2.213	0.765	0.368
DCM1	83	0.39	0.15	0.26	3.480	2.679	0.525	0.276
DCM2	74	0.44	0.30	0.23	3.523	2.781	0.494	0.245
DCM3	88	0.42	0.25	0.23	3.889	3.156	0.554	0.177
DCM4	86	0.39	0.16	0.22	3.565	2.764	0.571	0.231
MLD1	82	0.44	0.28	0.24	3.956	2.979	0.690	0.288
EQ1	70	0.43	0.28	0.21	3.392	2.605	0.464	0.322
EQ2	50	0.47	0.37	0.19	2.627	2.051	0.349	0.225
IC1	73	0.42	0.27	0.21	3.324	2.669	0.471	0.185

ACKNOWLEDGMENTS

Our thanks to R. Armstrong and S. Carson for comments on the manuscript. Also, thanks to H. Ducklow, J. R. Toggweiler and G. Evans for many discussions during the course of this work. J. Olszewski prepared Figures 14 and 16. R. D. S. and J. L. S. were supported by the National Science Foundation Joint Global Ocean Flux Program (OCE 90-12333) and the U. S. Department of Energy uner contract DEFG 02-90ER61052. M. J. R. F. was supported by U. S. Department of Energy contract DEFG 02-90ER61052. The support of GFDL/NOAA through the generosity of K. Bryan and J. Mahlman is gratefully acknowledged.

REFERENCES

Altabet MA (1989) Particulate new nitrogen fluxes in the Sargasso Sea. J Geophys Res 94:12771–12779

Ducklow HW, Fasham MJR (1992) Bacteria in the greenhouse: modeling the role of oceanic plankton in the global carbon cycle. In: Mitchell R (ed) Environmental Microbiology. Wiley-Liss, New York. pp 1–31

Evans GT, Parslow JS (1985) A model of annual plankton cycles. Biol Oceanogr 3:327–347

Eppley RW (1972) Temperature and phytoplankton growth in the sea. Fish Bull 70:1063–1085

Fasham MJR, Ducklow HW, McKelvie DS (1990) A nitrogen-based model of plankton dynamics in the oceanic mixed layer. J Mar Res 48:591–639

Fasham MJR, Denman K, Brewer PG (1991) The Joint Global Ocean Flux Study: goals and objectives. In: Corell RW, Anderson, PA (eds) Global Environmental Change. Springer-Verlag, Berlin. pp 245–257

Fasham MJR (1992) Modelling the marine biota. In: Heimann M (ed) The Global Carbon Cycle. Springer-Verlag, Berlin. in press

Fasham MJR, Sarmiento JL, Slater RD, Ducklow HW, Williams R (1993) A seasonal three-dimensional ecosystem model of nitrogen cycling in the North Atlantic euphotic zone: a comparison of the model with observations from Bermuda Station "S" and OWS "India". Global Biogeochem Cycles, in press

Feldman G, Kuring N, Ng C, Esaias W, McClain C, Elrod J, Maynard N, Endres D, Evans R, Brown J, Walsh S, Carle M, Podesta G (1989) Ocean color, availability of the global data set. Eos Trans AGU 70:633–648

Harrison DE (1989) On climatological monthly mean wind stress and wind stress curl fields over the world ocean. J Climate 2:57–79

Hellerman S, Rosenstein M (1983) Normal monthly wind stress over the world ocean with error estimates. J Phys Oceanogr 13:1093–1104

Laws EA, Bannister TT (1980) Nutrient- and light-limited growth of *Thalassiosira fluviatilis* in continuos culture, with implications for phytoplankton growth in the ocean. Limnol Oceanogr 25:457–473

Kawase M, Sarmiento JL (1986) Nutrients in the Atlantic thermocline. J Geophys Res 90:8961-8979

Levitus S (1982) Climatological atlas of the world ocean. NOAA Prof Pap 13, US Govt Print Office, Washington, DC 173 pp

Levitus S, Reid J, Conkright ME, Najjar RG, Mantyla A (1992) The distribution of phosphate, nitrate and silicate in the world ocean. Prog Oceanogr, in press

Martin JH, Knauer GA, Karl DM, Broenkow WM (1987) VERTEX: carbon cycling in the Northeast Pacific. Deep-Sea Res 34:267–285

Menzel DW, Ryther JH (1960) The annual cycle of primary production in the Sargasso Sea off Bermuda. Deep-Sea Res 6:351–367

Sakshaug E, Andersen K, Keifer DA (1989) A steady state description of growth and light absorption in the marine planktonic diatom *Skeletonema costatum*. Limnol Oceanogr 34:198–205

Sarmiento JL (1986) On the North and Tropical Atlantic heat balance. J Geophys Res 91:11677–11689

Sarmiento JL, Slater RD, Fasham MJR, Ducklow HW, Toggweiler JR, Evans GT (1993) A seasonal three-dimensional ecosystem model of nitrogen cycling in the North Atlantic Euphotic Zone. Global Biogeochem Cycles, in press

Steele JH, Henderson EW (1992) The role of predation in plankton models. J Plankton Res 14:157–172

Wroblewski JS (1977) A model of phytoplankton plume formation during variable Oregon upwelling. J Mar Res 35:357–394

DATA ASSIMILATION FOR BIOGEOCHEMICAL MODELS

Joji Ishizaka
National Institute for Resources and Environment
16-3 Onogawa, Tsukuba, Ibaraki, 305
Japan

1. Introduction

Following early modelling attempts by Sverdrup, Riley, Steele and others during the 1940s to 1960s, many ecosystem models for a variety of marine environments have been developed during the last 4 decades. These ecosystem models represent an evolution from simple models with 1 or 2 dimensions to more complex models with many dimensions. Recently three-dimensional, time-dependent ecosystem models have been developed (Walsh, 1988; Sarmiento et al., 1989). To some extent ecosystem model development has been hindered because biological and chemical data from coastal and ocean ecosystems are usually limited in time and space coverage. Continuous monitoring systems (Armstrong et al., 1967) supply more complete coverage of variables, such as temperature, salinity, chlorophyll fluorescence, and nutrients; however, these data are usually still limited by ship location.

The development of satellite-borne ocean colour sensors allows nearly synoptic measurement of a biological variable, surface chlorophyll concentration, with a minimum resolution of about 1 km and 1 day with 3-day coverage of the global ocean (Hovis et al., 1980; Gordon et al, 1980). The vast amount of data obtained from this sensor has revealed for the first time the complex nature of the space and time dependence of the surface phytoplankton pigment distributions. The availability of ocean colour data has opened the possibility of quantitative evaluation of the results of ecosystem models and introduced the possibility of improving model results through assimilation of these measurements into the models (Abbott, 1992). Also, ecosystem models that assimilate ocean colour

NATO ASI Series, Vol. I 10
Towards a Model of Ocean
Biogeochemical Processes
Edited by G. T Evans and M. J R Fasham

measurements can be used as dynamical interpolators to fill gaps between satellite observations. The general use of satellite-derived pigment fields with ecosystem models is discussed by Ishizaka and Hofmann (In press).

Data assimilation techniques have been successfully applied in meteorology and are routinely used with operative weather forecast models. More recently these techniques have been used with physical oceanographic circulation models. However, for biogeochemical models, data assimilation techniques have been only rarely applied, and there is uncertainty as to how appropriate these techniques are for these models. At this time, it might be argued that biogeochemical models are not mature enough to incorporate data assimilation techniques to predict biological or chemical variables. However, attempting data assimilation with the present models is a useful exercise that will expand our understanding of the models and of the ecosystem.

There are many areas of biogeochemical modelling that could benefit from the development of data assimilation techniques. For example, predicting the pCO_2 fields at the surface of the ocean (which is a quantity closely related to biological activity in the ocean) is needed because the CO_2 budget in the ocean is important to understand the global warming problem. Estimating global primary production is also important, and many studies suggest that it can be estimated from satellite ocean colour data (e.g. Platt & Sathyendranath, 1988; Sathyendranath et al., 1991). However, satellite phytoplankton pigment concentration data alone is not sufficient for this purpose for many reasons (e.g. cloud cover). An ecosystem model capable of assimilations satellite ocean colour data could be used to dynamically interpolate gappy satellite data. Predictions of red tides and low oxygen waters in coastal areas may also be possible with data assimilative ecosystem models.

In this article, data assimilation approaches that are used in meteorology and physical oceanography are described briefly. Next, two examples (Ishizaka, 1990a; Ishizaka, in prep.) that illustrate the possibilities and questions associated with applying data assimilation methods to ecosystem models are given.

2. Data Assimilation in Meteorology and Physical Oceanography

It is not my intention to review the entire history and details of the data assimilation techniques used in meteorology and physical oceanography circulation models. These subjects are covered in detail by others (e.g. Haidvogel & Robinson, 1989; Anderson & Willebrand, 1989). However, it is useful to briefly describe the data assimilation approaches that are used in meteorology and physical oceanography so that differences between the models used in these fields and biogeochemical models can be highlighted.

One straightforward definition which comes from the final purpose of data assimilation, is improvement of model predictions by using real data (cf. weather prediction). There are three different conditions which have to be specified for numerical model; initial conditions, boundary conditions, and internal parameter values. It is possible to specify all of these conditions with real observation data. Real observation data can be "assimilated" into models implemented for limited domains as "realistic" boundary conditions. For many ecosystem models, most of the parameters are "assimilated" from experimental data, and unknown

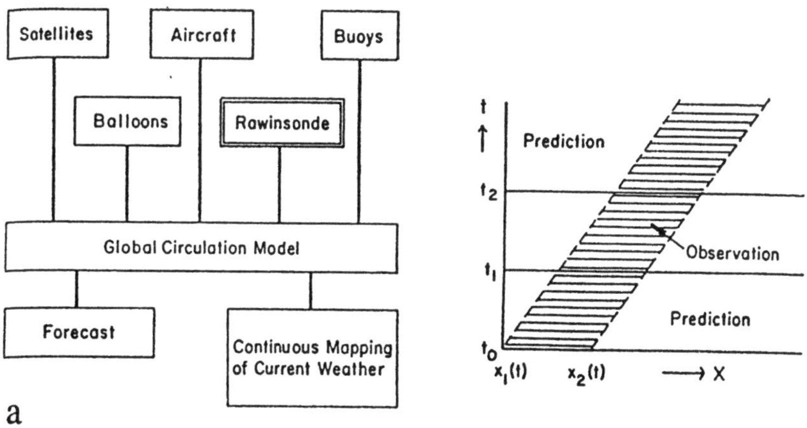

Fig. 1. Concepts of data assimilation (Kasahara, 1972). a) The prediction model (global circulation model) serves as an integrator of various observed data. b) Data from different locations are mixed with prediction by the model.

parameters are often estimated by adjusting the parameters so that the simulated distributions match field distributions (e.g. Hofmann and Ambler, 1988). However, the methods for estimating the initial conditions with real observation data are usually called "data assimilation." Model can be reinitialized whenever the data for the initialization are available; in other words, the model results can be upgraded by real data. This process seems to be easy, but as will be clear in the later description, it is not as easy as we think at first.

Techniques for using data to improve model prediction were first given serious attention by meteorologists during the Global Atmospheric Research Program (GARP) / First GARP Global Experiment (FGGE), which took place in the late 1960s-70s. This experiment resulted in vast amounts of meteorological data, satellite data in particular, becoming available. Although the data from this program were extensive, they had different accuracies and characteristics, and even the satellite data were not completely synoptic. Thus, techniques for 4-dimensional data assimilation were needed in order to combine the various data with time-dependent 3-dimensional global circulation models. Thus, data assimilation techniques have their beginning in attempts to integrate various data sets (with different accuracies and characteristics) with models to construct an overall distribution of properties or circulation (Charney et al. 1969; Kasahara, 1972; Fig. 1). Furthermore, McPherson (1975) defined data assimilation as the absorption of real data into a model after adjusting the data so that the data and model match each other (mutual accommodation of observations and a model).

Obviously techniques for mapping were important in weather prediction. Thus, various mapping techniques (objective analysis) were developed. Initially, mapping was just a mechanical interpolation; however, statistical interpolation methods that used statistical constraints, such as optimal interpolation techniques (e.g. Gandin, 1965) were soon developed. Furthermore, diagnostic relationships from physical dynamics (e.g. geostrophic balance) were used to combine different data types. Finally, a prognostic model was used to create the maps.

The essence of mapping is the extraction of "real (important) information" from noisy and gappy data. The question is then what is the real (important) information in the measurements for meteorology. Besides measurement errors, meteorological data include phenomena that span a range of scales. Whether the data are information or noise depends on the scale of interest. For example, if the large global scale phenomena are of interest, then small scale turbulence or gravity waves are noise.

Prognostic models can be used for dynamical mapping of meteorological data, but at the same time, the model needs data for verification and improvement. Similar problems are associated with using data with the model. If the measurements include noise which is not consistent with the model dynamics, then such data may not be useful for model verification and improvement. For example, if measurements that include small scale gravity waves are used to initialize a large scale circulation model, then the gravity waves will propagate in the model and induce large errors in the model (initialization shock).

Thus, meteorological data assimilation techniques are focussed on extracting information from data which is consistent with the model dynamics and then using the data for further integration of the dynamical model. The easiest method for upgrading a predictive model with data simply replaces model variables with observations. However, this method usually does not work because of the initialization shock mentioned previously. Optimal interpolation can be used to smooth noisy 2-dimensional data with a certain length scale. For 4-dimensional data assimilation, time sequence analysis (e.g. Kalman filter), which is equivalent to the optimal interpolation in 4-dimensions with a error function based on the predictive model and observation, and variational assimilation techniques, such as adjoint methods, have been developed. The space and time frequency at which data are assimilated is an important aspect of these data assimilation techniques.

In physical oceanography, assimilation of satellite altimeter data into circulation model has been the most studied. However, assimilation of XBT/CTD has also been studied. Although the techniques used for assimilation of physical oceanographic data are similar to those used for meteorological data, one large difference in the two is that much less physical oceanographic data is available. Most oceanographic data assimilation studies have been focussed on the equatorial area, not just because of ENSO events, but because the time scales of the physical phenomenon in this region are more clearly separated into "information" and "noise" than those at the mid-latitude (Leetma & Ming, 1989). This is another difference between meteorology and physical oceanography; the separation of time scales is easier in meteorology than in physical oceanography.

3. Real Data (Satellite Pigment) Experiments

Presently there are few examples of data assimilation, particularly reinitialization or upgrade of model results with observed data, in biogeochemical models. Najjar et al. (1992) recently used objectively mapped surface phosphate distributions to force the 3-dimensional global phosphate model coupled with a general circulation model. This model includes a simple phosphate uptake term which corresponds to new production (i.e. downward flux). They used the difference between the predicted model results and field data to estimate this simple biological term in the model.

Ishizaka (1990a) described a modelling study in which satellite-derived pigment data were assimilated into a 2-dimensional ecosystem model developed for the southeastern U.S. continental shelf. The physical portion of this model was, in a sense, a use of data assimilation because current meter measurements were optimally interpolated (Ishizaka, 1990b) to obtain the flow and temperature fields used in the model. The biological part of the model (Ishizaka, 1990c) contained four components; N (nutrient), P (phytoplankton), Z (zooplankton) and D (detritus). The detrital component is needed for mass balance, hence it does not represent true detritus. The model formulation is standard;

$$\frac{\partial B}{\partial t} + u\,\frac{\partial B}{\partial x} + v\,\frac{\partial B}{\partial y} - Kx\,\frac{\partial^2 B}{\partial x^2} - Ky\,\frac{\partial^2 B}{\partial y^2} = S + V \qquad (1)$$

where B is the concentration of each biological component, and x and y are the along- and across-shelf directions, respectively. The horizontal advection velocities in the x and y directions, u and v, were specified using the optimally interpolated fields. The horizontal eddy diffusivities, Kx and Ky, are assumed to be constant. The time rate of change of B by biological processes is given by S. This terms was formulated using saturating functions (Monod's equation for nutrient uptake and Ivlev's equation for grazing) and linear death terms. The second term on the right side, V, represents the effects of vertical advection and diffusion. This term was estimated from the difference between simulated and interpolated temperature distributions by assuming that the total nitrogen mass could be estimated from a conservative relationship between nitrogen and temperature. Detailed descriptions of the above model are given in Ishizaka (1990b, c).

Pigment distributions from the Coastal Zone Color Scanner (CZCS) were available for the southeastern U.S. continental shelf for 9 days in April 1980, and they were suitable for direct comparison with the model results and for assimilation into the model. Time integration of the model was started from constant initial conditions from 15 days before the first CZCS data, and boundary conditions for the model were specified as constant or temperature dependent. Model parameters were chosen from the field observation data.

Ishizaka (1990c) quantitatively compared the model results with satellite pigment data using root mean square errors and correlation coefficients. These statistics gave a measure of similarity between the magnitude and patterns in the model and those in the satellite data. Many simulations with different parameter sets and different boundary conditions were done. The simple statistics showed clearly the effect of

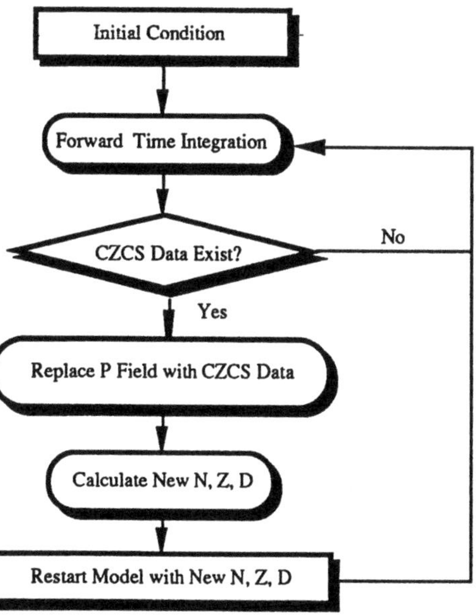

Fig. 2. Schematic of the procedure used for assimilation of Coastal Zone Color Scanner data into the physical-biological model described in Ishizaka (1990a). Phytoplankton, nutrient, zooplankton and detritus fields are indicated as P, N, Z and D, respectively.

variations in parameters on the ability to simulate phytoplankton fields observed in ocean colour measurements. One conclusion from these numerical experiments was that the phytoplankton patterns are controlled primarily by horizontal advection, but that the magnitude of the phytoplankton concentration is affected by biological processes and in enhanced by the shelf-edge upwelling on the southeastern U.S. continental shelf.

The above simulation model was used with ocean colour measurements to do a series of numerical experiments to test data assimilation techniques. Figure 2 shows a schematic of the data assimilation procedure used by Ishizaka (1990a). For this modelling study, the data assimilation technique consisted of a simple replacement of the simulated phytoplankton fields with the satellite-derived phytoplankton fields. The effects of the data assimilation procedure were evaluated by comparing the model results with a sequence of satellite observation that followed the day used for assimilation (April 10). In addition to simple replacement experiments, Ishizaka (1990a) also

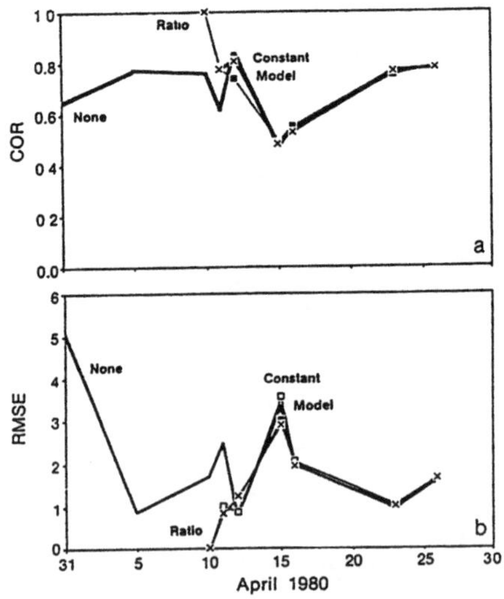

Fig. 3. a) Correlation and b) root mean square error (RMSE) between the model and satellite phytoplankton fields with data assimilation on April 10 (Ishizaka, 1990a). None indicates the simulation of phytoplankton distribution obtained with no data assimilation. The three different approaches used for upgrading of the ecosystem components other than phytoplankton are shown as ratio (cross), model (filled square), and constant (white square).

considered methods for updating other components of the ecosystem and assimilation of only partial phytoplankton fields.

In general, simulated phytoplankton fields became similar to the ocean colour measurements for several model days after the assimilation; however, the simulated phytoplankton fields quickly returned to the values obtained from the model with no assimilation (Fig. 3). The other components of the model ecosystem, i.e. nutrient, zooplankton, and detritus, also need updating to be consistent with the new phytoplankton fields. Three methods were tested for estimating values of the other three components that are consistent with the updated phytoplankton field. In the first method, only the phytoplankton fields were upgraded and the other fields were specified using the simulated distribution. The second method used constant nutrient and zooplankton fields, which were specified using the initial condition values of the model. The detritus fields were specified by assuming that nitrogen mass was conserved with temperature. The third method assumed that a ratio between nutrient, zooplankton, and detritus and total mass was conserved with the upgrading. The results obtained with these methods (Ishizaka, 1990a) did not show widely different phytoplankton distributions (Fig. 3). However, for other models the effects of the different methods may be larger as discussed in the next section.

The effect of partial upgrading was also tested by replacing only half of the phytoplankton fields with the satellite data. The effect of partial upgrading was about half of that upgrading the entire phytoplankton field. Computationally, a partial upgrade did not cause serious problems. For physical models, partial upgrading sometimes induces shock because gravity waves are induced by the discontinuity resulting from the data assimilation process. For an ecosystem model, it is unlikely that the discontinuity associated with partial upgrades will create spatial waves that will affect ecosystem behavior, although the possibility of biologically-induced waves exists (Okubo, 1980). In this sense, data assimilation with a biogeochemical model may be relatively easier than with a physical model.

Ishizaka (1990a) next assimilated a sequence of 8 satellite phytoplankton fields into the ecosystem model and used the resulting simulated phytoplankton distributions to estimate phytoplankton carbon and nutrient nitrogen fluxes across the continental shelf. The time change of flux based on the non-upgraded (free from observed data) and upgraded (forced by observed data) models showed similar patterns (Fig. 4). However, the phytoplankton carbon flux was reduced by about half, and the nutrient nitrogen flux was increased by a factor of two.

The reduction of phytoplankton flux occurred because the phytoplankton concentration was overestimated by the original model, and assimilation of satellite phytoplankton fields reduced this overestimation. Assimilation of phytoplankton concentrations successfully forced the model phytoplankton concentration close to the satellite value. This result indicated that the ecosystem model can be used for the

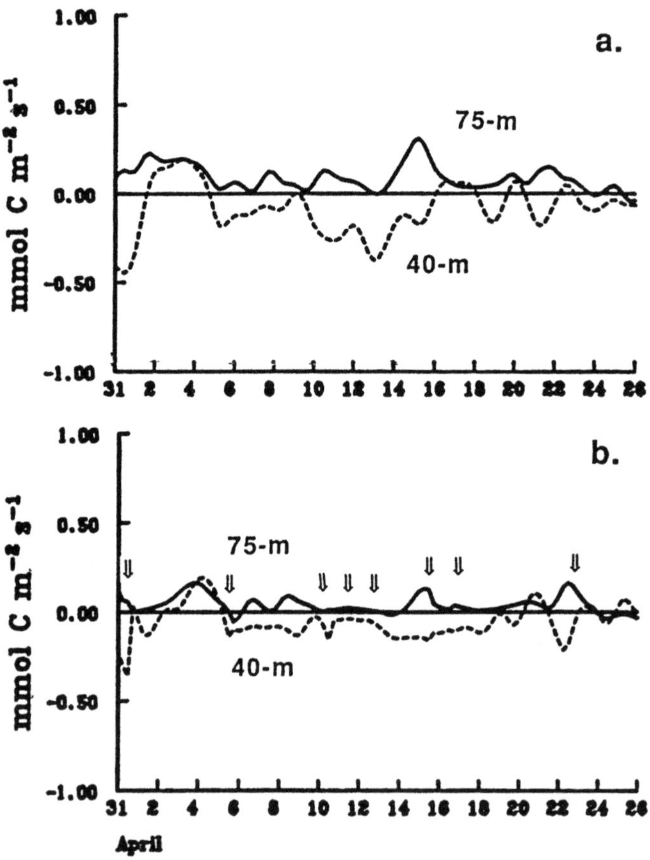

Fig. 4. Time change of phytoplankton carbon flux across the 40-m and 75-m isobaths of the southeastern U.S. continental shelf (Ishizaka, 1990a). Positive and negative values indicate the offshore and onshore flux, respectively. Model results with non-upgrading (a) and with upgrading by 8 CZCS data fields (b) are shown. Arrows indicate the time of assimilation.

dynamical interpolation of satellite phytoplankton data. On the other hand, the estimate of the nutrient concentration was unrealistically high after the data assimilation, and the nutrient nitrogen flux probably became worse than the results of non-upgrading model. This is because the crude estimation of upwelling term, which is the input of nutrient to the system, induced larger errors with the data assimilation.

It is important to notice that the simulated nutrient fields were worse after data assimilation of phytoplankton than without assimilation. Treatment of the other ecosystem components in the data assimilation should be also important, although the three methods tested by Ishizaka (1990a) for estimating other components did not show large differences in the final phytoplankton results. This point is addressed in the next section with updating experiments that use a simple time dependent biological model.

4. Spatially Uniform Model

As shown in the previous section, space- and time-dependent ecosystem models can be improved by assimilation of ocean colour data. However, uncertainty exists as to the effects of assimilation of real data into complicated simulation models. Simple time-dependent ecosystem models with no spatial heterogeneity provide a simple framework for investigating the effects of data assimilation. Also, time-dependent ecosystem models include the processes responsible for material transformations, which are essential in biogeochemical models, but are not usually included in atmospheric and physical oceanographic models.

One approach for examining the effects of data assimilation on a certain model is "identical twin experiments". Twin experiments have been used extensively to study data assimilation effects in atmosphere and ocean circulation models. In this study, a simple ecosystem model was implemented to test data assimilation with a twin experiment approach.

This method used one model as the "real" condition and two other models to simulate the "reality". First, the model was run with a standard parameter set and initial conditions; this model run can be thought as representing 'the real world'. Next, the model was run with a small amount of error included in the parameters or initial conditions; this model run can be thought as 'model of the world'. Finally, the 'model of the world' was run until a certain time step, at which 'real world' was observed and assimilated the observed data into 'model of the world' according to different possible rules and after which the simulation continued; this run can be called 'upgraded model of the world'. The question is how the accuracy (deviation from 'the real world') of the 'upgraded model' compares with that of the 'model of the world'.

The assimilation method used in this numerical experiment was simple replacement of the phytoplankton concentration with the value obtained from the first model. A simple model, with little representation to the real ocean, was used in this study because less complicated ecosystems make it easier to understand the basic response of the ecosystem model to data assimilation. The parameter values used in the model were also chosen for simplicity and ease of understanding.

Detail of the twin experiments and the results will be given in elsewhere (Ishizaka, in prep.), and these are briefly described here. Most simple data assimilation experiments were conducted with a batch culture model of phytoplankton in which nutrient-limited phytoplankton growth was modeled without any grazer for the phytoplankton and without additional nutrient source except the initial condition.

The batch culture model is formulated by well-known Michaelis-Menten type as:

$$\frac{dP}{dt} = \mu P = \mu_m \frac{N}{K_s+N} P \qquad (2)$$

$$\frac{dN}{dt} = -\mu P = -\mu_m \frac{N}{K_s+N} P \qquad (3)$$

where P and N are phytoplankton and nutrient concentrations, μ is the phytoplankton growth and nutrient uptake, μ_m is the maximum value of μ and K_s is the half saturation constant.

Figure 5 shows one of the results of upgrading of phytoplankton for the simulation of phytoplankton batch culture. In this case, error was included in the value of P_0, initial phytoplankton concentration. The model was upgraded at time=0.2, when phytoplankton growth is about half of the final condition, by resetting the phytoplankton concentration of model 3 (upgraded model) to the value from model 1 (real world) at the same time. At this time, the error of the phytoplankton concentration became zero. After the upgrade, phytoplankton concentration had little error, relative to the control case, during the exponential growth phase and reached a steady state concentration with small error. The nutrient concentration was not upgraded in this experiment, and the steady state nutrient condition (0) did not depend on error in the initial condition. However, the error of nutrient concentration relative to the control during the growth phase was reduced by data assimilation.

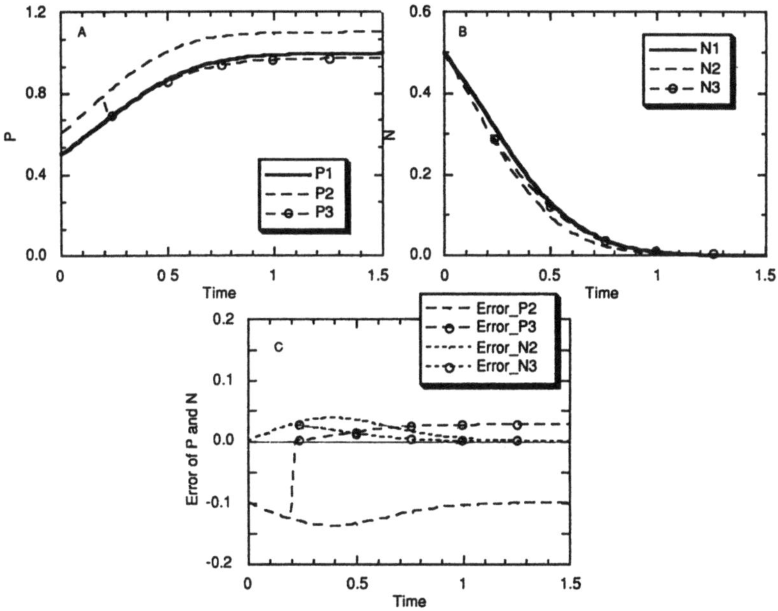

Fig. 5. Time change of the phytoplankton (a), nutrient (b), and error (c) from the batch culture simulation. Control case (no error) is shown as P1 and N1 (real world). Results of model with error in the phytoplankton initial condition are shown as P2 and N2 (model of the world). Results of the model with data assimilation are shown as P3 and N3 (upgraded model).

Table 1 summarizes the results of some numerical experiments with the batch culture model with different types of error sources and different types of data assimilation methods. Those cases in which the model simulations were improved by assimilation and resulted in a perfect match between the assimilated and control model are indicated by p. The cases in data assimilation gave better results than no assimilation is denoted by b. The cases in which the model results were worse or unchanged by data assimilation are shown by w and u, respectively. In some cases, errors in parameters or initial conditions did not cause errors in the phytoplankton or nutrient concentrations, and therefore data assimilation had no effect with model solution. These cases are indicated in Table 1 by (p). The error in the model solution was evaluated when the data were assimilated, after assimilation and before reaching a steady state, and at the steady state condition.

Table 1. Results of the data assimilation for the batch culture simulations. Numbers for the error sources are parameter values used for the simulations, and the error values for the parameters are in parenthesis. The assimilation methods are indicated by P_C and P_C=-N_C for upgrading of only phytoplankton concentration and for upgrading of both nutrient and phytoplankton, respectively. The time periods for evaluation were indicated by A, T, and S for the time of assimilation, the transient solution, and the steady state solution, respectively. Symbols are explained in the text.

Error Sources	Assimilation Methods		P			N	
		A	T	S	A	T	S
1. μ_m=3(+3)	P_C	p	b	w	u	b	(p)
	P_C=-N_C	p	b	(p)	p	b	(p)
2. K_s=0.5(+0.1)	P_C	p	b	w	u	b	(p)
	P_C=-N_C	p	b	(p)	p	b	(p)
3. P_0=0.5(-0.1)	P_C	p	b	w	u	b	(p)
N_0=0.5(+0.1)	P_C=-N_C	p	p	(p)	p	p	(p)
4. P_0=0.5(+0.1)	P_C	p	b	b	u	b	(p)
	P_C=-N_C	p	b	u	w	w	(p)
5. N_0=0.5(+0.1)	P_C	p	b	b	u	b	(p)
	P_C=-N_C	p	b	u	w	w	(p)

Two different types of data assimilation were used; upgrades of the phytoplankton concentration only (P_C) and upgrades of both the phytoplankton and nutrient concentrations (P_C=-N_C). When the nutrient concentration was also upgraded, it was assumed that the total mass of the phytoplankton and nutrient was not changed with the upgrading. This type of upgrading is an example of assimilation respecting the known dynamics of the system, an analogue of respecting geostrophic balance for physical circulation model. The two different assimilation approaches gave different results.

The effects of the upgrade were also dependent on the error sources of the 'model of the world'. Five cases of the different error sources were examined. For the cases 1 and 2, model parameters (μ_m and K_s) were in error. For the case 3, the partition (P_0:N_0) of the initial condition was in error. For the cases 4 and 5, the initial condition of phytoplankton and nutrient concentrations were in error (P_0, N_0). These five cases showed different but consistent effects of upgrade.

Upgrading the phytoplankton concentration resulted in a solution with relatively less error in this component after the assimilation for all the cases with both upgrading method. However, upgrading of cases 1, 2 and 3, of which the error sources were model parameters and partition of initial conditions, showed worse results when only the phytoplankton concentrations were upgraded (Pc). There were no errors for the total mass of these three cases, and the upgrading of only phytoplankton concentration caused the error in the total mass and resulted in the error of the steady state condition. Upgrading with mass conservation (Pc=-Nc) for these three cases resulted perfect matching between the 'real world' and 'model of the real world' which was also expected even without upgrading. When there were errors in the total mass of the 'model of the world' (cases 4 and 5), upgrading with total mass conservation (Pc=-Nc) resulted no improvement for the steady state. However, upgrading of only phytoplankton concentration (Pc) improved the steady state solution.

Upgrading did not change the steady state solutions of nutrient concentration which were zero for all the cases with and without upgrading. For the cases with no error of total mass (cases 1, 2 and 3), the upgrading with mass conservation (Pc=-Nc) improved the results, and the upgrading without mass conservation (Pc) also improved even when the upgrading did not directly change the nutrient concentration. The upgrading without mass conservation (Pc) for the cases with error in total mass (cases 4 and 5) similarly improved the nutrient concentration after the upgrading. However, the upgrading with mass conservation (Pc=-Nc) for these two cases made the solution slower to converge to the steady state. This is because the error of total mass in the initial conditions accumulated into the nutrient concentration by correcting the phytoplankton concentration.

The effects of data assimilation were also examined in a chemostat-type phytoplankton-nutrient model of the form:

$$\frac{dP}{dt} = (\mu\text{-}D)\,P \qquad (4)$$

$$\frac{dN}{dt} = -\,\mu P + D(U\text{-}N), \qquad (5)$$

where D is the dilution rate and U is the nutrient concentration in the reservoir. The phytoplankton growth rate, μ, is a Monod type expression which is shown in equations (2) and (3).

Overall, the results obtained from upgrading the chemostat model were similar to those obtained with the batch culture model. However, the primary difference was in the value obtained for the steady state solution. For the chemostat simulation, the steady state depends only on the parameter values and not the initial conditions. Thus, the steady state solutions were not affected by any upgrading technique used in this study. The results of upgrading when errors existed in the new parameter, D, were similar to those obtained for the other phytoplankton parameters.

The effect of upgrades on the other new parameter, U, were similar to those obtained for the phytoplankton parameters and the initial conditions at the steady state and at the transient solutions, respectively. In particular, the nutrient concentration became worse, after the upgrading of phytoplankton concentration by the methods with or without conservation of total mass. This parameter, U, is the source of nutrient in the system and similar to the upwelling term of the model of Ishizaka (1990a). The behaviors after the upgrades were also similar between this spatially uniform model and the 2-D model of Ishizaka (1990a). When the nutrient source parameter (or term) was overestimated, upgrading (reduction) of phytoplankton concentration slowed the transformation of nutrient to phytoplankton and resulted the accumulation of error (overestimation) in nutrient concentration.

The results of model upgrade became more complex as more components of the ecosystem were added. Inclusion of a simple zooplankton response added the complication that the timing of the data assimilation has an effect on the model results. This is because the error in the model moves from one component to other components through the food chain. Thus, one expects a more complex model response to the upgrading when there are more components of the food web.

These results show that assimilation of phytoplankton data into a simple time-dependent model is not straightforward. Differences in error sources, methods of assimilation, and the timing of the assimilation all result in different model solutions when data assimilation is used. From the batch culture experiments, it is clear that the suitable methods of assimilation depends on the error sources. In the simulation of the real world, what causes the error is not certain and often there are many sources of errors, including observation itself. Furthermore, much complex ecosystem structure is preferable for the real simulation and increases uncertainty of the effects of the data assimilation.

5. Discussion

This study gives a brief introduction to data assimilation techniques used in meteorology and physical oceanography and provides two examples of data assimilation for biogeochemical models. The application of assimilation techniques to biogeochemical models is not well developed and many issues on this area remain to be resolved.

Correct representation of spatial and temporal scales is an important aspect of the data assimilation procedures used in meteorological and oceanographic circulation models. The success of a particular data assimilation procedure depends largely on information at the correct scale from the data. For biogeochemical problems, events (e.g. mixing) that occur at short space and time scales are important (e.g. Goldman, 1988). How such events can be treated in data assimilation techniques for biogeochemical models is uncertain.

There are other problems specific to the assimilation of data into biogeochemical models. For example, the formulations for expressions of the biological model are to a large extent uncertain. No standard formulations exist for mathematical expressions of interactions in multi-species communities or for closure terms, such as mortality and grazing. Some approaches to these problems are given in other chapters of this book. Each parameter value in a biogeochemical model has some uncertainty associated with it. It is almost impossible to select a set of biological parameter values that can apply to the global ocean because of uncertainties in understanding of multi-species responses and of physiological factors. Because of these uncertainties, one might think that it is impossible to use data assimilation with models to make predictions of biogeochemical cycles. However, these uncertainties may be a strong reason to investigate data assimilation techniques for biogeochemical models because data assimilation studies provide a framework for determining the "best" set of model and field data that do not contradict each other.

The sample models given in this study indicate that assimilation of satellite derived phytoplankton distributions can significantly improve predictions of those fields from model. The three day coverage of the global phytoplankton distribution that is provided for sensors such as SeaWiFS and OCTS is probably sufficient for assimilation into a global ecosystem model. This opens the possibility of predicting the global primary production fields in the ocean. However, the simple models used

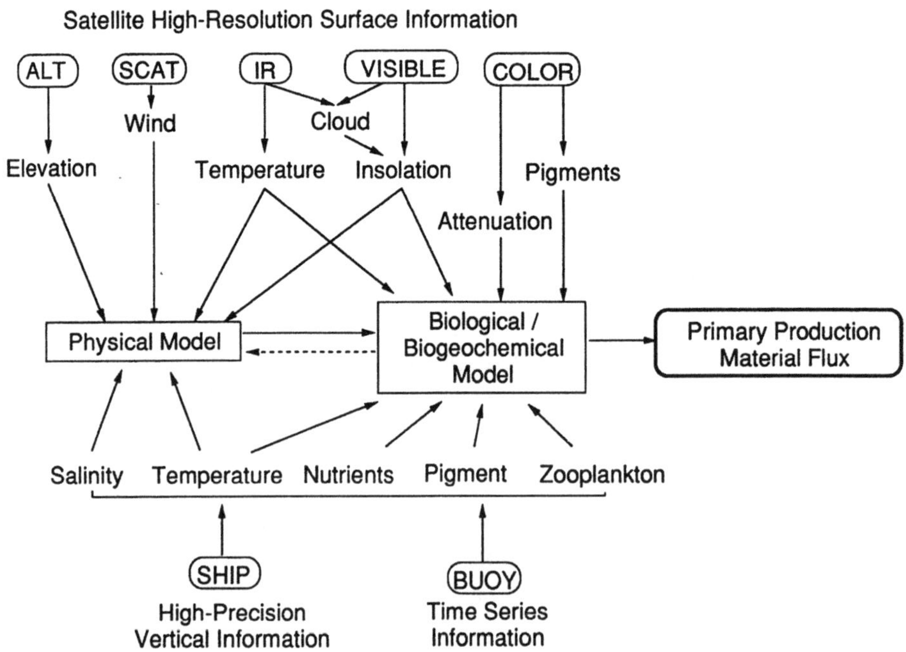

Fig. 6. Possible future data assimilation of various data with biogeochemical models.

in this study also showed the difficulty and necessity of improving the other components of the ecosystem model, such as nutrients and zooplankton.

An additional problem arises in assimilation of satellite ocean colour data into a three-dimensional biogeochemical model. Satellite ocean colour data contain information averaged from the surface to one optical depth. The problem then arises as to how to extract vertical information from the averaged satellite data. Platt and Sathyendranath (1988) used empirical (diagnostic) relationship vertical profiles to estimate primary production from satellite-derived ocean colour data. Empirical relationships also exist between nutrients and temperature (Zentara & Kamykowski, 1976). These empirical relationships may be useful for constraining data assimilation methods and for projecting information to deeper parts of the euphotic zone.

As shown, the effects of data assimilation depend on the sources of error, the assimilation methods, and the timing of the assimilation. Assimilation of only phytoplankton measurements is probably not enough to adjust all of the model ecosystem components. This is a discouraging result for the prediction of quantities other than phytoplankton, such as pCO_2. It is likely that prediction of such quantities will require assimilation of other data sets into biogeochemical models. These other data may not be available on a global scale as will be the ocean colour data. Hence, methods that can combine different data sets with different type and frequency will need to be developed.

Although spatial coverage is poor, moored buoy data such as bio-optical measurements, acoustic zooplankton counts (Flagg & Smith, 1989), and nutrient and oxygen sensor measurements may be useful for assimilation into models because of their time continuity (Dickey, 1988). Other variables from fixed stations, like Station P, are also useful. In terms of model input, many different types of physical data, such as temperature and light data, including satellite-derived SST and irradiance, are available and could potentially be assimilated into a model. Furthermore the physical part of a coupled physical-biological model may be improved by assimilation of various type of data, which in turn will improve the biological portion of the model. Future possibilities of data assimilation to biogeochemical models with various data sets are shown in Figure 6.

Acknowledgements
I would like to acknowledge Drs. Geoffrey Evans and Mike Fasham for giving me the opportunity to attend the workshop. I thank Drs. Geoffrey Evans, Eileen E. Hofmann and Donald Eliason for their comments on the manuscript. I also thank Dr. Uli Wolf and Aida Starke for their help during the workshop.

References

Abbott MR (ed) (1992) US JGOFS Report:Workshop on modelling and satellite data assimilation. US JGOFS Planning Office WHOI Woods Hole MA

Anderson DLT, Willebrand J (eds) (1989) Oceanic circulation models: Combining data and dynamics. Kluwer

Armstrong FAJ, Stearns CR, Strickland DH (1967) The measurement of upwelling and subsequent biological processes by means of the Technicon Autoanalyzer and associated equipment. Deep-Sea Res 14: 381-389

Charney J, Halem M, Jastrow R (1969) Use of incomplete historical data to infer the present state of the atmosphere. J Atmos Sci 26: 1160-1163

Dickey, TD (1988) Recent advances and future directions in multi-disciplinary in situ oceanographic measurement system. In: Rothschild BJ (ed) Toward a Theory on Biological-Physical Interactions in the World Ocean. Kluwer. p 555-598

Flagg CN, Smith SL (1989) On the use of the acoustic Doppler current profiler to measure zooplankton abundance. Deep-Sea Res 36: 255-274

Gandin LS (1965) Objective analysis of meteorological fields. Israel Program for Scientific Translations Jerusalem

Goldman JC (1988) Spatial and temporal discontinuities of biological processes in pelagic surface waters. In: Rothschild BJ (ed) Toward a Theory on Biological-Physical Interactions in the World Ocean. Kluwer. p 273-296

Gordon HR, Clark DK, Mueller JL, Hovis WA (1980) Phytoplankton pigments from the Nimbus-7 Coastal Zone Color Scanner: Comparisons with surface measurements. Science 210: 63-66

Haidvogel DB, Robinson A (1989) Special issue: data assimilation. Dynam Atmos Oceans 13: 171-513

Hofmann, E. E. and J. Ambler (1988) Plankton dynamics on the outer southeastern U.S. continental shelf. Part II: A time-dependent biological model. J Mar Res 46: 883-917

Hovis WA, Clark DK, Anderson F, Austin RW, Wilson WH, Baker ET, Ball D, Gordon HR, Mueller JL, El-Sayed SZ, Sturm B, Wrigley RC, Yentsch CS (1980) Nimbus-7 Coastal Zone Color Scanner: system description and initial imagery. Science 210: 60-63

Ishizaka J (1990a) Coupling of Coastal Zone Color Scanner data to a physical-biological model of the southeastern U.S. continental shelf ecosystem. 3. Nutrient and phytoplankton fluxes and CZCS data assimilation. J Geophys Res 95: 20201-20212

Ishizaka J (1990b). Coupling of Coastal Zone Color Scanner data to a physical-biological model of the southeastern U.S. continental shelf ecosystem. 1. CZCS data description and Lagrangian particle tracing experiments. J Geophys Res 95: 20167-20181

Ishizaka J (1990c). Coupling of Coastal Zone Color Scanner data to a physical-biological model of the southeastern U.S. continental shelf ecosystem. 2. An Eulerian model. J Geophys Res 95: 20183-20199

Ishizaka J, Hofmann EE (In press). Coupling of ocean colour data to physical-biological models. In: Ocean colour: Theory and applications in a decade of CZCS experience. Kluwer Academic

Kasahara A (1972) Simulation experiments for meteorological observing systems for GARP global experiment. Bull Amer Meteor Soc 54: 252-264

Leetma A, Ming J (1989) Operational hindcasting of the tropical Pacific. Dyn Atmos Oceans 13: 465-490

McPherson RD (1975) Progress, problems, and prospects in meteorological data assimilation. Bull Amer Meteor Soc 56: 1154-1165

Najjar, RG, Sarmiento JL and JR Toggweiler (1992) Downward transport and fate of organic matter in the ocean: simulations with a general circulation model. Global Biogeochem Cycle 6: 45-76

Okubo A (1980) Diffusion and ecological problems: Mathematical models. Springer-Verlag New York

Platt T, Sathyendranath S (1988) Oceanic primary production: Estimation by remote sensing at local and regional scales. Science 241: 1613-1620.

Sathyendranath S, Platt T, Horne EPW, Harrison WG, Ulloa S, Outerbridge R, Hoepffner N (1991) Estimation of new production in the ocean by compound remote sensing. Nature 353: 129-133

Sarmiento J, Fasham MJR, Siegenthaler U, Najjar R, Toggweiler JR (1989) Models of chemical cycling in the ocean: Progress Report II. Ocean Tracers Laboratory Technical Report No. 6. Princeton University Princeton

Walsh JJ (1988) On the Nature of Continental Shelves. Academic San Diego

Zentara SJ, Kamykowski D (1976) Latitudinal relationships among temperature and selected plant nutrients along the west coast of North and South America. J Mar Res 35: 321-337

AN ANNOTATED BIBLIOGRAPHY OF MARINE BIOLOGICAL MODELS

Ian J. Totterdell
James Rennell Centre for Ocean Circulation
Gamma House
Chilworth Research Centre
SOUTHAMPTON SO1 7NS
United Kingdom

1. Introduction

Many papers have been written about models of pelagic ecosystems. Over 100 of these papers, written over a period of more than 40 years, are included in a bibliography at the end of this appendix. Most introduce a model, normally of several trophic levels, discussing its aims and structure, plus the results obtained. Some papers consider only a particular subsystem of an ecosystem model, such as the uptake of nutrients or the relationship between the rate of photosynthesis and light intensity, or a certain aspect of the interaction of the physics of the ocean and the biota. A few papers are reviews of other models, or are conceptual models.

The ecosystem models that are presented in the listed papers have been examined and are discussed here in terms of their aims, their overall structure and their subsystems. The processes that are featured and the variety of functional forms used to represent them are described. The models are referenced in terms of the papers that present them, by means of the number assigned to that paper in the bibliography. Some papers present more than one variations on a certain model, in which case the different forms are called 'a', 'b', etc.

2. Overview of the models' structure

The models were constructed to study phenomena in a number of different temporal and/or physical regimes. As well as those listed below, [4, 6, 12] and [93] modelled the results of controlled ecosystem experiments.

NATO ASI Series, Vol I 10
Towards a Model of Ocean
Biogeochemical Processes
Edited by G T. Evans and M J.R Fasham

2.1. Time-scales

(i) Annual cycles. Models which are intended to produce repeating cycles cannot ignore processes which are important for nutrient regeneration. The following models ran simulations for at least a year, with seasonal variations in the external forcings:
[1, 17, 20, 21, 23, 28, 31, 43, 45, 59, 60, 68, 73, 81, 83, 88].
Two models published before the advent of suitable computing power ([70] and [76]) were not able to consider the time-variation of the system but examined steady-states in all seasons.

(ii) Particular periods. The majority of models ran for only parts of a year. Some were compared to data collected during a certain period, while others were intended to simulate phenomena of limited duration, such as a spring bloom. Models included in these categories were:
[3, 24, 26, 29, 30, 33, 34, 35, 36, 37, 39, 41, 46, 50, 51, 53, 55, 61, 65, 66, 71, 72, 74, 75, 77, 90, 91, 95, 96, 98, 100, 101, 102, 103, 105].

(iii) Theoretical response. A few papers examined the approach to steady-state or to limit cycles of a simple mathematical model of an ecosystem. Sometimes perturbations were introduced and the system's response studied. The time-scales for these changes were of the order of days. Such papers included:
[2, 13, 22, 27, 32, 52, 56, 58, 78, 85].

(iv) Steady-state. The following models allowed no time-variation, but considered only the steady-state:
[9, 18, 69, 86, 94, 106].

(v) Paper [8] describes a GCM which does not resolve seasonal cycles, but the annual primary production is simulated for a period of many years.

2.2. Physical environment

(i) Deep ocean. The models listed in this section includes both those that are explicitly intended for areas where the ocean is deep (i.e. there is a substantial depth of water below the seasonal thermocline) as well as those

whose geographical location is not defined but which do not consider the ocean bottom:
[2, 8, 17, 20, 21, 23, 28, 41, 43, 69, 72, 73, 86, 88, 95, 96, 101, 102, 106].

(ii) Shelf sea. Models which are designed to describe ecosystems on continental shelves are included here, along with those which feature the sea-floor. However models defined as coastal (see the next section) are excluded. The following are shelf sea models:
[1, 3, 11, 24, 33, 34, 37, 51, 53, 59, 60, 66, 68, 70, 71, 74, 75, 76, 77, 85, 90, 91].

(iii) Coastal. This section lists models designed for areas near the coast or in bays, inlets or inland seas. The ecosystems found in these areas are usually strongly influenced by effects such as run-off of freshwater from the land or upwelling near the shore which can be crucial for the supply of nutrients: however not all the models include these effects explicitly. Coastal models include:
[31, 35, 36, 39, 46, 55, 58, 61, 81, 82, 98, 100, 103].

(iv) Lakes. A few of the papers in the bibliography describe freshwater ecosystems in lakes. The following models are included because they feature similar processes to oceanic systems:
[9, 13, 26, 45, 50, 56].

2.3. Physical structure

Most of the models have only one physical dimension, that of depth. However some papers, mainly those not considering light limitation, describe models that are zero-dimensional, being vertically as well as horizontally averaged. There are also models which feature one or two horizontal dimensions, including [8, 29, 31, 35, 37b, 55, 74b, 89, 90, 91, 98] and [103]. The inclusion of vertical variation allows the light profile and stratification effects to be modelled. The following sections consider how the variation in depth is featured.

(i) One layer only. Most of the models in this category consider only mixed layer processes, with no transfer of material to or from any other layer. In some, the depth of the mixed layer is allowed to vary with time. Single-layer models include:
[6, 9, 13, 26, 27, 31, 32, 34, 45, 46, 50, 51, 59, 68, 69, 71, 76, 83].

(ii) Two layers, one implicit. These models feature a second layer below the first, in which nutrient and other values are prescribed (and usually constant). No biological activity is allowed in this lower, implicit layer, but by mixing across the interface, or by entrainment, nutrients are supplied to the active layer. Models with this structure are:
[17, 20, 21, 23, 28, 36, 37, 39, 61, 75, 78, 86, 89].

(iii) Two or more explicit layers. The following models have two or more biologically active layers:
[1, 18, 73, 88].

(iv) Continuous profile. Many models calculate profiles of the chemical and biological components. Since these must be calculated numerically, the profiles cannot be continuous, but instead values are found at frequent intervals down the water column, usually in steps of one to ten metres. Some of the models featuring this structure impose a mixed layer and demand that the points corresponding to it have the same values of.concentrations, etc.:
[3, 11, 12, 22, 29, 33, 43, 60, 72, 81, 85, 101, 102].
Others do not impose a mixed layer, although the eddy diffusivity might be much greater near the surface than at depth, producing a well-mixed but not perfectly-mixed layer:
[2, 4, 24, 30, 35, 41, 55, 65, 66, 70, 74, 77, 90, 91, 95, 96, 103, 105].

3. Forcings

The models featuring time variation usually prescribe the values of some of the environmental factors such as insolation, temperature, etc., and occasionally some of the variables (e.g. grazing rate). This section lists the models which force each factor or variable. Models are not included however if they set a value which is constant in time and (where applicable) depth.

3.1 Physical forcings

(i) Solar radiation. All the models specify the time variation of the insolation, except those which do not consider light limitation of growth. The incoming radiation is either calculated from astronomical equations, specified

by a simple function or taken from measurements. The following models include diurnal variation:
[1, 4, 22, 24, 33, 34, 41, 60, 61, 65, 66, 85b, 88, 100, 101, 102, 103].
The following models feature no light limitation and hence insolation is not included:
[26, 27, 29, 30, 51, 52, 53, 56, 58, 59, 75, 93, 94].

(ii) Temperature. In some models, the rates of certain processes are dependent on the ambient temperature. The following models specify the time variation of the temperature, from measurements, simple functions, or the outputs of linked physical models:
[1, 3, 4, 6b, 8, 11, 26, 28, 31, 34, 36, 37, 45, 50, 60, 61, 70, 73, 74, 81, 83, 88, 89, 103].

(iii) Mixing or vertical diffusion. This process can be very important for the supply of nutrients to the euphotic zone. These models represent the process as a mixing rate, varying with time and/or depth:
[1, 11, 28, 75, 88].
Most models however prescribe the eddy diffusivities, either by specifying the values or by calculating from wind stress, temperature profiles or the output of physical models. Those which specify diffusivities are:
[3, 22, 24, 30, 41, 60, 65, 66, 77, 85, 95, 96, 98, 100].
Those which calculate diffusivities are:
[4, 29, 39, 43, 70, 73, 74, 81, 103, 105].

(iv) Winds and currents. Some models force the wind strengths, often to calculate the eddy diffusivities:
[1, 36, 39, 43, 71, 81, 103].
Models [1] and [74b], which both feature horizontal variation, force the currents.

(v) Mixed layer depth. The depth of the base of the mixed layer is forced to vary in many models, and the development of phytoplankton blooms can be quite sensitive to this variation. The depth is often calculated from measurements of temperature or chemical profiles. Models featuring this forcing include:
[17, 21, 22, 23, 28, 36, 37, 45, 60, 61, 72, 88, 101, 102, 105].

3.2 Biological forcings

(i) Nutrients. Some of the models, particularly those simulating only particular parts of the year, do not complete the nutrient cycle. Many set a constant nutrient concentration in a biologically-inactive layer, but models [28] and [68] set the nutrient levels in the biologically active layer(s) from observations, and model [33] forces the upwelling.

(ii) Grazing rates. The closure term is sometimes herbivory or (more rarely) carnivory, and the rate (usually per unit prey biomass) is often prescribed. Some models force the grazing pressure to follow a simple function of time while others set the herbivore or carnivore density from observations. Models [28d] and [33] prescribe the rate of carnivory at the highest trophic level,while a number of models prescribe the rate of herbivory:
[4, 9, 13, 24, 28b, 33, 37, 41, 55, 65, 68, 98, 100].

(iii) Other biological forcings. Model [6b] varies the maximum growth rate of phytoplankton with time to represent the changes in community composition. Model [88] forces the mortality rate (including predation, respiration and natural death) to vary with the time of year by means of a function which approximately follows the mixed layer temperature.

4. Model state variables

4.1. Chemical variables

(i) Nitrogen. Most models feature a single chemical nutrient. Sometimes this is explicitly stated to be nitrate, or all forms of nitrogen combined, but in a few cases it has to be deduced to be nitrogenous by its interactions. Unless 'nutrient' is stated to be another compound (e.g. phosphorous) or to represent a number of alternative chemical species (models [56] and [94]) models featuring that variable are listed here:
[2, 3, 4, 6, 11, 13, 18, 20, 21, 24, 27, 28, 29, 30, 32, 34, 35, 36, 37, 39, 51, 55, 58, 59, 61, 73, 74b, 77, 78, 83, 85, 88, 89, 90, 91, 95, 96, 100, 101, 105, 106].

(ii) Nitrate and ammonia. A number of models separate oxidised and reduced nitrogen (nitrate, or 'new' nitrogen, and ammonia, or 'regenerated' nitrogen, respectively). Ammonia is produced by excretion or remineralization of dead cells and waste products, whereas nitrate tends to be mixed up from depth. Very few models complete the nitrate budget explicitly: in the biologically-active layers there is a net flux of nitrogen from nitrate to ammonia, and this is usually balanced by an implicit oxidation of nitrogen in the unmodelled deep layers. Separation of the two nutrients allows inhibition by ammonia of nitrate uptake to be modelled, and for there to be discrimination in the rates of growth on the two substrates. Models with both nitrate and ammonia are:
[1, 17, 23, 41, 43, 52, 60, 74a, 81, 86, 98, 103].

(iii) Phosphorous. This is sometimes the limiting nutrient, especially in the earlier models. A few include it as an additional element to compare to experimental data. The following models include phosphorous or its compounds:
[8, 11, 45, 65, 66, 68, 69, 70, 75, 76, 93, 98].

(iv) Silicate. This is usually included when diatoms are present, because it is used to form their shells. In some models it can be the limiting factor for diatom growth, in others it is used as a tracer to be compared to experimental data. These models feature silicate:
[4, 6b, 74, 98].

(v) Dissolved organic matter. This pool of carbon and nitrogen is included in models under different names, either as DOM, DON or DOC (dissolved organic matter, nitrogen and carbon respectively). Only model [52] uses more than one of these terms, since it keeps independent budgets of carbon and nitrogen. Models which feature a variable called DOM are [11] and [59]. Those which feature DOC are:
[1, 52, 73].
Those which feature DON are:
[17, 23, 31, 46, 52, 86, 93].

4.2. Biotic variables

(i) No explicit biology. A minority of the models include no explicit variable representing a biological component. These models ([8, 29, 30] and [35]) are primarily concerned with the distributions of chemical species, and consider biological processes only in that light. Primary production for example is modelled as a flux from one chemical component to another, the rate of which is governed by the levels of nutrient and light available at that time and depth.

(ii) Phytoplankton. Nearly all models include one or more variables which represent photosynthesizing plankton. Most consider a generic phytoplankton (sometimes represented by 'chlorophyll'), though parameter values may be taken from a particular species. A few models feature two or more phytoplankton variables, and these often represent individual species. Models with a single phytoplankton variable are:
[1, 2, 3, 6a, 9, 11, 13, 18, 21, 22, 23, 24, 26, 27, 28, 31, 32, 33, 36, 37, 39, 41, 43, 45, 55, 56, 58, 61, 65, 66, 68, 69, 70, 71, 72, 73, 74b, 75, 77, 78, 81, 82, 83, 85, 88, 89, 90, 91, 93, 95, 96, 98, 100, 101, 102, 103, 105, 106].
Models with two or more phytoplankton variables are:
[4, 6b, 17, 20, 34, 50, 52, 59, 60, 74a, 86, 94].

(iii) Herbivores. Although many models have only one variable representing herbivores, in a large proportion the variable is explicitly stated to represent the particular group of organisms that is most important in the modelled ecosystem. Models with two or more herbivore variables usually specify the groups these represent. It should be noted that a group of organisms which a model listed here includes as a herbivorous group may be included in another model as an omnivorous group, depending on the models' structure. Models which feature a single herbivore variable are:
[1, 11, 18, 20, 21, 23, 27, 28cd, 31, 32, 34, 36, 45, 46, 50, 55, 58, 60, 61, 69, 70, 71, 73, 75, 77, 78, 86, 93, 94, 95, 103, 105, 106].
Models which feature two or more herbivore variables are:
[3, 6b, 33, 59, 96, 98].

(iv) Omnivores. The following models have one variable representing omnivores:
[17, 28d, 46, 55, 61, 94, 96].
The following models have two or more variables representing omnivores:
[51, 52, 59, 95].

(v) Carnivores. Models featuring a single carnivore are:
[6, 70, 94, 96].
Models featuring more than one carnivore are:
[11, 31, 59, 61, 95].

(vi) Bacteria. Generally this term refers to heterotrophic bacteria, with

photosynthesizing bacteria being represented by the phytoplankton variable: only model [60] explicitly includes autotrophic bacteria. The bacteria are usually agents for the remineralization of DOM and detritus. Models including heterotrophic bacteria are:
[17, 23, 31, 46, 51, 52, 59, 61, 86, 93, 95, 96].

(vii) Detritus. This compartment can include fecal pellets, dead cells, etc. Models featuring detritus as an explicit state variable are:
[3, 6, 17, 23, 52, 55, 58, 59, 61, 74a, 86, 95, 96, 101, 102].

5. Chemical Cycles

5.1. Nutrient uptake and autotrophic growth

Most models incorporate some form of limitation of growth by nutrients, or rather their lack. The majority assume that as soon as nutrient is taken up it is used for growth. An alternative to this assumption, involving the use of extra state variables, is to include internal pools of substrate. This formulation is used by models [16, 50, 90, 93b] and [101]: the other models listed in this section do not discriminate nutrient uptake and growth.

(i) Functional forms. The most common form of the function relating the rate of uptake of a particular nutrient to that nutrient's concentration in the water is the Monod or Michaelis-Menten term:

$$\text{Uptake rate} = V_{max} \frac{N}{N+K_N}$$

where V_{max} is the maximum rate, N is the nutrient concentration and K_N is the concentration at which the rate is half the maximum (the half-saturation). This form is used (for at least one nutrient, if there are several) by models:
[1, 2, 3, 4, 6, 8, 13, 17, 18, 20, 21, 23, 24, 27, 28, 29, 30, 31, 34, 35, 36, 39, 41, 45, 46, 50, 52, 55, 56, 58, 60, 61, 65, 66, 73, 74, 81, 83, 85, 88, 90, 91, 93, 98, 101, 103, 105, 106].
Some other models, particularly the earliest ones, modelled the rate of uptake as proportional to the nutrient concentration:
[32, 68, 69, 70, 94].

The following models used a variety of other functions, some of them complicated:
[11, 37, 43, 51, 75, 76, 89, 95, 98, 100].

(ii) Multiple limitation. The rate of growth is often limited by more than one nutrient, or by one or more nutrients and the insolation. There are two cases to be considered, that where two nutrients can be used as alternative substrates (usually nitrate and ammonia) and that where two (or more) nutrients are each needed for growth. In the former case a single modified Michaelis-Menten function is often used, frequently with inhibition:

$$\text{Uptake rate} = V_{max} \left(\frac{N}{N+K_N} \exp(-UA) + \frac{A}{A+K_A} \right)$$

where A is the concentration of ammonia, K_A the half-saturation and U the inhibition constant. This form is often referenced to paper [103]. Models which include the inhibition of nitrate uptake by ammonia are:
[17, 23, 34, 74a, 98, 103].

For the case of two non-substitutable nutrients or limiting factors (e.g. nitrogen and silicate, or nitrogen and illumination) two different formulations are used. The most common calculates the degree of limitation due to each nutrient or factor and multiplies them together to give the overall limitation: this is used by the following models for the combination of at least two factors (especially light and nutrient):
[1, 2, 3, 4, 6, 8, 17, 20, 21, 24, 28, 34, 36, 37, 39, 41, 45, 50, 55, 56, 60, 66, 68, 70, 73, 76, 78, 81, 83, 85, 86, 88, 93ab, 95, 98, 100, 103, 106].
The other, less common, formulation is often referred to as Liebig's law, but should be called Blackman's law: the total rate is governed by whichever of the alternative factors is most limiting. This formulation is used for the combination of at least two factors in these models:
[6, 11, 43, 61, 65, 69, 74a, 90, 91, 93c, 96, 98].
Models [6] and [98] appear in both lists because different formulations are used for different combinations of limiting factors.

Temperature affects the rate of growth in some models, but it is not a limiting resource in the same way as those considered above, and its limitation factor (see section 8) is always combined with the others by multiplication.

5.2. Release and regeneration of nitrogen

There are several processes by which nitrogen and nitrogenous compounds are returned to solution. Excretion, egestion and respiration are processes associated with healthy, viable organisms, while natural mortality and cell lysis are not. Some papers use the term 'excretion' to mean only the release of ammonia, but others use it to mean the release of all waste products, having no separate 'egestion' process. In this discussion the processes have mainly been listed by the terms used by the papers' authors.

(i) Excretion. The following models feature a rate of excretion which is biomass-specific (usually for phytoplankton):
[3, 4, 6, 11, 17, 23, 51, 59, 77, 93b, 95, 96, 103].
The following models feature a rate of excretion which is determined by the rate of feeding (zooplankton) or nutrient-uptake (phytoplankton):
[13, 55, 60, 61, 76].

(ii) Egestion. Applied exclusively to zooplankton, this term refers mainly to the production of fecal pellets, which in some models are part of detritus and can have a sinking rate. In a few models the product of egestion is assumed to be remineralized immediately to ammonia. Only model [58] has a biomass-specific rate of egestion: in all other cases the rate is determined by the rate of feeding. Models [11, 27] and [86] include processes explicitly called 'sloppy feeding' and refering to the release of some parts of the prey into the water without going through the gut: some others include this in egestion. Models featuring a rate of egestion determined by the rate of feeding are:
[1, 6, 17, 23, 51, 52, 58, 60, 61, 73, 78, 94b, 95, 96, 98, 103, 105, 106].

(iii) Respiration. Technically this process involves only carbon compounds, but many models have set carbon:chlorophyll and carbon:nitrogen ratios for the biotic variables and so it is often modelled as a loss of biomass, whatever the units used to measure that component. Some models include the process as part of natural mortality, but the following feature it as a separate process:
[1, 9, 20, 21, 28, 31, 33, 36, 37, 41, 45, 52, 59, 60, 65, 66, 68, 70, 72, 73, 74a, 75, 81, 82, 90, 91, 93, 100, 101].

(iv) Cell lysis. Only models [52] and [103] include the release of cell contents by lysis, in the former to DOC and DON pools and in the latter to detritus.

(v) Natural mortality.This term is used in a large number of models: in some it represents the processes by which viable cells die (including, for example, viral infection and predation by unmodelled predators higher in the food chain), while in others it is meant to represent a number of processes by which biomass is lost, including respiration and excretion. These models include natural mortality to represent only processes leading to cell death:
[6, 17, 23, 34, 36, 45, 46, 50, 52, 55, 56, 59, 65, 73, 74a, 95, 96, 101, 105].
The following models include the term to represent also processes by which biomass is lost:
[1, 2, 3, 4, 11, 20, 21, 27, 39, 43, 55, 58, 66, 71, 74b, 85, 86, 88, 89, 94, 98, 106].

Some of the above processes involve the release of nitrogen in the form of complex compounds rather than simple molecules. In models where this is the case, the regeneration of ammonia and nitrate (or 'nutrient') is often included as an explicit process, usually as a constant (specific) rate. Regeneration is featured in these models:
[1, 8, 11, 58, 81, 86, 90, 91, 95, 98, 103].

The following models constrain the total amount of nitrogen (or phosphorous) in the system, in the form of nutrient, phytoplankton and zooplankton (for example), to be constant:
[1, 3, 8, 27, 58, 94a, 105, 106].

6. Light and photosynthesis

The calculation of the rate of photosynthesis at a given depth is an important part of most of the models. Several different functional forms are used, as shown below. A number of the papers in the bibliography are reviews and studies of simple and commonly-used formulations:
[14, 15, 19, 25, 40, 42, 49, 64, 96].

6.1. Light as a function of depth

Beer's law, whereby the intensity of a given wavelength of light decreases exponentially with depth, is used to describe the profile of illumination in almost every model. Most also take the further simplifying step that the extinction coefficient is independent of wavelength (in the visible band) so that the intensity of the total photosynthetically available radiation (PAR) varies as a simple exponential:

$$I(z) = I(0)\ \exp(-kz)$$

where $I(z)$ is the intensity at depth z, and k is the extinction coefficient. Models [24] and [88] assume to the contrary that visible light is better described as two sub-bands with different extinction coefficients. The bands usually correspond approximately to red light and green light, but are split so that a given proportion of the intensity is in each band at the surface, rather than at a certain wavelength.

In some models the extinction coefficient is a function of the concentration of phytoplankton. This function is usually either linear, or of the form:

$$k = k_0 + aP + bP^{2/3}$$

where P is the phytoplankton concentration, a and b are constants and k_0 is the extinction coefficient of the water in the absence of phytoplankton. The following models use the linear function to parameterize self-shading:
[9, 11, 17, 20, 23, 24, 28, 33, 39, 41, 60, 61, 73, 81, 85, 88, 95, 96, 100, 103].
These models use the more complicated function:
[1, 36, 37, 43, 66, 69, 74, 98].
Models [50] and [72] include self-shading but do not specify the function.

6.2. Photosynthesis-light functions

Once the light profile has been calculated the rate of photosynthesis can be determined. The general features that models try to include are that at low light intensities the rate of photosynthesis is proportional to the light, but as the intensity increases the rate becomes constant (saturation) and then may begin to decrease as chlorophyll is broken down (inhibition). This section shows the various forms used for $f(I)$, where I is the light intensity and

$$\text{Rate of photosynthesis} = V_{max} f(I)$$

The parameters used here are often not the same as in the relevant papers, but in each case there is a unique correspondence.

(i) Proportional. Some models, particularly the earlier ones, make the rate of photosynthesis proportional to the light (this formulation does not however display saturation or inhibition):
[68, 70, 71, 82, 90, 91, 100, 102a, 105].

(ii) Michaelis-Menten. The same form that is frequently used for the rate of nutrient-uptake is sometimes applied to light (once again, inhibition is not displayed):
[2, 6a, 22, 34, 43, 86, 88, 98, 102b].

(iii) Steele's function. This has the form:

$$f(I) = \frac{I}{I_S} \exp(1-\frac{I}{I_S})$$

where I_S is the intensity giving the maximum rate of photosynthesis. Both saturation and inhibition are included. Models using this form are:
[1, 36, 37, 41, 50, 60, 73, 76, 77, 81, 83, 102c].

(iv) Exponential saturation. This was first used by Platt *et al.* (1980; paper [64]), and has the form:

$$f(I) = (1-\exp(-\frac{I}{I_S}))\ \exp(-b\frac{I}{I_S})$$

although it usually does not feature inhibition (i.e. b=0). I_S is a constant determined from photosynthetic parameters. Models [24, 72] and [74] use this function.

(v) Hyperbolic saturation (Smith's function). This also can feature inhibition, but is often used without (b=0):

$$f(I) = \frac{I/I_H}{(1+(I/I_H)^2)^{1/2}} \frac{1}{(1+b(I/I_H)^2)^{n/2}}$$

n is a constant, often given the value 1. If there is no inhibition, I_H is the intensity at which the rate of photosynthesis is half the maximum. The following models use this function:
[9, 17, 20, 21, 23, 26, 39, 61, 103].

(vi) Tanh function. This is used by models [28, 33] and [85] and has the form:

$$f(I) = \tanh \frac{I}{I_C}$$

where I_C is a constant, determined by photosynthetic parameters.

Other functions are used by the following models:
[3, 4, 6b, 11, 45, 55, 60, 65, 66, 69, 89, 95, 101, 106].

Most models set constant values for the parameters, but a few allow them to change according to the light history of cells (photo adaption). Models [9, 22, 74, 101] and [102] feature this process.

7. Grazing functions

This section lists the most common functions relating the rate of feeding (in terms of biomass of prey lost per unit volume per unit grazer density) to the concentration or density of the prey.

(i) Michaelis-Menten. This form is used by models [28d, 52] and [60] to describe carnivory, and by the following models to describe herbivory:
[1, 20, 21, 24, 28cd, 36, 37, 45, 46, 52, 58, 61, 65, 73, 77, 93, 98].

(ii) Ivlev function. This has the form:

$$\text{Rate of grazing} = G_{max}\,(1\text{-}\exp(-kP))$$

where P is the concentration or density of the prey and k is a constant. It is used by model [78] to describe carnivory, and by these models to describe herbivory:
[3, 4, 6, 27a, 34, 41, 51, 55, 78, 96, 100, 103, 105a, 106].

Other functions (including a simple linear relation) are used by these models for herbivory and/or carnivory:
[3, 11, 13, 18, 23, 27b, 32, 33, 55c, 59, 60, 69, 70, 75, 77, 78, 86, 95, 105b].

Grazing thresholds are featured by some of the models: if the prey density P falls below this level grazing stops. Also, when the density is above threshold value P_0, P is normally replaced by $(P\text{-}P_0)$ in the grazing function. Including grazing thresholds can help the survival of prey populations at very low densities. The following models feature thresholds:
[3, 4, 6, 20, 21, 24, 28c, 34, 41, 46, 52, 59, 61, 65, 73, 77, 93, 98, 103].

Preferences are used by several models in which a grazer feeds on more than one type of prey. Often these are expressed simply by having a different maximum rate of feeding for each type of food, but sometimes they are functions of the densities of the possible prey.The models which feature preferences are:
[11, 13, 17, 20, 23, 31, 34, 37, 46, 59, 95].

8. Other model features

(i) Temperature. In many of the models the rates of some processes (especially photosynthesis and respiration) vary with the ambient temperature, usually according to an exponential function. The following models have at least some temperature-dependent rates:
[1, 3, 4, 6b, 28, 31, 34, 36, 37, 45, 50, 60, 70, 73, 74, 81, 83, 88, 103].

(ii) Sinking. The falling under gravity of some living and dead cells and fecal pellets is a fundamental part of the biological pump whereby carbon is exported to deep water from the euphotic zone. The models which feature a sinking velocity (usually constant) for some cells or detritus are:
[3, 4, 6, 9, 17, 23, 41, 45, 50, 52, 60, 61, 65, 69, 70, 74a, 77, 78, 81, 85, 86, 88, 89, 95, 96, 100, 101, 102].

(iii) Size-dependency. Most models express the amount of each component in terms of a continuum, e.g. biomass or chlorophyll per unit volume. However a few ([13, 33, 34] and [36]) consider the number of organisms per unit volume and allow their sizes and weights to vary.

(iv) Vertical migration. Some zooplankton move vertically in the water column on a daily cycle, coming up to the surface layers at night to feed and spending the day at depth. This behaviour is featured in models [11, 33, 95, 96] and [97], and is studied in detail in paper [5].

(v) Cohorts. Several species of zooplankton (e.g. copepods) pass through a number of distinct stages before becoming mature adults. Juvenile stages are included in models [34, 36] and [77].

(vi) Closure. At some point the food chain has to be terminated, and this closure term is often a mortality function. A number of different forms are used, but the most fall into one of two categories. These models use a constant specific rate of mortality:
[1, 2, 3, 17, 18, 20, 21, 23, 43, 45, 73, 85, 89, 90, 91, 106].

These models have a specific rate of mortality which is dependent on the density of the relevant component:
[24, 28c, 34, 36, 60, 69, 75, 78].
Other forms of the closure term are used by models [6b, 39, 50, 77, 86] and [88].

Acknowledgement

This bibliography and discussion were compiled under (The United Kingdom) Department of the Environment contract PECD 7/12/37.

Bibliography

[1] Agoumi A, Gosse P, Khalanski M (1985) Numerical modelling of the influence of the vertical thermal structure on phytoplankton growth in the English Channel. In: Gibbs PE (ed) Proceedings of the 19th European Marine Biology Symposium, Plymouth, Devon, 16-21 September 1984. Cambridge University Press, Cambridge, p 23

[2] Altabet MA, Robinson AR, Walstad LJ (1986) A model for the vertical flux of nitrogen in the upper ocean: Simulating the alteration of isotopic ratios. J Mar Res 44:203-225

[3] Andersen V, Nival P (1988) A pelagic ecosystem model simulating production and sedimentation of biogenic particles: role of salps and copepods. Mar Ecol Prog Ser 44:37-50

[4] Andersen V, Nival P (1989) Modelling of phytoplankton population dynamics in an enclosed water column. J mar biol Ass UK 69:625-646

[5] Andersen V, Nival P (1991) A model of the diel vertical migration of zooplankton based on euphausiids. J Mar Res 49:153-175

[6] Andersen V, Nival P, Harris RP (1987) Modelling of a plankton ecosystem in an enclosed water column. J mar biol Ass UK 67:407-430

[7] Anderson TR (1992) Modelling the influence of food C:N ratio and respiration on growth and nitrogen excretion in marine zooplankton and bacteria. J Plankton Res 14:1645-1671

[8] Bacastow R, Maier-Reimer E (1990) Ocean-circulation model of the carbon cycle. Climate Dynamics 4:95-125

[9] Bannister TT (1974) A general theory of steady-state phytoplankton growth in a nutrient saturated mixed layer. Limnol Oceanogr 19:13-30

[10] Bartell SM, Brenkert AL, Carpenter SL (1988) Parameter Uncertainty and the behaviour of a size-dependant plankton model. Ecol Modelling 40:85-95

[11] Belyaev VI, Lenin AI, Petipa PS (1975) Mathematical model of a pelagic ecosystem. Mar Biol 31:1-6

[12] Bienfang P, Szyper J, Laws E (1983) Sinking rate and pigment responses to light-limitation of a marine diatom: implications to dynamics of chlorophyll maximum layers. Oceanol Acta 6:55-62

[13] Carpenter SR, Kitchell JF (1984) Plankton community structure and limnetic primary production. Am Nat 124:159-172

[14] Chalker BE (1980) Modelling light saturation curves for photosynthesis: an exponential function. J Theor Biol 84:205-215

[15] Cullen JJ (1990) On models of growth and photosynthesis in phytoplankton. Deep-Sea Res 37:667-683

[16] Droop MR (1968) Vitamin B12 and marine ecology. IV. The kinetics of uptake, growth and inhibition in Monochrysis Lutheri. J mar biol Ass UK 48:689-733

[17] Ducklow HW, Fasham MJR (1992) Bacteria in the greenhouse: modeling the role of oceanic plankton in the global carbon cycle. In: Mitchell R (ed) Environmental Microbiology. Wiley-Liss, New York, p 1

[18] Dugdale. RC (1967) Nutrient limitation in the sea: dynamics, identification, and significance. Limnol Oceanogr 12:685-695

[19] Eilers PHC, Peeters JHC (1988) A model for the relationship between light intensity and the rate of photosynthesis in phytoplankton. Ecol Modelling 42:199-216

[20] Evans GT (1988) A framework for discussing seasonal succession and coexistence of phytoplankton species. Limnol Oceanogr 33:1027-1036

[21] Evans GT, Parslow JS (1985) A model of annual plankton cycles. Biol Oceanogr 3:327-347

[22] Falkowski PG, Wirick CD (1981) A simulation model of the effects of vertical mixing on primary production. Mar Biol 65:69-75

[23] Fasham MJR, Ducklow HW, McKelvie SM (1990) A nitrogen-based model of plankton dynamics in the oceanic mixed layer. J Mar Res 48:591-639

[24] Fasham MJR, Holligan PM, Pugh PR (1983) The spatial and temporal development of the spring phytoplankton bloom in the Celtic Sea, April 1979. Prog Oceanogr 12:87-145

[25] Fee EJ (1969) A numerical model for the estimation of photosynthetic production, integrated over time and depth, in natural waters. Limnol Oceanogr 14:906-911

[26] Fee EJ (1973) A numerical model for determining integral primary production and its application to Lake Michigan. J Fish Res Board Can 30:1447-1468

[27] Franks PJS, Wroblewski JS, Flierl GR (1986) Behavior of a simple plankton model with food-level acclimation by herbivores. Mar Biol 91:121-129

[28] Frost BW (1987) Grazing control of phytoplankton stock in the open subarctic Pacific Ocean: a model assessing the role of mesozooplankton, particularly the large calanoid copepods *Neocalanus* ssp. Mar Ecol Prog Ser 39:49-68
[29] Garcon VC, Martinon L, Andrie C, Andrich P, Minster J-F (1989) Kinematics of CO2 fluxes in the tropical Atlantic Ocean during the 1983 northern summer. J Geophys Res 94:855-870
[30] Garside C (1985) The vertical distribution of nitrate in open ocean surface water. Deep-Sea Res 32:723-732
[31] Gordon DC Jr, Keizer PD, Daborn GR, Schwinghamer P, Silvert WL (1986) Adventures in holistic ecosystem modelling: the Cumberland Basin ecosystem model. Neth J Sea Res 20:325-335
[32] Hendrickson WH, Karri SBR, Shar AD (1985) Simulation in a plankton model. Int Rev Ges Hydrobiol 70:547-559
[33] Herman AN, Platt T (1983) Numerical modelling of diel carbon production and zooplankton grazing on the Scotian Shelf based on observational data. Ecol Modelling 18:55-72
[34] Hofmann EE, Ambler JW (1988) Plankton Dynamics on the outer southeastern U.S. continental shelf. Part II: A time-dependent biological model. J Mar Res 46:883-917
[35] Hofmann EE, Pietrafesa LJ, Klinck JM, Atkinson LP (1980) A time-dependent model of nutrient distribution in continental shelf waters. Ecol Modelling 10:193-214
[36] Horwood JW (1976) A model of primary and secondary production in Loch Storien, and its stability. In: Perscone P, Jaspers E (eds) Proceedings of the 10th European Symposium on Marine Biology. University Press, Wetteren Belgium, p 297
[37] Horwood J (1982) Algal production in the west-central North Sea. J Plankton Res 4:103-124
[38] Isaacs JP (1973) Potential trophic biomasses and trace-substance concentrations in unstructured marine food webs. Mar Biol 22:97-104
[39] Iverson RL, Curl HC Jr, Saugen JL (1974) Simulation model for wind-driven summer phytoplankton dynamics in Auke Bay, Alaska. Mar Biol 28:169-177
[40] Iwakuma T, Yasuno M (1983) A comparison of several mathematical equations describing photosynthesis-light curves for natural phytoplankton populations. Arch Hydrobiol 97:208-226
[41] Jamart BM, Winter DF, Banse K, Anderson GC, Lam RK (1977) A theoretical study of phytoplankton growth and nutrient distribution in the Pacific Ocean off the northwestern U.S. coast. Deep-Sea Res 24:753-773
[42] Jassby AD, Platt T (1976) Mathematical formulation of the relationship between photosynthesis and light for phytoplankton. Limnol Oceanogr 21:540-547
[43] Kiefer DA, Kremer JN (1981) Origins of vertical patterns of

phytoplankton and nutrients in the temperate, open ocean: a stratigraphic hypothesis. Deep-Sea Res 28A:1087-1105

[44] King FD (1987) Nitrogen recycling efficiency in steady-state oceanic environments. Deep-Sea Res 34A:843-856

[45] Kmet T, Straskraba M (1989) Global behaviour of a generalised aquatic ecosystem model. Ecol Modelling 45:95-110

[46] Laake M, Dahle AB, Eberlein K, Rein K (1983) A modelling approach to the interplay of carbohydrates, bacteria and non-pigmented flagellates in a controlled ecosystem experiment with Skeletonema Costatum. Mar Ecol Prog Ser 14:71-79

[47] Lange GD, Hurley AC (1975) A theoretical treatment of unstructured food webs. Fish Bull 73:378-381

[48] Laws EA, Chalup M.S (1990) A microalgal growth model. Limnol Oceanogr 35:599-608

[49] Lederman TC, Tett P (1981) Problems in modelling the photosynthesis-light relationship for phytoplankton. Bot Mar 24:125-134

[50] Lehman JT, Botkin DR, Likens GE (1975) The assumptions and rationales of a computer model of phytoplankton population dynamics. Limnol Oceanogr 20:343-364

[51] Moloney CL, Bergh MO, Field JG, Newell RC (1986) The effect of sedimentation and microbial nitrogen regeneration in a plankton community: a simulation investigation. J Plankton Res 8:427-445

[52] Moloney CL, Field JG (1991) The size-based dynamics of plankton food webs. I. A simulation model of carbon and nitrogen flows. J Plankton Res 13:1003-1038

[53] Moloney CL, Field JG, Lucas MI (1991) The size-based dynamics of plankton food webs. II. Simulations of three contrasting southern Benguela food webs. J Plankton Res 13:1039-1091

[54] Morrison KA, Therien N, Marcus B (1987) Comparison of six models for nutrient limitations on phytoplankton. Can J Fish Aquat Sci 44:1278-1288

[55] O'Brien JJ, Wroblewski JS (1973) A simulation model of the mesoscale distribution of the lower marine trophic levels off west Florida. Inv Pesq 37:193-244

[56] O'Brien WJ (1974) The dynamics of nutrient limitation of phytoplankton algae: a model reconsidered. Ecology 55:135-141

[57] O'Neill RV, DeAngelis DL, Pastor JJ, Jackson BJ, Post WM (1989) Multiple nutrient limitation in ecological models. Ecol Modelling 46:147-164

[58] Ohuchi A, Fukuoka J, Miyakoshi A, Suzuki M (1986) Modelling of the lower trophic levels of a marine ecosystem and its example of short-period variations of chlorophyll and nutrient in Harima-nada. Ecol Modelling 32:149-163

[59] Pace ML, Glasser JE, Pomeroy LR (1984) A simulation analysis of continental shelf food webs. Mar Biol 82:47-63

[60] Parker RA (1986) Simulating the develpment of chlorophyll maxima in the Celtic Sea. Ecol Modelling 33:1-11
[61] Parsons TR, Kessler TA (1987) An ecosystem model for the assessment of plankton production in relation to the survival of young fish. J Plankton Res 9:125-137
[62] Patten BC (1968) Mathematical models of plankton production. Int Rev Ges Hydrobiol 53:357-408
[63] Platt T, Denman KL, Jassby AD (1977) Modelling the productivity of phytoplankton. In: Goldberg ED, McCave IN, O'Brien JJ, Steele JH (eds) The Sea, Volume 6. J. Wiley and Sons, Chichester, p 807
[64] Platt T, Gallegos CL, Harrison WG (1980) Photoinhibition of photosynthesis in natural assemblages of marine phytoplankton. J Mar Res 83:687-701
[65] Radach G (1980) Preliminary simulations of the phytoplankton and phosphate dynamics during FLEX '76 with a simple two-component model. 'Meteor' Forsch-Ergebnisse Reihe A 22:151-163
[66] Radach G, Maier-Reimer E (1975) The vertical structure of phytoplankton growth dynamics: a mathematical model. Mem Soc R Sci Liege 6e serie 7:113-146
[67] Radford PJ (1979) Some aspects of an estuarine ecosystem model - GEMBASE. In: Jorgensen SE (ed) State-of-the-Art in Ecological Modelling, Vol. 7: Proceedings of the Conference on Ecological Modelling, Copenhagen, 28th August to 2nd September 1978. Pergamon Press, Oxford, p 301
[68] Riley GA (1946) Factors controlling phytoplankton populations on Georges Bank. J Mar Res 6:54-73
[69] Riley GA (1965) A mathematical model of regional variations in plankton. Limnol Oceanogr 10(suppl):R202-R215
[70] Riley GA, Stommel H, Bumpus DF (1949) Quantitative ecology of the plankton of the western North Atlantic. Bull Bingham Oceanogr Collect. 12:1-169
[71] Ross GG, Nival P (1976) Plankton Modelling in the Bay of Villefranche. J Theor Biol 56:381-399
[72] Sakshaug E, Slagstad D, Holm-Hansen O (1991) Factors controlling the development of phytoplankton blooms in the Antarctic Ocean - a mathematical model. Mar Chem 35:259-271
[73] Simonot J-Y, Dollinger E, Le Treut H (1988) Thermodynamic-biological-optical coupling in the oceanic mixed layer. J Geophys Res 93C:8193-8202
[74] Slagstad D, Stole-Hansen K (1991) Dynamics of plankton growth in the Barents Sea. Model studies. Polar Res 10:173-186
[75] Steele JH (1958) Plant production in the northern North Sea. Marine Res (Scot Home Dept) 1958(7):1-36
[76] Steele JH (1962) Environmental control of photosynthesis in the sea. Limnol Oceanogr 7:137-150

[77] Steele JH, Henderson EW (1976) Simulation of vertical structure in a planktonic ecosystem. Scottish Fisheries Res Rep No 5:1-27.

[78] Steele JH, Henderson EW (1981) A simple plankton model. Am Nat 117:676-691

[79] Steele JH, Henderson EW (1992) The role of predation in plankton models. J Plankton Res 14:157-172

[80] Steele JH, Mullin MM (1977) Zooplankton Dynamics. In: Goldberg ED, McCave IN, O'Brien JJ, Steele JH (eds) The Sea, Volume 6. J. Wiley and Sons, Chichester, p 857

[81] Stigebrandt A, Wulff F (1987) A model for the dynamics of nutrients and oxygen in the Baltic proper. J Mar Res 45:729-759

[82] Sverdrup HA (1953) On conditions for the vernal blooming of phytoplankton. J cons int explor Mer 18:287-295

[83] Takahashi M, Fujii K, Parsons TR (1973) Simulation study of phytoplankton photosynthesis and growth in the Fraser River estuary. Mar Biol 19:102-116

[84] Taylor AH (1988) Characteristic properties of models for the vertical distribution of phytoplankton under stratification. Ecol Modelling 40:175-199

[85] Taylor AH, Harris JRW, Aiken J (1986) The interaction of physical and biological processes in a model of the vertical distribution of phytoplankton under stratification. In: Nihoul JC (ed) Marine Interfaces Ecohydrodynamics. Elsevier Oceanography Series, 42:313-330

[86] Taylor AH, Joint I (1990) A steady-state analysis of the 'microbial loop' in stratified systems. Mar Ecol Prog Ser 59:1-17

[87] Taylor AH, Stephens JA (1993) Diurnal variations of convective mixing and the spring bloom of phytoplankton. Deep-Sea Res II 40:389-408

[88] Taylor AH, Watson AJ, Ainsworth M, Robertson JE, Turner DR (1991) A modelling investigation of the role of phytoplankton in the balance of carbon at the surface of the North Atlantic. Global Biogeochem. Cycles 5:151-171

[89] Taylor AH, Watson AJ, Robertson JE (1992) The influence of the spring phytoplankton bloom on carbon dioxide and oxygen concentrations in the surface waters of the North-East Atlantic during 1989. Deep-Sea Res 39:137-152

[90] Tett P (1981) Modelling phytoplankton production at shelf-sea fronts. Phil Trans R Soc Lond A 302:605-615

[91] Tett P, Edwards A, Jones K (1986) A model for the growth of shelf-sea phytoplankton in summer. Est Coast Shelf Sci 23:641-672

[92] Thingstad TF (1987) Utilization of N, P and organic C by heterotrophic bacteria. I Outline of a chemostat theory with a consistent concept of 'maintenance' metabolism. Mar Ecol Prog Ser 35:99-109

[93] Thingstad TF, Pengerud B (1985) Fate and effect of allochthonous organic material in aquatic microbial systems. An analysis based on chemostat theory. Mar Ecol Prog Ser 21:47-62

[94] Thingstad TF, Sakshaug E (1990) Control of phytoplankton growth in nutrient recycling ecosystems. Theory and terminology. Mar Ecol Prog Ser 63:261-272

[95] Vinogradov MYe, Krapivin VF, Menshutkin VV, Fleyshman BS, Shushkina EA (1973) Mathematical model of the functions of the pelagic ecosystem in tropical regions (from the 50th voyage of the R/V Vityaz'). Oceanology USSR 13:704-717

[96] Vinogradov ME, Menshutkin VV, Shushkina EA (1972) On mathematical simulation of a pelagic ecosystem in tropical waters of the ocean. Mar Biol 16:261-268

[97] Vollenweider RA (1965) Calculation models of photosynthesis-depth curves and some implications regarding day rate estimates in primary production measurements. Mem Ist Ital Idrobiol 18(suppl):425-457

[98] Walsh JJ (1975) A spatial simulation model of the Peru upwelling ecosystem. Deep-Sea Res 22:201-236

[99] Williams PJLeB (1981) Incorporation of microheterotrophic processes into the classical paradigm of the planktonic food web. Kieler Meeresforsch Sonderh 5:1-28

[100] Winter DF, Banse K, Anderson GC (1975) The dynamics of phytoplankton blooms in Puget Sound, a fjord in the north-western United States. Mar Biol 29:139-176

[101] Wolf KU, Woods JD (1988) Lagrangian simulation of primary production in the physical environment - the deep chlorophyll maximum and nutricline. In: Rothschild BJ (ed) Towards a Theory of Biological-Physical Interactions in the World Ocean. Reidel, Dordrecht, p 51

[102] Woods JD, Onken R (1982) Diurnal variation and primary production in the ocean - preliminary results of a Lagrangian ensemble model. J Plankton Res 4:735-756

[103] Wroblewski JS (1977) A model of phytoplankton plume formation during variable Oregon upwelling. J Mar Res 35:357-394

[104] Wroblewski JS (1984) Formulation of growth and mortality of larval northern anchovy in a turbulent feeding environment. Mar Ecol Prog Ser 20:13-22

[105] Wroblewski JS, Richman JG (1987) The non-linear response of plankton to wind mixing events - implications for the survival of larval northern anchovy. J Plankton Res 9:103-123

[106] Wroblewski JS, Sarmiento JL, Flierl GR (1988) An ocean basin scale model of plankton dynamics in the North Atlantic. I. Solutions for the climatological oceanographic conditions in May. Global Biogeochem Cycles 2:199-218

LIST OF PARTICIPANTS

Dr. Valerie Andersen
Station Zoologique
BP 28
06230 Villefranche sur Mer
France

Mr. Tom Anderson
James Rennell Centre for Ocean Circulation
Chilworth Research Centre
Chilworth, Southampton
SO1 7NS
United Kingdom
OMNET: via RENNELL.CENTRE
E-mail: tra@ub.nso.ac.uk

Dr. Robert Armstrong
AOS Program
Princeton University
PO Box CN 710
Princeton, NJ 08544-0710
USA
OMNET: via J.SARMIENTO
E-mail: raa@splash.princeton.edu

Dr. Helmut Baumert
Universitat Hamburg
Institut fur Meereskunde
Troplowitzstr. 7
2000 Hamburg 54
Germany
OMNET: via IFM.HAMBURG
E-mail: baumert@ifmsun2.ifm-hamburg.de

Dr. John Cullen
Dept. of Oceanography
Dalhousie University
Halifax, NS
Canada B3H 4J1
OMNET: J.CULLEN
E-mail: jcullen@ac.dal.ca

Mr. Helge Drange
Nansen Environmental Centre
Edvard Griegsvei 3A
N 5037 Solheimsviken
Bergen
Norway
OMNET: via O.JOHANNESSEN
E-mail: helge@fram.nrsc.no

Dr. Geoffrey T. Evans
Science Branch
Dept. of Fisheries + Oceans
PO Box 5667
St. John's, NF
Canada A1C 5X1
OMNET: G.Evans.DFO
E-mail: evans@mrspock.nwafc.nf.ca

Dr. Mike Fasham
James Rennell Centre
Gamma House
Chilworth Research Centre
Chilworth, Southampton SO1 7NS
United Kingdom
OMNET: via RENNELL.CENTRE
E-mail: mjf@ub.nso.ac.uk

Prof. John G. Field
University of Cape Town
Zoology Dept.
7700 Rondebosch, Cape Town
South Africa
OMNET: J.FIELD
E-mail: jgfield@ucthpx.uct.ac.za

Dr. George Fransz
NIOZ
P.O. Box 59
1790 Ab den Burg
Texel
The Netherlands
OMNET: via NIOZ.TEXEL

Dr. Bruce Frost
School of Oceanography, WB-10
University of Washington
Seattle, Wash. 98195
USA
OMNET: B.FROST

Ms. Veronique Garcon
CNES/GRGS
18 av Edouard Belin
31055 Toulouse Cedex
France
E-mail: Garcon@gaia.cnes.fr

Dr. Richard Geider
Univ. of Delaware
College of Marine Studies
Lewes Complex
Lewes, DE 19958-1298
USA
OMNET: R.GEIDER

Dr. Joji Ishizaka
National Research Institute
for Resources and Environment
16-3 Onogawa, Tsukuba
Ibaraki 305
Japan
OMNET: J.ISHIZAKA

Dr. George Jackson
Dept. of Oceanography
Texas A&M University
College Station, TX 77843
USA
OMNET: G.JACKSON
E-mail: jackson@benthos.tamu.edu

Mr. Fortunat Joos
Physics Institut
University of Bern
Sidlerstr. 5
3012 Bern
Switzerland
E-mail: joos@phil.unibe.ch

Prof. Boris Kagan
P.P. Shirshov Inst. of Oceanology
Russian Academy of Sciences
30 Pervaya Liniya
199053 St. Petersburg
Russia

Dr. Dale Kiefer
Dept. of Biological Science
U.S.C. University Park
Los Angeles
California 90089-0371
USA
OMNET: D.KIEFER

Dr. David Kirchman
University of Delaware
College of Marine Studies
Lewes, DE 19958
USA
OMNET: D.KIRCHMAN

Dr. Olivier Klepper
Centre for Mathematical Methods
National Institute for Public
Health and Environmental Protection
P.O. Box 1
3720 BA Bilthoven
The Netherlands
E-mail: cwmok@rivm.nl

Dr. Christiane Lancelot
Universite Libre de Bruxelles
GMMA
Campus de la Plaine CP 221
Boulevard du Triomphe
B-1050 Bruxelles
Belgium

Dr. Louis Legendre
Departement de Biologie
Universite Laval
Cite Universitaire
Pavillon vachon
Ste-Foy
Canada, QC G1K 7P4
OMNET: L.LEGENDRE
E-mail: 3602 jpbo@lavalvl1

Mr. Charles Lin
Dept. of Atmospheric and
Ocean Sciences
McGill University
805 Sherbrooke St. West
Montreal, QC
Canada, H3A 2K6
OMNET: C.LIN

Dr. John Marra
Lamont-Doherty Geological
Observatory
Palisades, N.Y. 10964
USA
OMNET: J.MARRA
E-mail:
marra@lamont.ldgo.columbia.edu

Mr. Eugene Murphy
British Antarctic Survey
High Cross
Madingley Rd.
Cambridge CB3 0ET
United Kingdom
OMNET: via BAS.UK

Prof. Paul Nival
Station Zoologique
BP 28
06230 Villefranche sur Mer
France

Dr. John Parslow
CSIRO
Division of Fisheries
GPO Box 1538
Hobart, Tas. 7001
Australia
OMNET: J.PARSLOW
E-mail: parslow@aqucous.ml.csiro.au

Dr. Tom Powell
University of California-Davis
Division of Environmental Studies
Davis, California 95616
USA
OMNET: T.POWELL

Dr. Guenther Radach
Universitat Hamburg
Institut fur Meereskunde
Troplowitzstr. 7
2000 Hamburg 54
Germany
OMNET: via IFM.HAMBURG

Mr. Fereidun Rassoulzadegan
Station Zoologique
BP 28
06230 Villefranche sur Mer
France

Prof. John Raven
University of Dundee
Biological Sciences
Dundee DD1 4HN
United Kingdom

Dr. Jaime Rodriguez
Dept. of Ecology
Faculty of Sciences
Univ. of Malaga
Campus Universitario de teatinos
29071 Malaga
Spain

Dr. Vladimir Ryabchenko
PP Shirshov Inst. of Oceanology
Russian Academy of Sciences
30 Pervaya Liniya
199053 St. Petersburg
Russia

Prof. Egil Sakshaug
Inst. for Marine Biochemistry
University of Trondheim
N-7034 Trondheim-Nth
Norway

Prof. Jorge L. Sarmiento
AOS Program
Princeton University
PO Box CN 710
Princeton, N.J. 08544-0710
USA
OMNET: J.SARMIENTO
E-MAIL: jls@splash.princeton.edu

Dr.-Ing. Wolfram Schrimpf
Joint Research Centre of CEC
Institute for Remote Sensing Applications
Building 69 - T.P. 690
I-21020 Ispra (VA)
Italy
E-mail: w.schrimpf@jrc.it

Dr. Marian Scott
University of Glasgow
Dept. of Statistics
Mathematics Bldg.
University Gardens
Glasgow G12 8QW
United Kingdom
E-mail: e.m.scott@uk.ac.glasgow

Dr. John Steele
Woods Hole Oceanographic Inst.
Woods Hole, Mass. 02543
USA
OMNET: J:STEELE

Mr. Arnold Taylor
Plymouth Marine Laboratory
Prospect Place, West Hoe
Plymouth PL1 3DH
United Kingdom
OMNET: via PML.UK

Mr. Paul Tett
School of Ocean Sciences
University of Wales
Menai Bridge
Gwynedd LL59 5EY
United Kingdom
E-MAIL: ossoo1@vaxa.bangor.ac.uk

Dr. T. Frede Thingstad
Universitetet i Bergen
Dept. of Microbiology and
Plant Physiology
Jahnebakken 5
N - 5007 Bergen
Norway

Dr. Ian J. Totterdell
James Rennell Centre
Chilworth Research Centre
Chilworth, Southampton
SO1 7NS
United Kingdom
OMNET: via RENNELL.CENTRE
E-mail: ijt@ub.nso.ac.uk

Dr. Uli Wolf
Institut fur Meereskunde
JGOFS Buro
Dusternbrooker Weg 20
2300 Kiel 1
Germany
OMNET: via JGOFS.KIEL
E-MAIL:
kwolf@meereskunde.uni-kiel.dbp.de

Dr. John Woods
Director Marine and Atmospheric
Sciences
NERC
North Star Avenue
Swindon SN2 1EU
United Kingdom
OMNET: J.WOODS

Dr. Fréd Wulff
Dept. of Systems Ecology
University of Stockholm
S 10691 Stockholm
Sweden
E-mail: fred@ecology.su.se

SUBJECT INDEX

MIX
Papier aus verantwortungsvollen Quellen
Paper from responsible sources
FSC® C105338

If you have any concerns about our products,
you can contact us on
ProductSafety@springernature.com

In case Publisher is established outside the EU,
the EU authorized representative is:
Springer Nature Customer Service Center GmbH
Europaplatz 3, 69115 Heidelberg, Germany

Printed by Libri Plureos GmbH
in Hamburg, Germany